Submicron Porous Materials

Paolo Bettotti
Editor

Submicron Porous Materials

Springer

Editor
Paolo Bettotti
Department of Physics
University of Trento
Trento
Italy

ISBN 978-3-319-85044-3 ISBN 978-3-319-53035-2 (eBook)
DOI 10.1007/978-3-319-53035-2

Softcover reprint of the hardcover 1st edition 2017

Printed on acid-free paper

This Springer imprint is published by Springer Nature
The registered company is Springer International Publishing AG
The registered company address is: Gewerbestrasse 11, 6330 Cham, Switzerland

Preface

> Wherever he saw a hole he always wanted to know the depth of it.
>
> To him this was important.
>
> J. Verne, *Journey to the Center of the Earth*

Interfaces mediate our daily way to interact with things and at macroscopic surfaces they are the most common day-to-day experience. Flat surfaces are fundamental in many applications, but the mastering of material's porosity is even more fascinating as it is an effective way to tune both type and strength of materials interactions. This is the main reason why porous materials are very important in science and technology nowadays.

Any material is porous if we consider that its crystallographic unit cells might act as cages for smaller molecular species. More generally, and with the exception of ceramics and metals annealed at high temperature, nearly all materials show a porous structure with a typical length scale much larger than their unit cell. Porosity induces peculiar characteristics to surfaces and largely modifies their properties compared to those of the bulk states: wetting, tribology, chemical reactivity, and optoeletronic properties are only a few significant examples of characteristics modified by the presence of porosity.

Very often, the interest in controlling material's porosity arises because of the new surface-mediated properties created in dense materials upon their porosification.

As stated above, nearly any kind of material can be rendered porous. Generally, organic-based materials allow large flexibility in their synthesis conditions and a good control over their porous structure. In this broad class of materials, porosity is introduced either by adding porogens or by phase separation. Porogens are molecular species (or colloids) that are inserted during the material preparation and successively removed (often) exploiting their thermal degradation and removed as gas species. Otherwise, the pore can be formed by exploiting phase separation reactions, followed by the removal of one of the species to form the voids. On the other hand, some inorganic materials (mainly silicon, III-V semiconductors, alumina and titania) show peculiar corrosion processes that enable the formation of well-controlled porous structure: under controlled etching conditions these materials spontaneously form self-arranged lattices of pores, sometimes with rather good periodicity. Compared to random 3D porous structures, their highly ordered pores permit a better control on applications that require a well-known pore length or highly homogeneous volumetric porosity (for example in some separation techniques). Moreover, they show a greater resistance to thermally and chemically harsh environments. Hybrid materials sit in between the organic and inorganic families. They are formed starting from organic precursors that polymerize to form an inorganic network.

Porous materials might have an extremely high specific area (up to thousands of $\frac{m^2}{g}$) and this fact is exploited to increase their performance in a number of different applications:

- in sensors, a large specific area increases the density of the binding site and thus decreases the minimum detectable signal;
- in thermoelectrics, porosity is a way to decrease thermal conductivity and increase the potential difference created across the junction;
- in catalysis, a porous support maximizes the interaction between the catalyst and the reagent, thus increasing the overall efficiency of the reaction;
- in energy storage, a large porosity is required to maximize the storage capacity;
- in photovoltaics, porous materials are used to decrease reflectance and increase photon absorption probability and thus cell efficiency.

But the simple value of the specific area is not sufficient to completely qualify the properties of a porous material. In fact, it is of fundamental importance that porosity has to be accessible to the species of interest (e.g., they have to be able to reach the active sites). Thus, a fine control over pore size distribution is needed to optimize the material properties, and the design of hierarchically structured porous materials is an important step to achieve an optimal use of the entire material surface.

IUPAC nomenclature defines nanopores as smaller than 2 nm, mesopores in the range 2–20 nm, and macropores greater than 20 nm. Such classification is quite limited because how a porous material behaves largely depends on the ratio between the pore size and the dimension of the species that interact with it, rather than on the absolute pore size. Thus, in this monograph we use the term submicron to underline the importance of the mesoscale, intended as the length scale where the property is not determined by molecular scale details, yet it is different from the one shown by the bulk state. Mesoscale physics addresses such an intermediate scale where the material's properties cannot be described by effective medium theories, yet they are not determined by the fine details of molecular species involved.

This monograph treats a selected series of topics about fabrication, characterization, and applications of topical porous materials. The first two chapters deal with the use of carbon-based porous materials in catalysis and water treatment of silica- and carbon-based materials. Chapters 3 and 4 cover two types of organic-based porous materials. Chapter 5 introduces porous anodic alumina as an inorganic material with highly ordered pores. Chapters 6 and 7 treat silica and titania based sol-gel materials. Chapter 8 describes the basics of porous silicon fabrication, its use in optical sensing, and drug delivery applications. Chapter 9 shows how thermal transport is affected in porous materials. Chapter 10 describes how to model diffusion in porous materials. Chapter 11 describes a peculiar spectroscopic technique to characterize material porosity.

A final note about the etymology of the word “pore”. It comes from the Greek and means passage, to go beyond. I hope this monograph to be a useful document to go beyond our current mastering of material’s porosity at the submicron scale and to help both nonspecialists and scientists to understand and develop innovative porous materials.

Trento, Italy
2016

Paolo Bettotti

Contents

Contributors

A. Arenillas Microwaves and Carbon for Technological Applications (MCAT-INCAR-CSIC), Oviedo, Spain

E. Blanco Dep. Física de la Materia Condensada and Instituto de Microscopía Electrónica y Materiales (IMEYMAT), Universidad de Cadiz, Puerto Real, Cádiz, Spain

Paolo Bettotti Nanoscience Laboratory, Department of Physics, University of Trento, Trento, Italy

Anna Brudzisz Faculty of Chemistry, Department of Physical Chemistry & Electrochemistry, Jagiellonian University in Krakow, Krakow, Poland

Agnieszka Brzózka Faculty of Chemistry, Department of Physical Chemistry & Electrochemistry, Jagiellonian University in Krakow, Krakow, Poland

Lei Caihong Guangdong Provincial Key Laboratory of Functional Soft Condensed Matter, School of Materials and Energy, Guangdong University of Technology, Guangzhou, People's Republic of China

Carmen Ciotonea Univ. Lille, CNRS, Centrale Lille, ENSCL, Univ. Artois, UMR 8181 - UCCS - Unité de Catalyse et Chimie du Solide, Lille, France

Frank A. Coutelieris Department of Environmental & Natural Resources Management, University of Patras, Agrinio, Greece

Jean-Philippe Dacquin Univ. Lille, CNRS, Centrale Lille, ENSCL, Univ. Artois, UMR 8181 - UCCS - Unité de Catalyse et Chimie du Solide, Lille, France

Xin Gao Center for Applied Energy Research, University of Kentucky, Lexington, KY, USA

Gavriil Gavriil Department of Environmental & Natural Resources Management, University of Patras, Agrinio, Greece

D. Lacroix LEMTA, Université de Lorraine, Vandœuvre-lès-Nancy, France; LEMTA, CNRS, Vandœuvre-lès-Nancy, France

James Landon Center for Applied Energy Research, University of Kentucky, Lexington, KY, USA

Kunlei Liu Center for Applied Energy Research, University of Kentucky, Lexington, KY, USA

S. Mariazzi Stefan Meyer Institute for Subatomic Physics, Austrian Academy of Sciences, Vienna, Austria

J.A. Menéndez Microwaves and Carbon for Technological Applications (MCAT-INCAR-CSIC), Oviedo, Spain

Ayokunle Omosebi Center for Applied Energy Research, University of Kentucky, Lexington, KY, USA

Chiara Piotto Nanoscience Laboratory, Department of Physics, University of Trento, Trento, Italy

M. Ramirez-del-Solar Dep. Física de la Materia Condensada and Instituto de Microscopía Electrónica y Materiales (IMEYMAT), Universidad de Cadiz, Puerto Real, Cádiz, Spain

N. Rey-Raap Department of Chemical Engineering, University of Porto, Porto, Portugal

Sébastien Royer Univ. Lille, CNRS, Centrale Lille, ENSCL, Univ. Artois, UMR 8181 - UCCS - Unité de Catalyse et Chimie du Solide, Lille, France

Xu Ruijie Guangdong Provincial Key Laboratory of Functional Soft Condensed Matter, School of Materials and Energy, Guangdong University of Technology, Guangzhou, People's Republic of China

Myungeun Seo Graduate School of Nanoscience and Technology, Korea Advanced Institute of Science and Technology (KAIST), Daejeon, Republic of Korea

Grzegorz D. Sulka Faculty of Chemistry, Department of Physical Chemistry & Electrochemistry, Jagiellonian University in Krakow, Krakow, Poland

K. Termentzidis LEMTA, Université de Lorraine, Vandœuvre-lès-Nancy, France; LEMTA, CNRS, Vandœuvre-lès-Nancy, France

M. Verdier LEMTA, Université de Lorraine, Vandœuvre-lès-Nancy, France; LEMTA, CNRS, Vandœuvre-lès-Nancy, France

Ewa Wierzbicka Faculty of Chemistry, Department of Physical Chemistry & Electrochemistry, Jagiellonian University in Krakow, Krakow, Poland

Leszek Zaraska Faculty of Chemistry, Department of Physical Chemistry & Electrochemistry, Jagiellonian University in Krakow, Krakow, Poland

Chapter 1
Structured and Surface-Modified Carbon Xerogel Electrodes for Capacitive Deionization

James Landon, Xin Gao, Ayokunle Omosebi and Kunlei Liu

1.1 Introduction

Porous materials are critically important for a wide variety of applications including liquid and gas-phase separations, catalyst supports, and energy storage/generation applications, to name a few [1–4]. In this chapter, we will review the impact of submicron porous materials on the progress made in the field of capacitive deionization (CDI) also known as a form of electrostatic ion separation. CDI uses porous carbon electrodes for the separation and concentration of ionic content from liquid streams with the use of an applied voltage [5–8]. While CDI was first demonstrated in the 1960s, the development of more advanced and structured submicron porous carbon materials has been crucial to the advancement of this technology beyond laboratory-scale demonstrations [9, 10]. Initial CDI experiments focused on the use of carbon particles and applied potentials to remove dissolved ionic species in a flow-through deionization cell. This initial configuration was often called a flow-through capacitor as opposed to capacitive deionization, which is now the more widely accepted name. These studies defined the possibility of using carbon materials to reversibly separate ionic content with an applied electrical

J. Landon · X. Gao · A. Omosebi · K. Liu (✉)
Center for Applied Energy Research, University of Kentucky,
2540 Research Park Drive, Lexington, KY 40511, USA
e-mail: kunlei.liu@uky.edu

J. Landon
e-mail: james.landon@uky.edu

X. Gao
e-mail: xin.gao1@uky.edu

A. Omosebi
e-mail: ayokunle.omosebi@uky.edu

P. Bettotti (ed.), *Submicron Porous Materials*,
DOI 10.1007/978-3-319-53035-2_1

voltage, but further advancements were needed to the porous carbon materials themselves before the field truly gained traction, including the proper use of advanced carbon materials.

1.2 Development of Porous Carbon Materials for Capacitive Deionization

In the late 1980s and early 1990s, a new carbon material was developed by Pekala and others at Lawrence Livermore National Laboratory called the carbon aerogel [11–13]. The first carbon aerogels were synthesized through a relatively simple polymerization reaction between resorcinol and formaldehyde followed by a gelation period, solvent exchange steps, supercritical drying, and finally carbonization of the material into a conductive carbon network. A simplified procedure for this synthesis process is shown in Fig. 1.1.

This synthesis process was subsequently modified through adjustments to the solvent exchange and drying stages to simplify the synthesis. Shown below in Fig. 1.2 is the synthesis process of carbon xerogel materials, which provides for optional solvent exchange steps and removes the supercritical drying stage, instead using vacuum or air drying prior to the carbonization process to form the final conductive material.

While the carbon xerogel synthesis process is simpler to carry out than the aerogel route, which requires a supercritical drying chamber, the loss of surface area due to pore collapse is detrimental to surface area critical applications [14]. During drying, capillary forces in the microporous space may cause pores to collapse, resulting in lowered surface area and limiting the total number of adsorption sites.

Therefore, a third synthesis method was developed for the production of porous carbon gel structures using a freeze drying process in the production of carbon cryogels to mitigate pore collapse prior to the drying process. Instead of supercritical or air drying in combination with solvent exchange steps, freeze drying is used directly after the gelation period to maintain the porosity of the polymerized structure. Subsequently, pressure is released from the freeze dryer to allow

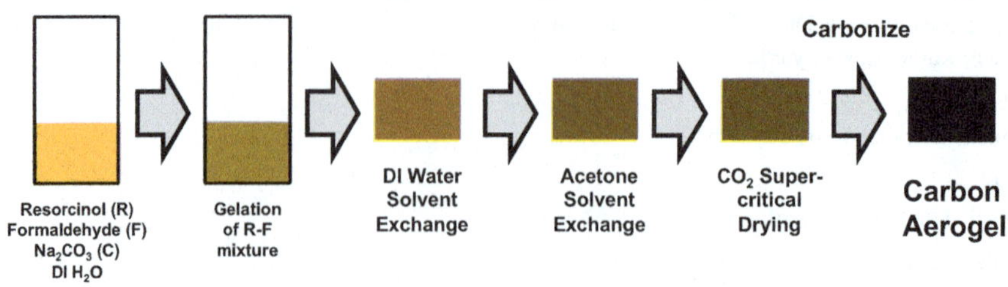

Fig. 1.1 Synthesis procedure for the production of carbon aerogel structures

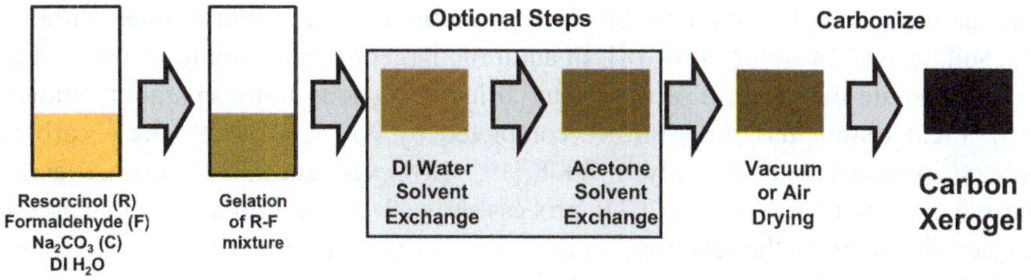

Fig. 1.2 Synthesis procedure for the production of carbon xerogel structures with optional solvent exchange steps and the use of vacuum or air drying prior to carbonizing

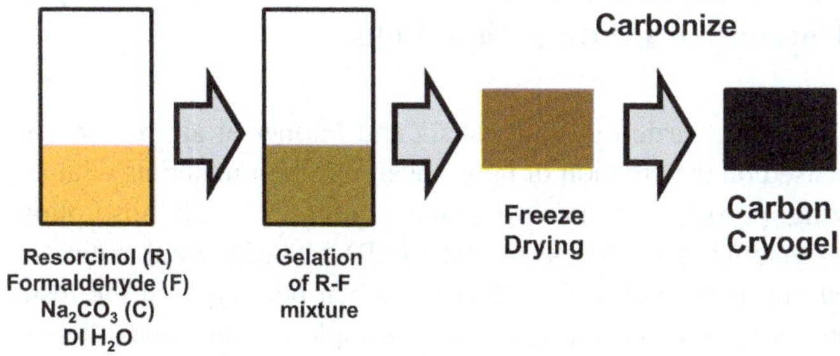

Fig. 1.3 Synthesis procedure for the production of carbon cryogel structures using a freeze drying process prior to carbonization

sublimation of the solvent, leaving behind a porous structure. This synthesis process is shown in Fig. 1.3.

All three synthesis processes, leading to the production of carbon aerogels, xerogels, and cryogels, result in the creation of a hierarchical network of carbon that contains substantial amounts of mesoporous and microporous adsorption sites that are necessary for significant separations to take place in CDI devices. Each synthesis process can be modified based on multiple factors, but three keys modifications are used

1. Initial resorcinol (R) and formaldehyde (F) ratio
2. Concentration of catalyst, e.g., Na_2CO_3, for pH adjustment
3. Dilution factor with water

These three factors can be used to alter the type of pore space (microporous, mesoporous, macroporous) and the amount of total carbon porosity in the resulting structure. The advent of this new carbon material and its easily modified pore structure helped to advance the CDI field, and substantial research into this possibly energy efficient separation process resumed with Farmer et al. publishing significantly in the 1990s.

Briefly, Farmer et al. demonstrated the promise of CDI as both a capable and efficient separation technology [15]. CDI in conjunction with the carbon aerogel material was used in the separation of conventional sodium chloride electrolytes as well as more complex mixtures involving ammonium, chromium, nitrate, carbonate, sulfate, and phosphate [16–18]. In addition, larger deionization trials were used to demonstrate the promise of this material including long-term separation studies [15]. Field testing has been further completed by Xu et al. using these carbon aerogel materials more recently in 2008 [19]. However, questions remained as to the effect of pore size on the CDI process as well as the importance of carbon surface chemistry to the resulting separation. These two points will be reviewed in detail in this chapter in the following sections.

1.3 Effect of Carbon Xerogel Porosity on Salt Adsorption in Capacitive Deionization Cells

Following the work carried out by Pekala and Farmer et al., research in the CDI field has focused on the creation of new, specific carbon materials with an emphasis on optimizing porosity to maximize the equilibrium salt adsorption capacity (SAC) and dynamic salt adsorption rate (SAR) [20–31]. Systematic studies have been carried out to determine the effect of carbon pore type on the resulting SAC with carbon xerogels being chosen as an example for this study. Resorcinol and formaldehyde were again chosen as chemical precursors in the synthesis process with deionized water and acetone used for solvent exchanges as well as an air-drying step, similar to the process shown in Fig. 1.2.

In this study, to modify the carbon pore space, the dilution factor was systematically changed to determine its effect on the resulting surface area, pore volume, and average pore size. Resorcinol and formaldehyde (RF) weight percentages between 37 and 50 were chosen in this study through variations in the amount of deionized water in the initial solution. Resorcinol and formaldehyde were mixed in a 1:2 mol ratio, and the Na_2CO_3 catalyst (pH adjustment) was added with a final molar ratio of 1500:1 resorcinol:catalyst. Deionized water was then added to produce the final RF weight percentages of 37, 40, 45, and 50. Solutions were mixed for a period of 30 min before the solution was infiltrated into a piece of conductive macroporous carbon cloth. Gelation was allowed to occur overnight before samples were heated at approximately 85 °C for a period of 24 h. Solvent exchange steps were carried out in deionized water for 2 h followed by acetone for 2 h. Finally, air drying was used for all samples for a period of 2 h before pyrolysis/carbonization. Samples were carbonized at 1000 °C using heating and cooling ramp rates of 1 °C/min, resulting in porous hierarchical carbon xerogel monolithic materials. A similar synthesis process has been carried out by Rolison et al. in the creation of their carbon nanofoams [3]. Scanning electron microscopy (SEM) images of these carbon xerogels are shown in Fig. 1.4 for the 37, 40, 45, and 50 RF weight percent electrodes.

The use of a dilution factor in the carbon xerogel synthesis process drastically affects the resulting external pore size of the resulting material, as evidenced by the

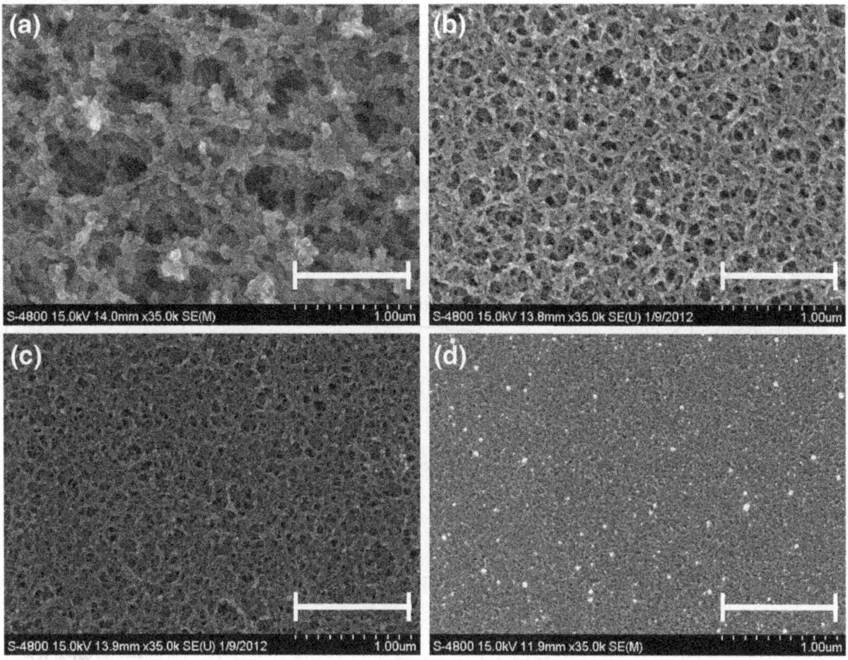

Fig. 1.4 SEM images of the external pore size for **a** 37 wt% RF, **b** 40 wt% RF, **c** 45 wt% RF, and **d** 50 wt% RF. External pore size is a strong function of the initial RF concentration with the pore size decreasing as the RF weight percent increases. Size bar in the figure is 1 μm. This material was first published by Landon et al. in 2012 in the Journal of the Electrochemical Society and is reproduced by permission of The Electrochemical Society [32]

difference in pore sizes shown in Fig. 1.4. With increases in the RF weight percent (decreasing the level of dilution), the resulting external pores range from >500 nm for 37 wt% RF to <50 nm for 50 wt% RF, thus providing an easily adaptable method to systematically modifying the pore space while maintaining the bulk of other materials properties. In addition to examining the external pore sizes, nitrogen adsorption/desorption experiments were carried out using a Micromeritics TriStar 3000 at 77 K. Nitrogen adsorption/desorption isotherms for all four of these materials are shown in Fig. 1.5 and differential pore volumes versus pore size, commonly called pore size distributions, are shown in Fig. 1.6. Classical type IV isotherms are seen for all of the RF carbon xerogel materials synthesized here as represented by an initial increase of adsorbed gas at low relative pressure followed by capillary condensation at higher relative pressures. Type IV isotherms are typically found for mesoporous materials, which is further shown by the pore size distribution in Fig. 1.6.

An increase in the RF weight percentage (via decreases in the level of dilution) increases the volume of adsorbed nitrogen per gram in these materials. Correspondingly, a decrease in the average pore size results from an increase in the RF

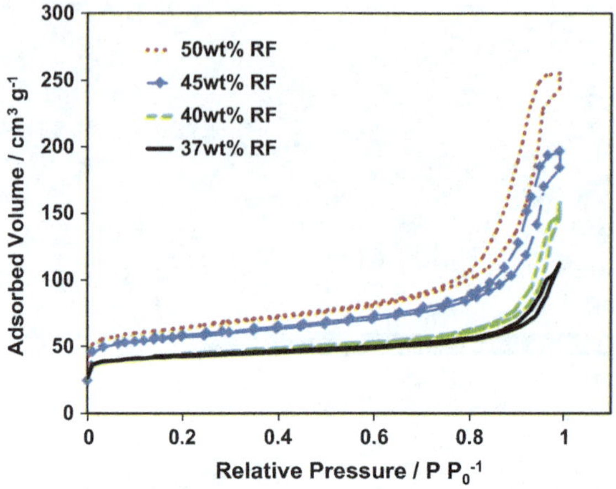

Fig. 1.5 Nitrogen adsorption/desorption isotherms for various RF weight percent electrodes. Type IV isotherms with a characteristic hysteresis at high relative pressure indicative of mesoporous materials are seen for all RF materials here. This material was first published by Landon et al. in 2012 in the Journal of the Electrochemical Society and is reproduced by permission of The Electrochemical Society [32]

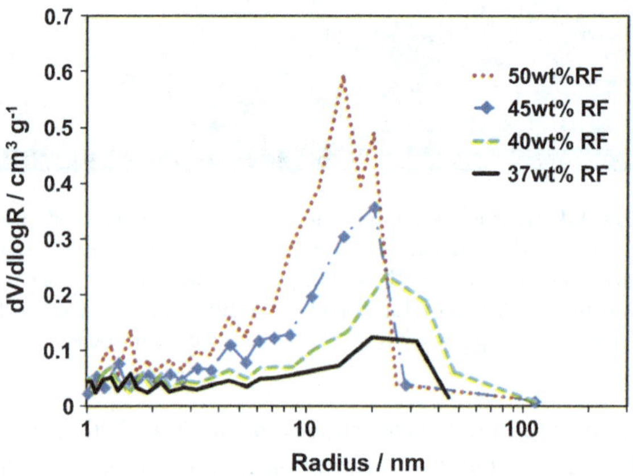

Fig. 1.6 BJH pore size distributions for various RF weight percent electrodes. Large increases in the pore volume are shown as the RF percentage increases while the pore size slightly decreases. This material was first published by Landon et al. in 2012 in the Journal of the Electrochemical Society and is reproduced by permission of The Electrochemical Society [32]

weight percent, as shown in Fig. 1.6, which correlates well with the decrease in the external pore size shown in Fig. 1.4. A summary of the physical properties including surface area (from Brunauer–Emmett–Teller, BET, method), total pore volume (from Barrett–Joyner–Halenda, BJH, method) and micropore volume (from the t-plot method) for each of these materials is shown in Table 1.1.

With the creation of these systematically modified carbon xerogel materials, CDI experiments were subsequently conducted to determine the impact of surface area and pore size (in the mesoporous range) on the resulting salt adsorption capacity for dilute sodium chloride solutions. A schematic of the CDI test system is shown in Fig. 1.7.

A constant volume of 500 mL of about 4.5 mM NaCl salt solution was circulated, without being deaerated, through the CDI cell at 5 mL min^{-1}. Electrodes were soaked in the NaCl solution overnight to equilibrate and sufficiently wet the electrode surface prior to CDI tests. Once conductivity measurements reached a steady state, current and conductivity data were recorded using a Graphtec midi

Table 1.1 BET surface area and BJH pore volume data from N_2 adsorption isotherm experiments with 37, 40, 45, and 50 wt% RF carbon xerogels. This material was first published by Landon et al. in 2012 in the Journal of the Electrochemical Society and is reproduced by permission of The Electrochemical Society [32]

Sample	Surface area/m^2 g^{-1}	Total pore volume/mL g^{-1}	Micropore volume/mL g^{-1}
37 wt% RF	144.3	0.13	0.05
40 wt% RF	147.1	0.23	0.04
45 wt% RF	194.3	0.25	0.05
50 wt% RF	239.1	0.42	0.06

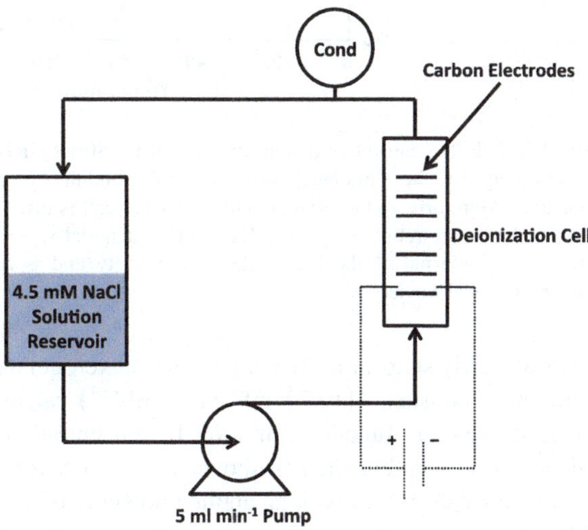

Fig. 1.7 Capacitive deionization setup composed of a deionization cell, conductivity sensor, 500 mL of about 4.5 mM NaCl solution, and a circulating pump set at approximately 5 mL min^{-1}. Total carbon electrode mass used was between 2.3–3.2 g for each experiment. This material was first published by Landon et al. in 2012 in the Journal of the Electrochemical Society and is reproduced by permission of The Electrochemical Society [32]

Logger GL220. A cell voltage of 1.2 V was applied using a DC regulated power source. Sixty minutes was taken for each adsorption (charging) and desorption (discharging) step. The salt adsorption capacity (SAC or Γ) of the synthesized carbon xerogel electrodes was determined from the change in concentration of the salt solution under potential divided by the total carbon electrode mass using Eq. (1). Total carbon used in each experiment with a pair of carbon electrodes ranged from 2.3 to 3.2 g.

$$\Gamma = (MV\sigma_{diff})/mA \tag{1}$$

where M is the molecular weight of NaCl (58.44 g mol^{-1}), V is the volume of solution, σ_{diff} is the conductivity difference between the adsorption and desorption

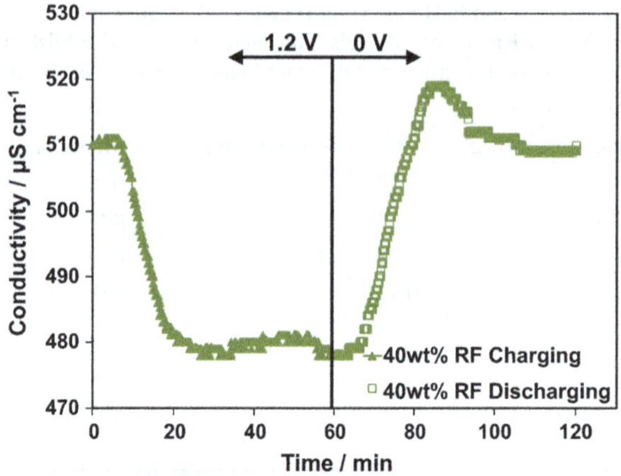

Fig. 1.8 Selected capacitive deionization result for 40 wt% RF electrodes showing a drop in the conductivity of the circulated solution corresponding to charging the CDI cell at 1.2 V. Conductivity returns to the original value after the cell is discharged at 0 V, meaning previously adsorbed ions are returned to the solution. This material was first published by Landon et al. in 2012 in the Journal of the Electrochemical Society and is reproduced by permission of The Electrochemical Society [32]

steps at steady state, m is the mass of carbon xerogel used in the CDI cell, and A is a calibration constant (112.77 mS cm^{-1} mM^{-1}) calculated from the slope of conductivity as a function of NaCl concentration. A typical plot of the adsorption/desorption data is shown in Fig. 1.8 for 40 wt% RF electrodes. The conductivity decreases with an applied potential of 1.2 V from approximately 510–480 µS/cm. When the cell is discharged at 0 V, the conductivity increases back to its original value of 510 µS/cm, which depicts the reversibility of salt adsorption/desorption on the carbon xerogel electrodes used for CDI applications.

CDI tests were carried out for each carbon xerogel electrode synthesized here (37–50 wt% RF). The SAC values are presented and compared in Fig. 1.9 to determine the effect of the surface area and porosity of these carbon xerogel electrodes. No substantive difference in the SAC is found using these different carbon xerogel electrodes even though the surface area nearly doubles from 144.3 m^2/g for the 37 wt% RF carbon xerogel electrodes to 239.1 m^2/g for those prepared with 50 wt% RF. Therefore, while porous carbon electrodes with higher surface areas are clearly important in providing adsorption sites for ionic species under an applied voltage, other important factors must be considered in designing and operating these CDI systems to generate higher SAC values. In fact, two important considerations have been discussed/proposed in recent years to (1) increase the SAC by enhancing microporosities and (2) targeting chemical surface charge of carbon electrodes to improve SAC and mitigate the degradation of the SAC via a new configuration of electrode pairs based upon the potential of zero charge. These two topics will be reviewed in detail in the following three sections to demonstrate the proper use of submicron porous materials for CDI applications.

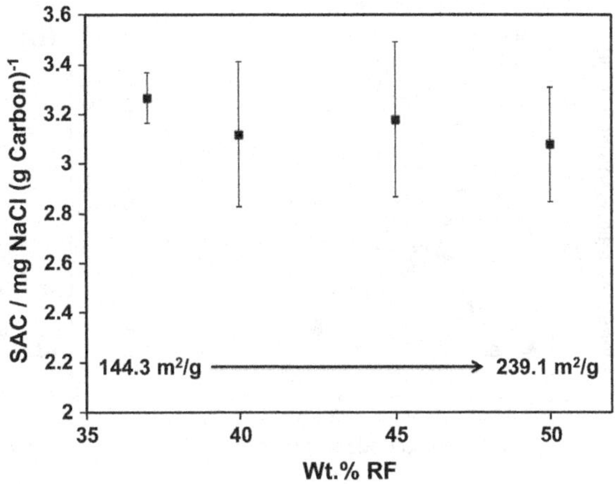

Fig. 1.9 Salt adsorption capacity in mg NaCl per g of carbon xerogel as a function of the RF wt%. No substantial difference in the capacity is found between each carbon xerogel electrode used in these CDI tests. This material was first published by Landon et al. in 2012 in the Journal of the Electrochemical Society and is reproduced by permission of The Electrochemical Society [32]

1.4 Development of Microporous Carbon and Salt Adsorption Capacity Analysis

While mesoporous carbon xerogel materials make quite effective electrodes for CDI applications, SAC values of less than 5 mg NaCl per gram of electrode remained relatively low compared to commercially available activated carbon electrodes. To understand the impact of the porosities on the SAC, Porada et al. extensively investigated and summarized the SAC values resulting from various carbon electrodes reported in the CDI literature, including commercially available activated carbon, carbon aerogel, carbon xerogel, carbide-derived carbon, carbon nanotubes, and so on [7, 22, 33, 34]. Based upon the salt adsorption capacity analysis (ζ-analysis) depicted in Ref. [33] combined with the modified Donnan model, Porada et al. concluded that pore sizes smaller than 1.1 nm contribute much more substantially to the SAC than larger pores [35, 36]. Therefore, this conclusion may support that the SAC values in Fig. 1.9 were similar, which is due to no substantive difference in the microporosities associated with each carbon xerogel electrode (See Table 1.1).

Furthermore, the ζ-analysis for the first time demonstrated that the SAC can be predicted when the pore size distribution of the carbon electrode is known. To perform the ζ-analysis, pore size distribution (PSD) must be taken into account to determine contributions for porosity metrics towards the overall SAC. For instance, Fig. 1.10 shows the measured and predicted SAC (a) without and (b) with considering the PSD. It can be seen that without the use of PSD in the ζ-analysis, the measured SAC deviates sharply from the predicted SAC for many carbon materials of interest in this study, but particularly for certain carbide-derived carbons (CDC) and activated carbons (AC). Once the PSD is accounted for, the measured SAC matches quite closely with the predicted SAC, demonstrating the importance of the PSD used in the ζ-analysis for the carbon electrodes used in a CDI device.

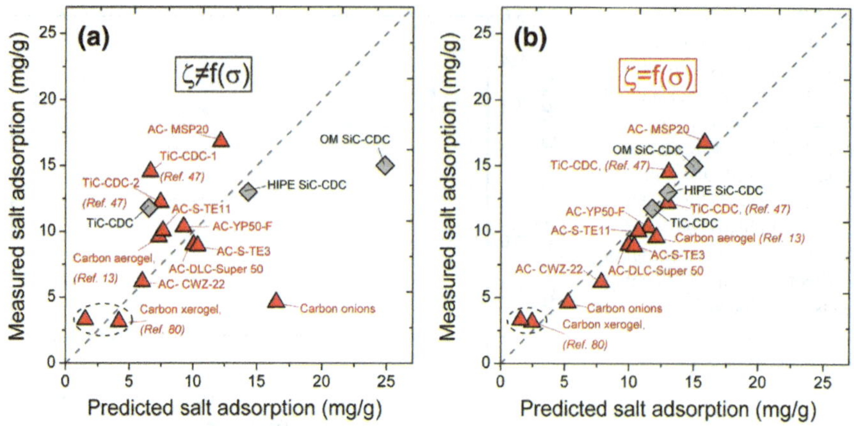

Fig. 1.10 Parity plots for salt adsorption for three carbide-derived carbons (*gray diamonds*) and twelve other materials (*red triangles*) per gram of carbon in both electrodes combined. **a** Salt adsorption capacity ζ assumed independent of pore size σ. **b** Optimized ζ(σ)-function. Reproduced from Ref. [33] with permission from the Royal Society of Chemistry

In addition to the ζ-analysis, work by Han et al. emphasized the importance of pore size on resulting salt separations [37]. In this case, an experimental investigation was conducted to determine the effect of microporosity and mesoporosity on the resulting salt adsorption capacities. Overlap of the double layer is mentioned as reasoning behind limited salt adsorption in microporous space while this overlap is considered negligible for adsorption in mesoporous regions. Ultimately, Han et al. drew similar conclusions to that of Porada et al. in that microporosity is extremely important for the resulting SAC. Furthermore, Han notes that mesoporosity also plays a role in adsorption, especially when considering selectivity of ion adsorption. Zou et al. have also carried out similar comparison experiments in assessing the importance of mesoporosity and microporosity for capacitive deionization [38, 39]. In summary, through these studies, in order to increase the mass of salt adsorbed per gram of carbon xerogel electrode, higher densities of micropores are required, which can be achieved by CO_2 and/or KOH treatments according to the CDI literature or using carbon aerogel materials that were previously discussed [40].

For all of the studies mentioned in this section of the chapter, and indeed for much of the CDI literature as a whole, the importance of carbon porosity has been stressed while the surface chemistry of the carbon electrode has been somewhat neglected. In the next section, we will review the importance of surface chemistry and demonstrate that while porous materials are important for CDI, including salt adsorption rate and salt adsorption capacity, incorrect or shifting surface chemistry can lead to negligible net salt adsorption and separation failure.

1.5 Importance of Surface Chemistry for Porous Carbon Materials in Capacitive Deionization

While the level of microporosity (<2 nm pores) has a profound impact on the salt adsorption capacity and needs to be an included parameter in future design studies for carbon electrodes in capacitive deionization, the effect of surface charge (or degree of basicity or acidity) is undeniable for CDI applications [41]. Studies on carbon surface chemistry have been carried out to demonstrate that the accumulation and presence of fixed chemical surface charge on carbon electrodes can affect the long-term cycle life of carbon electrodes in capacitive deionization cells as well as be used as a design parameter in the creation of more stable separation systems.

Previously in this chapter, the effect of porosity of carbon xerogel materials on the resulting salt adsorption capacity was found to be minimal, with no strong evidence that higher surface areas or smaller pore sizes (in the mesoporous range) could substantially increase the electrode's capability for salt removal. However, the presence of surface charge was not taken into account in these studies. Therefore, in this segment, the use of surface charge to both explain changes in the salt adsorption capacity and be leveraged for higher efficiency cells will be shown.

Long-term CDI cycling experiments were conducted with 50 wt% RF carbon xerogel electrodes to track separation performance over hundreds of hours. Two (2) liters of deaerated 4.3 mM NaCl was circulated at 75 mL min^{-1} through a stack of carbon xerogel electrodes with charge and discharge times of 30 min each. A charging potential of 1.2 V was used while discharging took place at short circuit. Salt adsorption capacity was calculated using Eq. (1) with a total mass of electrodes of approximately 4 g in these studies. The charge efficiency (Λ) for these experiments was calculated using the following equation:

$$\Lambda = (\Gamma F)/(\mathrm{M}Q_{\mathrm{ad}}), \tag{2}$$

where Γ is the salt adsorption capacity, calculated from Eq. (1), F is Faraday's constant, M is the molecular weight, and Q_{ad} is the electronic charge density passed during the adsorption step. Conductivity and current were tracked over the life of the experiments. A recirculating batch-mode operation similar to Fig. 1.7 was used.

The conductivity response of the CDI cell with carbon xerogel electrodes was examined after repeated adsorption/desorption steps. Shown in Fig. 1.11 are selected profiles from this investigation through 233 cycles (equivalent to 233 h of total run time). As can be seen, during the initial charging and discharging steps, the conductivity decreases with an applied potential, resulting in a lower output conductivity of the cell as highlighted by the adsorption arrow in Fig. 1.11a. Likewise, when the cell was short circuited, desorption of salt is seen as highlighted by a rise in conductivity shown by the subsequent desorption arrow. However, by the 128th cycle, desorption of salt is seen at the onset of charging at 1.2 V, and this desorption continues to grow with increased cycling. Once the 227th cycle is reached (Fig. 1.11c), the desorption event at the onset of charging actually reaches higher

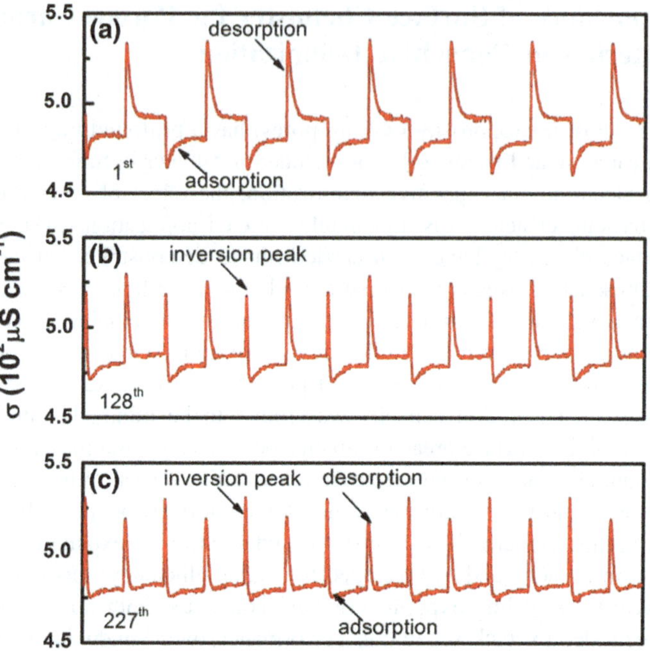

Fig. 1.11 Selected conductivity profiles using 2 L of 4.3 mM deaerated NaCl solution at 75 mL min^{-1}. The CDI cell was operated at 1.2 V for adsorption and at the short-circuit potential for desorption (1.2/0 V) for a total of 248 cycles. Time taken for each adsorption and desorption was 30 min, respectively. This material was first published by Gao et al. in 2014 in the Journal of the Electrochemical Society and is reproduced by permission of The Electrochemical Society [42]

conductivity values than observed during the discharging step. Indeed, the net adsorption of the cell becomes negligible by this time period, which was approximately 10 days in this study.

The salt adsorption capacities and charge efficiencies of this separation were also examined using Eqs. (1) and (2), respectively. Shown in Fig. 1.12 are the resulting performance parameters with similar initial SAC values to those seen in Sect. 1.3. While the salt adsorption capacity is similar to that seen in prior studies at the onset of cycling, this value steadily decreases over the lifetime of cycling until reaching negligible quantities after 227 cycles. The charge efficiency follows a similar path as the salt adsorption capacity decreases and the charged passed during adsorption remains constant.

These changes in the adsorption and desorption profiles as well as the conventional separation performance parameters signaled that either the structure of the carbon xerogel may have been altered or the surface chemistry of the carbon itself may have been affected by the applied potentials. Therefore, to rule out structural changes as a possibility, N_2 adsorption/desorption isotherms were carried out on the pristine carbon xerogel material as well as a used anode and used cathode from the

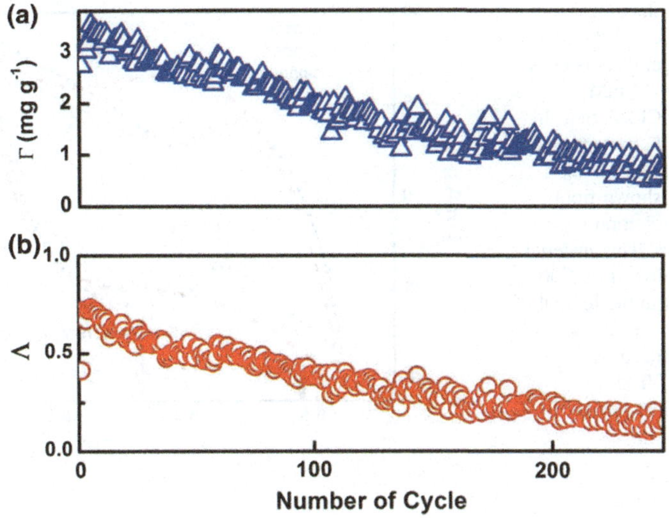

Fig. 1.12 Evaluation of deionization performance according to Fig. 1.11. This plot shows **a** the electrosorption capacity (Γ), which was calculated via Eq. (1), and **b** the charge efficiency (Λ), which was quantified by Eq. (2). This material was first published by Gao et al. in 2014 in the Journal of the Electrochemical Society and is reproduced by permission of The Electrochemical Society [42]

CDI cell. The pristine carbon xerogel had a pore volume of approximately 0.50 mL g^{-1} while the used anode and used cathode contained pore volumes of 0.46 mL g^{-1} and 0.48 mL g^{-1}, respectively. These negligible differences in the pore volumes could not account for the loss of separation seen in Figs. 1.11 and 1.12.

To study changes in the carbon surface chemistry, voltammetric studies were carried out using cyclic voltammograms (CV). These CV measurements can detect small changes in the surface chemistry of electrodes, thereby helping to determine if substantive changes have occurred over the course of the separation study. Recent work by Omosebi et al. has also shown that electrochemical impendence spectroscopy can be used to carry out similar carbon surface characterizations [43, 44]. Shown in Fig. 1.13 are the resulting CVs for the pristine carbon xerogel as well as the used anode and used cathode. All of the curves appear somewhat similar, but one particular parameter has significantly changed over the course of the cycling study, the location of the potential of zero charge or the E_{PZC} [45]. The E_{PZC} is the potential at which a material has a minimum in capacitance or adsorbed ions at the electrode–electrolyte interface. For both the pristine carbon xerogel and the used cathode, similar locations of the E_{PZC} are found. This region is highlighted in Fig. 1.13 as "E_{PZC} cathode". While the E_{PZC} did not shift for the cathode, a marked shift is seen at the anode where this region is now approximately 0.3–0.4 V more positive than the pristine electrode. It is this shift in potential, resulting from changes in the carbon surface chemistry, that results in the decreasing salt

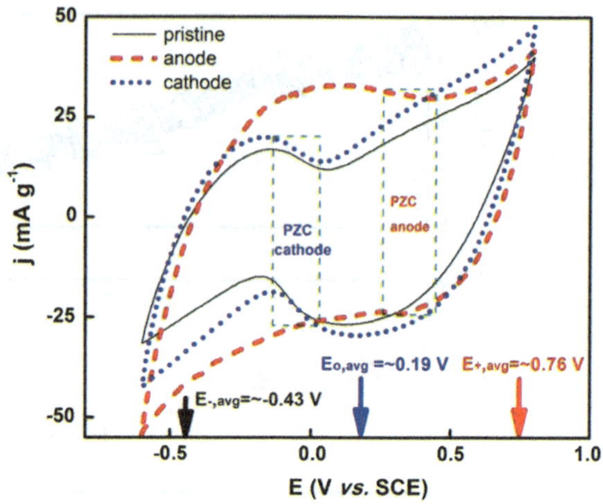

Fig. 1.13 Voltammetric characterization of the used carbon xerogel electrodes at 1 mV s^{-1} in 4.3 mM deaerated NaCl solution. In the plot, the E_+ potential, the E_- potential, and the E_o potential are shown from potential distribution measurements. This material was first published by Gao et al. in 2014 in the Journal of the Electrochemical Society and is reproduced by permission of The Electrochemical Society [42]

adsorption capacity and charge efficiency seen in Figs. 1.11 and 1.12, respectively. More detailed interpretation regarding CDI performance degradation can be found in Refs. [42, 46].

This shifting of the E_{PZC} has been verified to be caused by oxidation of the carbon electrode and the appearance of carboxylic groups that impart a negative charge to the surface as shown by recent work from Cohen et al. [47] and Gao et al. [48]. This oxidation/degradation has been shown to affect not only the carbon xerogel investigated here but also other carbon electrodes, meaning to create stable CDI separations, the issue of oxidation and shifting surface chemistry must also be resolved in addition to the development of highly porous carbon electrodes.

1.6 Creation of Surface-Functionalized Microporous and Mesoporous Materials

In order to combat oxidation/degradation issues that have affected submicron porous materials in prior CDI separations, efforts must be made to create carbon materials with stable surface chemistries under moderate applied electrical potentials of less than 2.0 V. Prior investigations by Cohen et al. and Gao et al. have shown that even voltages as low as 0.8 V can cause oxidation of the carbon electrode surface resulting in the previously demonstrated loss in salt adsorption capacity [42, 47–49]. Therefore, to truly make a stable electrode surface, oxidation must be fully considered in the as-synthesized carbon electrodes. Recent work by Gao et al. did just this in the creation of an inverted capacitive deionization (i-CDI) system that makes use of oxidized/silica-functionalized carbon xerogel anodes (Si-CX) paired with pristine carbon xerogel cathodes (CX). In this case, the anodes have already been oxidized with further oxidation only enhancing the electrode

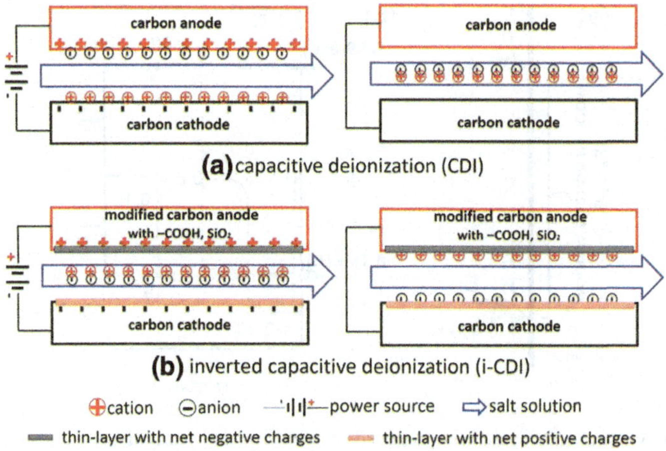

Fig. 1.14 Schematic of **a** capacitive deionization (CDI) and **b** inverted capacitive deionization (i-CDI) systems. Both systems have the same physical structure where carbon cathodes and anodes are separated by a flowing salt solution (it should be noted that the cathode and anode are defined as shown in the diagrams). However, because the net surface charges on the carbon surface can be managed by chemical and/or electrochemical modifications, the i-CDI system shows completely opposite adsorption–desorption behavior compared to the conventional CDI system. For instance, when the i-CDI system is charged using a power source, cations and anions are desorbed at the anode and cathode, respectively, and when the i-CDI system is discharged under a short-circuit condition, cations and anions are adsorbed at the anode and cathode, respectively. Reproduced from Ref. [47] with permission from the Royal Society of Chemistry

surface charge. However, unlike conventional CDI where ions are adsorbed under an applied potential and desorbed at short circuit, reduced potential, or potential reversal, ion adsorption in i-CDI takes place at short circuit while ion desorption occurs with an applied potential. Shown in Fig. 1.14 are schematics of this i-CDI process in comparison to conventional CDI systems.

The use of functionalized electrodes imbues the carbon xerogel electrodes in this i-CDI system with natural surface charge, meaning ion adsorption can occur at short circuit conditions to neutralize the inherent chemical charge on the carbon surface. Using the E_{PZC} parameter described earlier, where an electrode has a minimum in ion adsorption at a defined value, a potential can be applied to this cell to match the locations of the E_{PZC} values at the anode and cathode to cause desorption from the electrode surface. Shown in Fig. 1.15 are the resulting E_{PZC} values for the oxidized/silica-functionalized anode and the pristine carbon xerogel cathode (similar to the E_{PZC} values mentioned previously for carbon xerogel electrodes). These E_{PZC} values define the working voltage window at which adsorption and desorption can then be modulated in this i-CDI system. Additional oxidation of the carbon xerogel anode would only serve to further positively shift the E_{PZC} of the anode, thereby increasing the usable voltage difference between the anode E_{PZC} and cathode E_{PZC}, increasing the overall salt adsorption capacity of the separation cell.

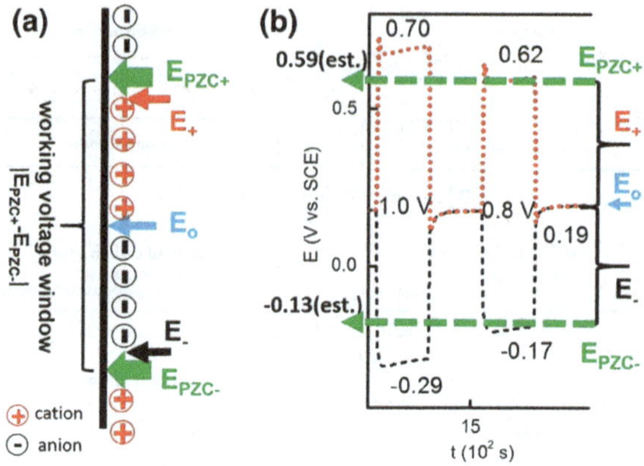

Fig. 1.15 **a** A novel CDI system is proposed, named inverted capacitive deionization (i-CDI). When the E_{PZC}s for both electrodes are dissimilarly located with respect to the short circuit voltage (E_o), the potential window between the E_{PZC}s at the anode and cathode can be used for the desalination process, but the adsorption–desorption behavior will be totally inverted. **b** Potentials distributed at the anode (E_+) and cathode (E_-) at a total cell potential of 0.8 and 1 V in 4.3 mM deaerated NaCl solution when a Si–CX anode and pristine CX cathode were used in a four-electrode cell. Reproduced from Ref. [47] with permission from the Royal Society of Chemistry

CDI was subsequently compared to i-CDI over substantially longer time periods than those shown in Fig. 1.11. Cycling tests were executed in a flow system similar to that shown in Fig. 1.7 with a conductivity sensor, a peristaltic pump, and a polyethylene tank. Experimental automation and timing were facilitated for this test with a relay box and computer control system. In this case, a single-pass CDI separation was carried out as opposed to the prior batch-mode tests performed. Therefore, 31 L of 4.3 mM deaerated NaCl solution was circulated at 75 mL min^{-1} through the separation cells. Each cell, CDI or i-CDI, was assembled with 16 pairs of electrodes separated by 0.15 cm thick silicon rubber spacers. The total mass of the CX electrodes including the carbon cloth substrates was approximately 16 g. Due to the large volume of salt solution used, the steady-state conductivity after the adsorption and desorption profiles is equivalent to the starting value for the bulk solution. Therefore, the salt adsorption capacity (Γ) was quantified by the solution flow rate (ϕ) multiplied by the integration of the conductivity curve with time (t) taken for both the adsorption and desorption steps, Eq. (3).

$$\Gamma = K\phi \int \sigma(\mathrm{t})\mathrm{d}t, K = \mathrm{M/mA}, \tag{3}$$

where M is the molecular weight of the salt, m is the mass of the carbon xerogel electrodes, and A (112.77 mS cm^{-1} mM^{-1}) is the slope calculated from a calibration plot of conductivity versus NaCl concentration. The charge efficiency, Λ, was calculated in the same manner as shown in Eq. (2). Shown in Fig. 1.16 is the resulting salt adsorption capacity for i-CDI and CDI separations over the course of

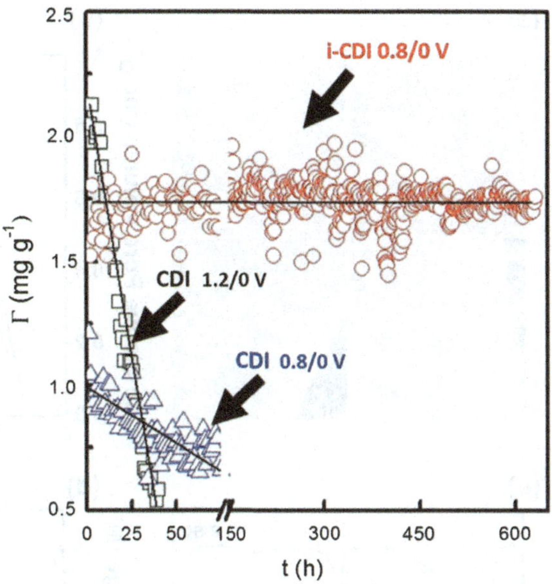

Fig. 1.16 Enhanced stability by the i-CDI system employed with a CX cathode and a Si–CX anode. This test was performed at 0.8/0 V in 31 L of 4.3 mM deaerated NaCl solution. Comparisons in performance to standard CDI operation under similar conditions with pristine CX electrodes are shown. In this plot, lines have been added to guide the eyes. Reproduced from Ref. [47] with permission from the Royal Society of Chemistry

600 h of operation, more than double the previous study. CDI separations were carried out at both 0.8 and 1.2 V charging potentials to demonstrate the effect of applied potential on the loss of salt adsorption capacity as well as to provide a similar basis for comparison to the i-CDI cell that was charged at 0.8 V. While both of the CDI separations see substantial losses in salt adsorption capacities by 50 h of operation, the i-CDI process is remarkably stable even after 600 h. This study demonstrated that by matching the physical properties of porous carbon materials with relevant and stable surface chemistries, significant gains can be made towards emerging efficient salt separations and cycle life time.

Therefore, with the advent of the importance of both microporous carbon materials and the use of targeted surface chemistries towards higher efficiency, higher salt adsorption capacity CDI systems, the combination of these two properties could further extend the impact and practicality of the CDI field. In recent work by Gao et al., modified carbon surfaces have been combined with microporous activated carbon cloth electrodes to demonstrate this concept [50].

Pristine Spectracarb (SC) activated carbon cloth electrodes were successfully modified by nitric acid and ethylenediamine solutions resulting in the chemical surface charge of the electrodes being negatively (N-SC) and positively enhanced (P-SC), respectively. More details concerning the exact surface modification procedures of these materials can be found in Ref. [50]. Changes in the physical characteristics of the SC electrodes were examined both before and after surface treatments, including the surface morphology, surface area, porosity, and resulting surface charge (as determined from chemical titrations). Surface morphologies for the SC electrodes are shown in Fig. 1.17a. The SC electrodes used here contain woven carbon fibers, and the overall structure of the SC material was maintained

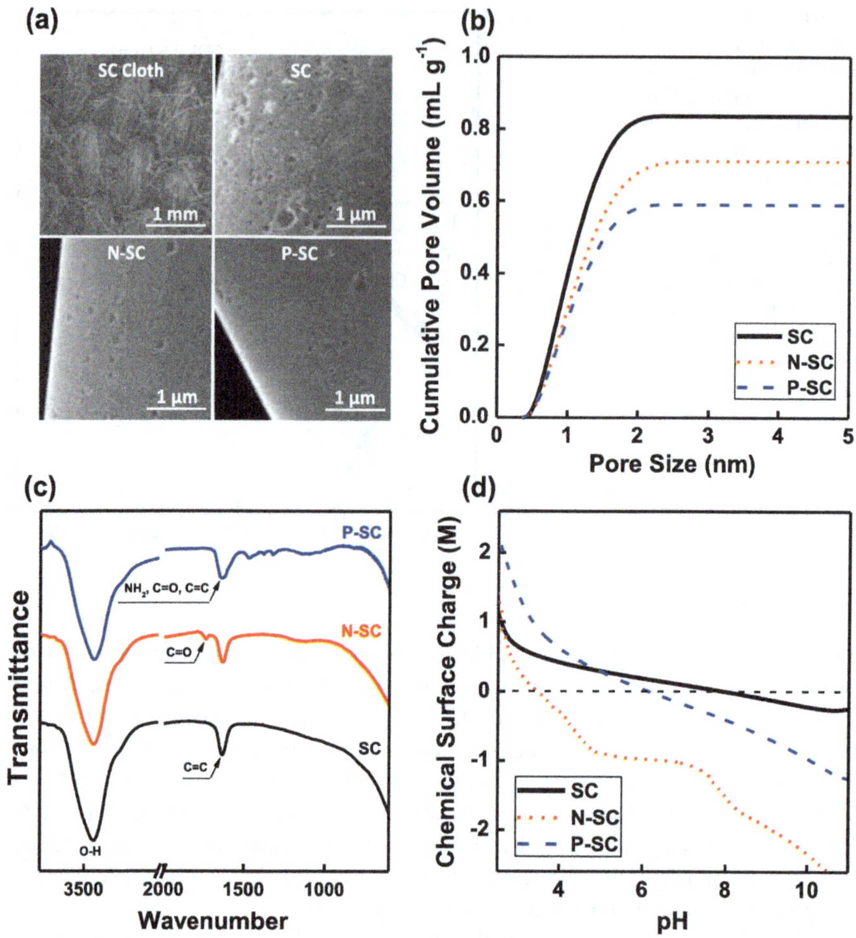

Fig. 1.17 Physical and chemical characterizations of the treated and untreated carbon electrodes. **a** Surface morphologies, **b** cumulative pore volumes, **c** surface chemical species, and **d** chemical surface charges of the SC, N-SC, and P-SC samples. Reproduced from Ref. [50] with permission from Elsevier

throughout the chemical treatments. However, the surface of the as-received SC fiber is somewhat rougher compared to that of the treated samples. Changes in the total pore volume of the treated electrodes were determined via gas adsorption-desorption tests and are shown in Fig. 1.17b. While microporosity is maintained throughout the nitric acid and ethylenediamine treatment process, there are changes in the cumulative pore volumes. Compared to the SC sample, the N-SC and P-SC pore volumes were reduced from around 0.82 mL g^{-1} to approximately 0.72 and 0.60 mL g^{-1}, respectively.

Shown in Fig. 1.17c are FTIR spectra for all three microporous carbon materials. In addition to O–H and C = C stretching vibrations found at 3430 and

1600 cm^{-1}, respectively, for all the samples, new bands are seen after the chemical treatments. Bands found at 1730 and 1150 cm^{-1} in the N-SC spectrum denote C=O and C–O stretching, respectively, indicating carboxylic groups (–COOH) on the N-SC carbon surface. For the P-SC material, a broad band between 1660 and 1580 cm^{-1} presents overlapped signatures for C=O stretching, NH_2 scissoring, and C=C stretching [51]. These characteristics indicate the formation of amine (–NH_2) and/or amide (–CO–NH–) functional groups, which has been documented elsewhere using similar treatments for carbon nanotubes and activated carbons [52–54]. Changes in the chemical surface charge are also observed in the P-SC and N-SC samples where the P-SC material displays a very positive chemical surface charge at lower pH values and the N-SC sample reveals a much more negative chemical surface charge over the tested pH range, especially at higher pH values.

These treated SC electrodes were subsequently employed in a CDI cell to determine their impact on adsorption of NaCl salt. To configure a CDI cell with these treated electrodes, the nitric acid modified SC with negatively enhanced surface charge was used as the cathode (N-SC) while the ethylenediamine modified SC with positively enhanced surface charge was employed as the anode (P-SC). CDI cells formed with the untreated and treated SC electrodes are designated as "cell I" and "cell II", respectively. Cell II was configured with P-SC anodes and N-SC cathodes. To demonstrate both CDI, enhanced CDI (e-CDI) and extended voltage CDI (eV-CDI) effects (see definitions for these two effects in Refs. [50, 55]), cycling experiments were performed using cell I and II where both CDI cells were charged at 1.0 V for 30 min and discharged from −1.0 to 0.6 V for another 30 min. Salt adsorption capacities and charge efficiencies for both untreated microporous carbon cloth electrodes (cell I) and with chemical treatments (cell II) are shown in Fig. 1.18. The use of surface treatments to enhance the chemical surface charge of these microporous carbon electrodes in cell II results in an increase in the working voltage window from 1.0 to 1.4 V and correspondingly increased the maximum SAC value from 9.6 to 17.1 mg g^{-1} when chemical charge is included. Therefore, even though the treated carbon electrodes contained lower overall microporous volumes, the SAC could be significantly enhanced through surface charge modifications and modulation of the discharge voltage. By combining materials with high concentrations of micropores and chemically modifying the surfaces of these electrodes, SAC values can be increased from ~3 mg g^{-1} for unmodified carbon xerogel electrodes described earlier to >15 mg g^{-1} when using surface charge enhanced microporous carbon materials. Additional descriptions concerning the e-CDI and eV-CDI effects as well as the amphoteric modified Donnan (amph-D) model used here to predict and interpret the results can be found in Ref. [50].

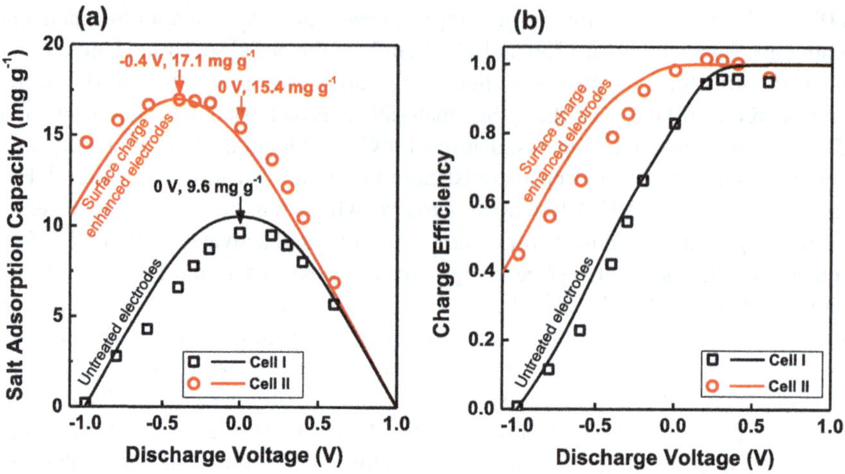

Fig. 1.18 **a** Salt adsorption capacity and **b** charge efficiency as a function of discharge voltage for untreated electrodes (cell I) and treated electrodes (cell II). In these experiments, the CDI cells were charged at 1.0 V for 30 min and discharged at different voltages for another 30 min. In these plots, *symbols* represent the experimental results obtained from cycling experiments, and *lines* denote the theoretical results calculated by the amph-D model, an advancement made on the modified Donnan model described earlier in this chapter [55]. Reproduced from Ref. [50] with permission from Elsevier

1.7 Conclusions

Porous carbon materials are crucial for a variety of applications including liquid and gas-phase separations and energy storage/generation devices. This chapter has focused on the use of submicron porous carbon materials for use as electrodes in capacitive deionization separation cells. Several advancements have been made to CDI electrodes over the last few decades, such as the development of porous, conductive carbon aerogels, theories and experiments demonstrating the importance of microporosity, and more recently the use of surface charge to enhance and modulate adsorption. Therefore, while porosity is clearly a primary design parameter in the creation of a salt separation system, surface chemistry/surface charge will need to be taken into account for future applications. The combination of surface charge and porosity are the pillars on which we must build to create more robust CDI separation systems with increased salt adsorption capacity and rate capable of contending with emerging separation requirements.

References

1. Pekala RW et al (1998) Carbon aerogels for electrochemical applications. J Non-Cryst Solids 225:74–80
2. Long JW et al (1999) Voltammetric characterization of ruthenium oxide-based aerogels and other RuO_2 solids: the nature of capacitance in nanostructured materials. Langmuir 15 (3):780–785
3. Lytle JC et al (2011) The right kind of interior for multifunctional electrode architectures: carbon nanofoam papers with aperiodic submicrometre pore networks interconnected in 3D. Energy Environ Sci 4(5):1913–1925
4. Bryning MB et al (2007) Carbon nanotube aerogels. Adv Mater 19(5):661–664
5. Oren Y (2008) Capacitive deionization (CDI) for desalination and water treatment—past, present and future (a review). Desalination 228(1):10–29
6. Anderson MA, Cudero AL, Palma J (2010) Capacitive deionization as an electrochemical means of saving energy and delivering clean water. Comparison to present desalination practices: will it compete? Electrochim Acta 55(12):3845–3856
7. Porada S et al (2013) Review on the science and technology of water desalination by capacitive deionization. Prog Mater Sci 58(8):1388–1442
8. Suss ME et al (2015) Water desalination via capacitive deionization: what is it and what can we expect from it? Energy Environ Sci 8(8):2296–2319
9. SALINE WATER CONVERSION (1960) Advances in chemistry, vol 27. American Chemical Society, p 264
10. Blair JW, Murphy GW (1960) Electrochemical demineralization of water with porous electrodes of large surface area, in SALINE WATER CONVERSION. American Chemical Society, pp 206–223
11. Pekala RW (1989) Organic aerogels from the polycondensation of resorcinol with formaldehyde. J Mater Sci 24(9):3221–3227
12. Wang J et al (1993) Carbon aerogel composite electrodes. Anal Chem 65(17):2300–2303
13. Fricke J et al (1992) Aerogels derived from multifunctional organic monomers. J Non-Cryst Solids 145:90–98
14. Job N et al (2005) Carbon aerogels, cryogels and xerogels: influence of the drying method on the textural properties of porous carbon materials. Carbon 43(12):2481–2494
15. Farmer JC et al (1996) Capacitive deionization of NaCl and $NaNO_3$ solutions with carbon aerogel electrodes. J Electrochem Soc 143(1):159–169
16. Farmer JC et al (1996) Capacitive deionization of NH_4ClO_4 solutions with carbon aerogel electrodes. J Appl Electrochem 26(10):1007–1018
17. Farmer J, Fix D, Mack G (1995) Capacitive deionization of water: an innovative new process. In: Conference: 5. international conference on radioactive waste management and environmental remediation, Berlin (Germany), 3–9 Sep 1995; Other Information: PBD: 9 Jan 1995. 1995; Lawrence Livermore National Laboratory, CA (United States). Medium: ED; Size: 15 p
18. Farmer JC et al (1997) Electrosorption of chromium ions on carbon aerogel electrodes as a means of remediating ground water. Energy Fuels 11(2):337–347
19. Xu P et al (2008) Treatment of brackish produced water using carbon aerogel-based capacitive deionization technology. Water Res 42(10–11):2605–2617
20. Kim T, Yoon J (2015) CDI ragone plot as a functional tool to evaluate desalination performance in capacitive deionization. RSC Adv 5(2):1456–1461
21. Zhao R et al (2013) Optimization of salt adsorption rate in membrane capacitive deionization. Water Res 47(5):1941–1952
22. Porada S et al (2012) Water desalination using capacitive deionization with microporous carbon electrodes. ACS Appl Mater Interf 4(3):1194–1199
23. Li H et al (2010) Novel graphene-like electrodes for capacitive deionization. Environ Sci Technol 44(22):8692–8697

24. Tsouris C et al (2011) Mesoporous carbon for capacitive deionization of saline water. Environ Sci Technol 45(23):10243–10249
25. Biesheuvel PM, van Limpt B, van der Wal A (2009) Dynamic adsorption/desorption process model for capacitive deionization. J Phys Chem C 113(14):5636–5640
26. Biesheuvel PM et al (2011) Theory of membrane capacitive deionization including the effect of the electrode pore space. J Colloid Interf Sci 360(1):239–248
27. Jeon S-I et al (2013) Desalination via a new membrane capacitive deionization process utilizing flow-electrodes. Energy Environ Sci 6(5):1471–1475
28. Wimalasiri Y, Zou L (2013) Carbon nanotube/graphene composite for enhanced capacitive deionization performance. Carbon 59:464–471
29. Yang J et al (2011) Development of novel MnO2/nanoporous carbon composite electrodes in capacitive deionization technology. Desalination 276(1–3):199–206
30. Walker PJ, Mauter MS, Whitacre JF (2015) Electrodeposited MnO_2 for pseudocapacitive deionization: relating deposition condition and electrode structure to performance. Electrochim Acta 182:1008–1018
31. Suss ME et al (2012) Capacitive desalination with flow-through electrodes. Energy Environ Sci 5(11):9511–9519
32. Landon J et al (2012) Impact of pore size characteristics on the electrosorption capacity of carbon xerogel electrodes for capacitive deionization. J Electrochem Soc 159(11):A1861–A1866
33. Porada S et al (2013) Direct prediction of the desalination performance of porous carbon electrodes for capacitive deionization. Energy Environ Sci 6(12):3700–3712
34. Porada S et al (2012) Effect of electrode thickness variation on operation of capacitive deionization. Electrochim Acta 75:148–156
35. Biesheuvel PM et al (2014) Attractive forces in microporous carbon electrodes for capacitive deionization. J Solid State Electrochem 18(5):1365–1376
36. Biesheuvel PM, Fu Y, Bazant MZ (2012) Electrochemistry and capacitive charging of porous electrodes in asymmetric multicomponent electrolytes. Russ J Electrochem 48(6):580–592
37. Han L et al (2014) Exploring the impact of pore size distribution on the performance of carbon electrodes for capacitive deionization. J Colloid Interf Sci 430:93–99
38. Li L et al (2009) Ordered mesoporous carbons synthesized by a modified sol–gel process for electrosorptive removal of sodium chloride. Carbon 47(3):775–781
39. Zou L et al (2008) Using mesoporous carbon electrodes for brackish water desalination. Water Res 42(8–9):2340–2348
40. Baumann TF et al (2008) High surface area carbon aerogel monoliths with hierarchical porosity. J Non-Cryst Solids 354(29):3513–3515
41. Arnold BB, Murphy GW (1961) Studies on the electrochemistry of carbon and chemically-modified carbon surfaces. J Phys Chem 65(1):135–138
42. Gao X et al Dependence of the capacitive deionization performance on potential of zero charge shifting of carbon xerogel
43. Electrodes during long-term operation (2014) J Electrochem Soc 161(12):E159–E166
44. Omosebi A et al (2014) Asymmetric electrode configuration for enhanced membrane capacitive deionization. ACS Appl Mater Interf 6(15):12640–12649
45. Omosebi A et al (2015) Continuous operation of membrane capacitive deionization cells assembled with dissimilar potential of zero charge electrode pairs. J Colloid Interf Sci 446:345–351
46. Bard AJ, Faulkner LR (2000) Electrochemical methods: fundamentals and applications. Wiley
47. Cohen I et al (2015) The effect of the flow-regime, reversal of polarization, and oxygen on the long term stability in capacitive de-ionization processes. Electrochim Acta 153:106–114
48. Gao X et al (2015) Surface charge enhanced carbon electrodes for stable and efficient capacitive deionization using inverted adsorption-desorption behavior. Energy Environ Sci 8 (3):897–909

49. Cohen I et al (2013) Long term stability of capacitive de-ionization processes for water desalination: the challenge of positive electrodes corrosion. Electrochim Acta 106:91–100
50. Cohen I et al (2013) Water desalination by capacitive deionization—advantages limitations and modification. ECS Trans 45(17):43–59
51. Gao X et al (2016) Complementary surface charge for enhanced capacitive deionization. Water Res 92:275–282
52. Gao X et al (2015) Enhanced salt removal in an inverted capacitive deionization cell using amine modified microporous carbon cathodes. Environ Sci Technol 49(18):10920–10926
53. Longhi M et al (2006) Preparation and characterization of aminated carbon from a single-step reaction. Chem Mater 18(17):4130–4136
54. Bifeng P et al (2006) Growth of multi-amine terminated poly(amidoamine) dendrimers on the surface of carbon nanotubes. Nanotechnology 17(10):2483
55. Vuković GD et al (2010) Removal of cadmium from aqueous solutions by oxidized and ethylenediamine-functionalized multi-walled carbon nanotubes. Chem Eng J 157(1):238–248

Chapter 2
Carbon Gels and Their Applications: A Review of Patents

N. Rey-Raap, A. Arenillas and J.A. Menéndez

2.1 Introduction

Scientific and technological advances over the past four decades have led to an increase in the demand for new technologies as well as improvements in the existing ones. In many cases these improvements involve the development of new materials. In particular, carbon has attached a great deal of interest, since natural carbon materials have extraordinary properties that confer on them a great potential for different applications. However, its impurities and a lack of control over its properties have led to its replacement by synthetic carbonaceous materials. These synthetic materials share the same structure as materials obtained from natural sources of carbon, with the added advantage that their chemical and porous properties can be adapted to the requirements of a particular application. Synthetic carbonaceous materials include carbon gels, porous materials composed of primary particles that are interconnected to create a three-dimensional network structure.

The formation of the polymeric structure depends to a great extent on the conditions established in the synthesis process, which in turn define the properties of the carbon gels. Therefore, the main advantage of carbon gels is that, by appropriately selecting the process conditions, the properties can be controlled and designed to achieve the optimum behaviour of carbon gels for a wide variety of

N. Rey-Raap
Department of Chemical Engineering, University of Porto,
Rua Dr. Roberto Frias s/n, 4200-465 Porto, Portugal
e-mail: nataliarey@fe.up.pt

A. Arenillas · J.A. Menéndez (✉)
Microwaves and Carbon for Technological Applications
(MCAT-INCAR-CSIC), Apartado 73, 33080 Oviedo, Spain
e-mail: angelmd@incar.csic.es

A. Arenillas
e-mail: aapuente@incar.csic.es

P. Bettotti (ed.), *Submicron Porous Materials*,
DOI 10.1007/978-3-319-53035-2_2

scientific fields and applications [1, 2]. Attracted by the great versatility of carbon gels, researchers around the world have focused on this type of material. However, despite its great commercial potential, the number of patents in relation to the amount of research being carried out and published is not very high (1 patent per nearly 100 research articles). These data are very revealing and must prompt the reader to ask himself the following questions concerning the low number of patents reported: is the synthesis process of carbon gels mature enough to be transferred to the industry? Have carbon gels really a wide sales market and can they be offered at competitive prices?

In this chapter, an attempt to write a review of patents on the synthesis and applications of carbon gels has been made in order to bring some enlightenment to these questions.

2.2 Carbon Gels

The first patented carbon gel was reported in 1989. Inspired by the synthesis of inorganic silica gels, Pekala and co-workers patented the methodology sol–gel for synthesizing organic gels, which are the precursors of carbon gels [3]. The methodology consists of provoking a polymerization reaction in which a liquid gradually condenses from a stable suspension of colloidal solid particles (sol) to create a three-dimensional porous network formed by polymeric chains in a liquid medium (gel), as shown in Fig. 2.1.

In this first patent, a polymerization reaction occurs between resorcinol and formaldehyde under alkaline conditions, i.e. in a reaction media composed of water and sodium hydroxide [3]. This reaction involves two main stages: (i) an addition reaction and (ii) a condensation reaction. Resorcinol is a trifunctional benzyl compound with two hydroxyl groups in positions 1 and 3 which allows

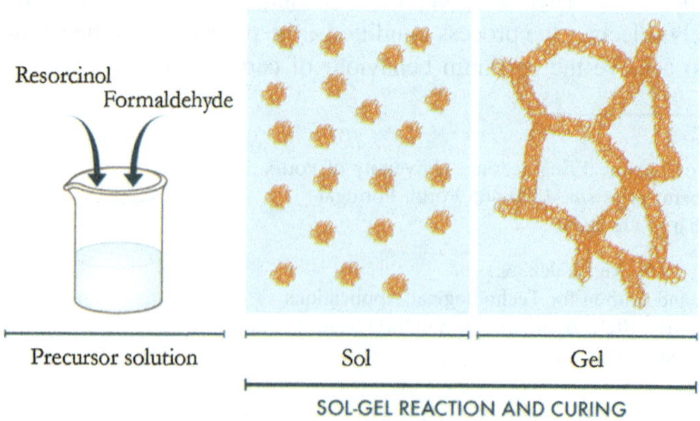

Fig. 2.1 Synthesis of organic gel by the sol–gel methodology

formaldehyde to be added in positions 2, 4 and 6 [3, 4]. In the presence of a basic catalyst, resorcinol anions are formed due to the abstraction of hydroxylic hydrogens. These anions are more reactive than resorcinol itself, which favours the addition of formaldehyde to form hydroxymethyl derivatives, as shown in Fig. 2.2.

At the same time as the addition reaction, condensation occurs, as shown in Fig. 2.3. The hydroxymethyl derivatives lose OH^{-} groups to form benzyl-type cations. These cations react with other benzene rings, giving rise to methylene and ether bonds [5]. Polymeric particles are then formed. These primary particles aggregate and cross-link with each other via polymer chains to form a three-dimensional porous network in a liquid medium.

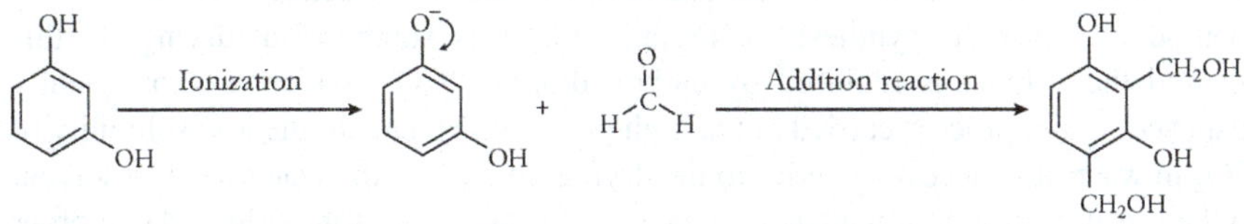

Fig. 2.2 Scheme of the addition reaction between resorcinol and formaldehyde

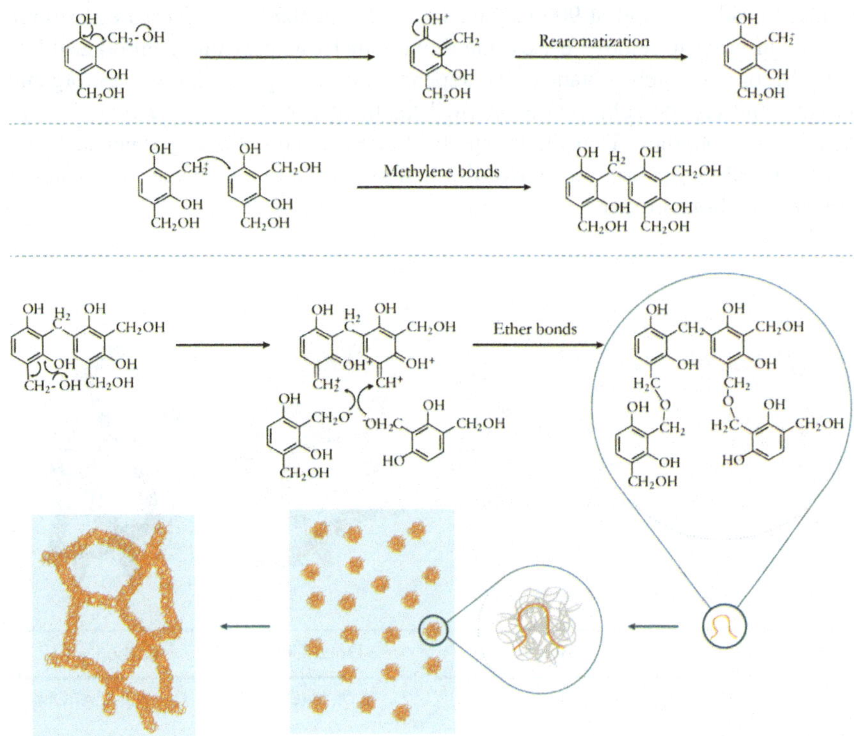

Fig. 2.3 Scheme of the condensation reaction between resorcinol and formaldehyde

The formation of primary particles causes the fluidity of the precursor solution to decrease. The time between the start of the reaction until the moment at which the solution loses fluidity is known as gelation time, while the exact moment at which the gel is obtained is called gelation point. After the gel point has been reached, the solid immersed in the liquid structure continues to evolve. This stage is called curing and during this stage the material hardens due to the simultaneous effect of cross-linkages between the polymer chains and to the partial evaporation of the solvent, which is mainly water. Once the sol–gel reaction and curing stages are completed, optional acid-washed steps can be performed in order to increase the degree of cross-linkages [3].

Organic gels obtained after the sol–gel reaction and curing stage must be dried to completely remove the solvent still embedded inside the structure. The first drying method used for the synthesis of organic gels was supercritical drying. In this method the solvent is replaced by carbon dioxide (CO_2) which is subsequently disposed of in a process carried out at high pressure [3]. Due to the low solubility of CO_2 in water it is necessary, prior to the drying, to replace the water used as solvent in the precursor solution with an organic solvent such as acetone (Fig. 2.4). In order to completely remove the residual water, multiple exchanges with fresh acetone are needed. After the solvent has been exchanged, the gel is dried by supercritical drying. First, the acetone is replaced with liquefied carbon dioxide and then the gel enclosed inside a vessel at 900 psi and 14 °C. Sequentially, the pressure is slowly reduced in order to remove the carbon dioxide that was previously introduced [6].

Dried organic gels obtained by supercritical drying are known as organic aerogels and can be subjected to thermal treatments to stabilize the material and modify its properties. Depending on the treatment conditions, oxygenated and hydrogenated groups can be removed, which gives rise to a carbonaceous material known as carbon aerogel.

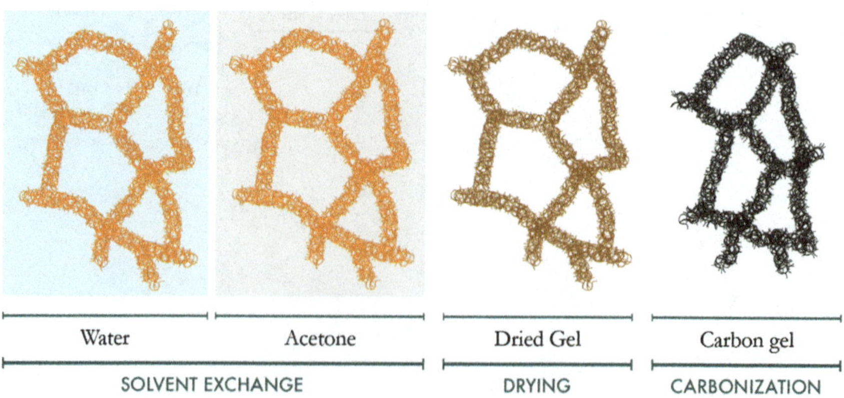

Fig. 2.4 Solvent exchange and drying stage in the synthesis of organic gel

Carbon aerogels obtained by this method are unique and have physicochemical properties, such as a high specific surface area and pore volume, a low density, an excellent electrical conductivity and a high purity. These properties can be controlled and designed to fit the requirements of a specific application. Furthermore, carbon gels can be synthesized in the form of monoliths, powders, microspheres or thin films, with high packing densities [1]. These characteristics go some way to explain why carbon gels have acquired a high commercial value and why the interest to synthesize them at industrial scale has increased over the last decades.

However, as explained above, the most common process for synthesizing carbon aerogels involves several steps: sol–gel reaction, curing, acid-wash, solvent exchange, supercritical drying and carbonization. Some of these stages are rather expensive and difficult to handle, which makes the process unfeasible at an industrial scale. Consequently, over the last 25 years great efforts have been made to overcome the main drawbacks of this process with the two-fold aim of: (i) achieving a feasible industrial synthesis process and (ii) improving the properties of carbon gels in order to increase the number of applications for which they can be used and increase their marketability.

2.3 Improvements in the Synthesis Process of Carbon Gels

The main stages involved in the most common method for synthesizing carbon gels are both costly and time-consuming. On the one hand, the polymerization reaction between resorcinol and formaldehyde is spontaneous but slow, and to promote it is essential to heat the precursor solution up. The most widely used process involves heating in electric furnaces to a temperature not exceeding 100 °C to prevent the precursor solution from boiling. This method requires several days for the sol–gel reaction and curing stage. The use of electric furnaces is very costly due to high level of energy consumption and long periods of time required to produce the organic gel. On the other hand, supercritical drying not only requires extreme drying conditions entailing a substantial increase in production costs, but usually several solvent exchange steps are also needed. These factors are some of the main obstacles to the implantation of this synthesis process at industrial scale. Consequently, research in this field is being carried out to develop easier, faster and cheaper methods of synthesizing carbon gels in order to make them more attractive and competitive than other commercial carbonaceous materials.

2.3.1 Drying Methods

The structure of carbon gel is reflected in the structure of dried gel which, in turn, is determined by the organic wet gel. These features can be created during the polymerization reaction but, if care is not taken in the removal of the solvent, the structure cannot be preserved. During drying, surface tensions occur in the liquid–vapour interface generated onto the surface of the pores [7]. As a result, capillary forces appear. If the mechanical strength of the polymer structure is not high enough to counteract these capillary forces, the structure could completely collapse, giving rise to non-porous materials [4]. From the equation of Young–Laplace shown in Fig. 2.5 and given that pores are spherical, it can be inferred that the degree of collapse will depend on (i) the specific surface energy, (ii) contact angle between the porous structure and the solvent and (iii) pore size. Accordingly, the drying must be conducted under conditions designed to avoid the collapse of the polymeric organic wet structure and to enable the dried gel to retain its original surface structure and porosity [8]. It has already been explained that this can be achieved by supercritical drying, which give rise to organic aerogels. However, the high level of energy, time and cost consumption involved have led researchers to seek other drying methods [9].

Freezing-drying has been widely patented as a drying method for organic gels that can avoid the collapse of the polymeric structure. Materials obtained by this method are known as organic cryogels. The concept of the freezing-drying method is to promote nucleation rather than crystal growth [8]. By freezing the wet gel a fine ice crystal structure within the primary particles is formed due to the widespread nucleation of ice crystals. As the solvent is cooled, the solvent undergoes crystallization until the eutectic point is reached. At this point, it undergoes the transition from a liquid to a solid until complete freezing is achieved [8]. After the material has been frozen, the solvent is removed by sublimation under low pressure for 3 days [7, 9, 10]. Sublimation is performed in a lyophilizer pre-cooled at a temperature that ranges from −10 to −50 °C, depending on the type of solvent employed [8, 9, 11]. Liquid nitrogen or ethanol cooled in dry ice are usually used to freeze the solid [8].

As inferred from several patents, freezing-drying has an advantage over supercritical drying as solvent exchange can be avoided if the proper solvent is used [7]. Instead, wet gels can be filtered or decanted in order to remove any excess dissolvent before freezing [11]. These simplifications have made it possible to commercialize carbon cryogels. An example of this is Energy2 Technologies Inc.,

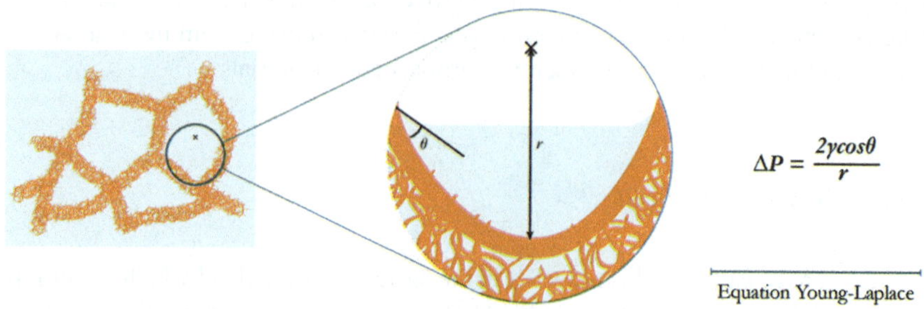

Fig. 2.5 Scheme of capillary forces on porous materials

which produces uniquely tailored carbon cryogels at costs that are competitive with the traditional carbons used in high performance applications.

Nevertheless, although supercritical and freezing drying are efficient methods for the fabrication of organic porous materials that remain porous after pyrolysis, both the methods are rather tedious, expensive and difficult to handle [12]. In order to further simplify and reduce the cost of these drying processes, new methods of producing carbon gel that do not require the step of supercritical or freezing drying have been reported. For example, wet gels can be dried in a vacuum by heating at a temperature between 50 and 100 °C for about from 2 to 20 h [7]. Centrifugation followed by evaporative drying can be also used as a drying method. Obviously, the structure of the gel must be able to withstand the centrifugation process. The pressure differential method can also be applied in order to force the liquid out of the structure by displacing it with gas, a vacuum [13, 14] or another liquid solvent [7]. Both centrifugation and the pressure differential method remove most of the liquid but not all and hence, post-evaporative drying methods are needed to completely remove the solvent [7].

These latter methods have led several researchers to look for a drying method based exclusively on the direct evaporation of the solvent. The first subcritical drying method, in which a simple air drying procedure was applied, also involved several solvent exchanges with different solvents such acetone, cyclohexane [13] or n-pentane [15] prior to the drying step. Subsequently, evaporative drying without a solvent exchange step was patented. First, drying was performed over a period of 40 days in order to avoid shrinkage [16]. However, it was observed with time that evaporative drying was also feasible in just 3 days [4, 17], without triggering the collapse of the structure. The materials obtained by this drying method are known as organic xerogels and can be obtained by a simpler, quicker and less expensive process making it more likely to be adopted at industrial scale.

It should be pointed out, however, that the synthesis of organic xerogels undergoes substantial shrinkage and even cracking of the structure upon drying at ambient conditions [18]. As indicated in Fig. 2.5, the degree of shrinkage depends on the specific surface energy, contact angle between the porous structure and the solvent and pore size. In the particular case of xerogels, the two first terms remain constant. Hence, shrinkage depends on the shape of the polymeric structure. Materials with a lack of sufficient mechanical strength to withstand the stresses arising during drying will tend to shrink and crack [4]. However, this drawback can be minimized by selecting the appropriate conditions for the previous sol–gel reaction and curing stages in order to obtain materials whose mechanical strength counteracts the capillary forces and avoids or reduces the risk of collapse.

2.3.2 Heating Methods

Typically, the precursor solution is sealed in a heat-resistant container, made of polypropylene or glass, which is placed in a thermostat-controlled oven, where the sol–gel reaction and curing stages are performed. The recent advances in drying processes now allow the drying to be carried out in the same electric oven. To this end, is necessary that the container enables the removal of the solvent [4]. However, the use of electric furnaces in all three main stages of the synthesis process is still very costly due to the high level of energy consumption and long times required to produce the organic xerogels (from 4 to 7 days). In order to obtain more competitive materials, some patents employ faster and cheaper heating sources.

Electrochemical processes can be used as methods to initiate the polymerization reaction [18]. Once the reagents involved in the sol–gel reaction have been selected and mixed in the appropriate proportions, two electrodes (cathode and anode) are immersed in the solution. A DC electric current is then supplied via the electrodes at room temperature for a period of 1–3 days. The electric current is in the range of 0–8 A with a voltage ranging from 0.01 to 60 V [18]. The current applied is essential for tailoring the final properties of the carbon gels. For example, an increase in the voltage results in carbon gels with a lower surface area [18].

Although to the best of our knowledge, there are no patents on the use of ultrasonic radiation as a heating method for the synthesis of carbon gels, this particular method has been reported in scientific journals [12, 19, 20]. This in itself makes it worthy of mention in this chapter. Ultrasonic radiation was first applied to perform drying. The use of this heating source avoids the solvent exchange stages and the need for high operating pressures. However, it is a long drying procedure which takes about 34 h. Years later, the polymerization reaction was also accelerated by ultrasonic radiation. This strategy was found to be very helpful for increasing reaction rates, products yields and for shortening the reaction time required.

Microwave radiation has been reported to be a very efficient heating source for synthesizing carbon xerogels [12]. At first, microwave heating was used only in the drying stages but it was soon realized that this technology was applicable to the whole process [21]. The general idea of this method is to apply a controlled microwave radiation to the precursor solution in order to promote polymerization, curing and drying. Carbon xerogels with similar characteristics to conventionally synthesized xerogels are obtained, but with a considerable saving of time, i.e. up to 90% (3–4 h under microwave radiation compared to several days by conventional heating) [21, 22]. In addition, there are several other advantages to be had by using microwave technology to fabricate carbon xerogels. In microwave heating it is possible to determine the different stages of the process: sol–gel reaction, curing and drying. This make it possible to allows control the viscosity of the reaction media which, in some cases, is essential to obtain organic xerogel in a specific form (i.e., sphere, monolith, film, etc.) [23].

Nowadays, microwave heating is presented as the most effective and economical method for synthesizing organic xerogels at an industrial scale. It involves a simple and fast method from which very competitive and high added value materials can be obtained. A good example is afforded by Xerolutions Ltd. which, since 2015, has been commercializing carbon xerogels synthesized by microwave radiation.

2.4 Improvements in the Properties of Carbon Gels

The main advantage of carbon gels is that is possible to tailor their properties to meet the requirements of the final application. The properties can be designed by selecting the appropriate synthesis conditions. This includes, but is not limited to: (i) the nature and concentration of the main reagents, (ii) the introduction of doping agents or additives at different stages of the synthesis process and (iii) the application of thermal treatments.

2.4.1 Nature and Concentration of the Main Reagents

The synthesis of carbon gels begins with the preparation of an aqueous solution containing hydroxylated benzenes, aldehydes, solvents and, when necessary, a fourth compound, known as the catalyst, which regulates the pH and catalyses the reaction. The nature and concentration of these reagents play a key role in tailoring the properties of organic gels.

As explained above, the first and also the most commonly used reagents for the synthesis of carbon gels are resorcinol and formaldehyde. Their concentration has a direct effect on the final porous properties of carbon gels. Formaldehyde is responsible for the formation of crosslinkages between primary particles. An increase in the concentration of formaldehyde leads to highly branched structures of high mechanical strength. These structures are able to counteract the surface tension generated during drying, so that shrinkage is minimized and the carbon gels obtained have higher pore volumes [2, 24].

The polymerization reaction between resorcinol and formaldehyde is a spontaneous but slow reaction. Therefore, a catalyst is often added to the precursor solution to initiate the sol–gel reaction and modify the final formation of the polymeric structure. Catalysts can be basic such alkali carbonates, acetates or hydroxides [3, 9, 10, 15, 25] or acidic, which includes, among others, hydrochloric acid, acetic acid, hydrobromic acid, uric acid, phosphoric acid or sulfuric acid [7, 11, 26, 27]. In general, regardless of its nature, increasing the amount of catalyst substantially reduces the time required for the polymerization reaction run its course [7]. There is, however, a maximum allowable amount of catalyst that can be added to each particular precursor solution, since an excessive reduction in the reaction time could give rise to non-porous materials.

The concentration of the catalyst is related to the pH of the precursor solution, one of the most important variables in the synthesis of carbon gels, as it has a strong influence on their final porous properties [4, 8]. Its effect can be explained in relation to the polymerization reaction between resorcinol and formaldehyde. The reaction starts with the abstraction of the hydroxyl hydrogen from the resorcinol, as previously shown in Fig. 2.2. The addition of a basic catalyst promotes the abstraction of hydroxyl hydrogens and in turn, the formation of resorcinol anions.

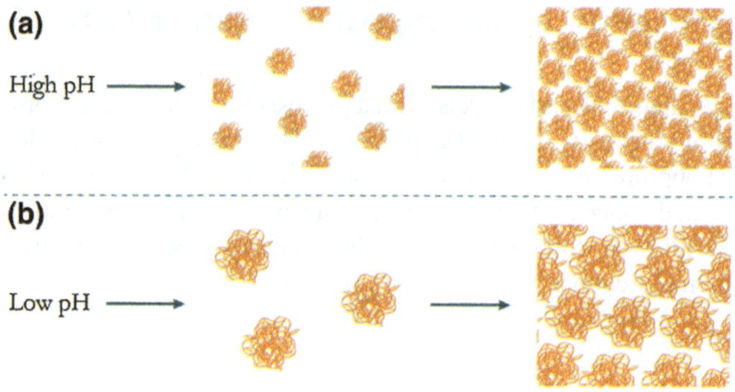

Fig. 2.6 Formation and growth of primary particles depending on the pH of the precursor solution

Therefore, increasing the concentration of catalyst (i.e., raising the pH value) causes an increase in the number of resorcinol anions to which the formaldehyde can be added. As the number of resorcinol anions increases, the addition reaction proceeds faster, giving rise to the formation of a large number of small primary particles [24]. Polymeric structures with a large amount of small primary particles lead to organic gels with narrow pores (Fig. 2.6a). Conversely, a low catalyst concentration (low pH value) generates a small number of resorcinol anions. The addition of formaldehyde takes less time than when adding more catalyst. As a result, a condensation reaction is promoted and the medium persists longer in the nucleation stage, favouring the formation of structures with a small number of large primary particles. These polymeric structures lead to organic gels with pores of larger size (Fig. 2.6b).

In contrast to the base-catalyzed route, which is based on activating the aromatic ring (Fig. 2.2), the acid-catalyzed route is based on accelerating the reaction by increasing the electrophilicity of formaldehyde [28]. The protonation of formaldehyde is followed by a nucleophilic attack on the resorcinol, leading to its hydroxymethylation. The hydroxymethyl groups formed are then protonated and react with other benzene rings giving rise to methylene and ether bonds. The acid-catalyzed process yields organic gels with a similar chemical composition to that obtained using basic catalysts. However, variations in morphology have been observed due to differences in the gelation mechanisms. With acid catalysts, the formation of the hydroxymethyl derivative occurs more slowly and, as a result, structures with a larger pore size are obtained [28, 29].

The reaction medium is also an important factor in the synthesis of carbon gels, particularly the specific surface energy, which is one of the factors governing the capillary forces, as previously explained (Fig. 2.5). Several solvents are available including deionized water, acetone, cyclohexane, methanol or ethanol. Each of these compounds has a different surface tension and therefore, the behaviour during the curing and drying stages will result in different pore structures of the final materials. Water exhibits the highest surface tension value (72.8 mN/m vs. 23.7,

25.5 and 22.6 mN/m for acetone, cyclohexane and methanol, respectively), causing a further collapse of the polymeric structure when the solvent is removed. Some patent authors have demonstrated that the degree of collapse when using water can be minimized by adding other solvents or mixtures of them to assist the polymerization reaction or enhance certain useful properties of the carbon gels. For example, small concentrations of alcohols such as ethylene glycol, propylene glycol or glycerine (between 3 and 13% by weight of the total solution [7]) can be added to reduce the average pore size and to increase the strength of the polymeric structure. However, adding alcohol or increasing alcohol content tends to increase the gelation time. In order to minimize this effect, the concentration of the other reagents (hydroxylated benzene, aldehyde and catalyst) can be modified to overcome the effect of the alcohol.

Despite its high surface tension value, water is the least expensive reaction medium and therefore, the most preferred for synthesizing carbon gels. The water concentration exercises a significant influence on the meso–macroporosity of organic gels, due to the number and size of the primary particles formed during the polymerization reaction [5, 30]. For the same volume of precursor solution, a decrease in the concentration of water entails an increase in the concentration of resorcinol and formaldehyde and the higher the concentration of resorcinol, the greater the number of anions generated. In this way a large number of small primary particles are formed, leading to materials with smaller pores (mesoporous materials). As the concentration of water increases, i.e. the resorcinol-formaldehyde content decreases, fewer primary particles of larger size are obtained, giving rise to materials with larger pores (macroporous materials). However, if the water concentration is too high, the number of primary particles will be so small that gelation will not occur [5].

Although resorcinol and formaldehyde are the most common reagents, other potential combinations have been also patented, including hydroxylated benzenes from the group comprising phenol, catechol, hydroquinone and phloroglucinol or a mixture of them [3, 7, 8]. Partially pre-polymerized forms or aromatic compounds such as phenolic resin [27] or melamine [26], respectively, can be also used. Alternatively, formaldehyde can been replaced with furfural or other saturated aldehydes [7, 8, 27].

The nature and concentration of all reagents plays a fundamental role in the synthesis of carbon gels and, particularly, in designing their final properties. This gives an idea of the complexity and at the same time the versatility of the process, since each variable can modify the effect of the others. Therefore, the choice of a suitable combination of reagents is essential for the sol–gel reaction to take place properly and for controlling the properties. Some patented examples that illustrate the importance of preparing an appropriate precursor solution are described in detailed below.

2.4.1.1 Melamine Formaldehyde Aerogels

As explained above, the synthesis process begins with heating the precursor solution that contains resorcinol (R), formaldehyde (F), water and a basic catalyst. As the reaction occurs, the precursor solution turns a different colour. As a result, dried RF organic aerogels become dark red in colour and have a large absorption coefficient within the visible spectrum. The colour of the RF aerogels is a result of oxidation products formed during the polymerization, such as quinones [26]. This fact limits the use of the RF aerogels for certain optical applications where the material needs to transmit light and be non-absorptive in the visible spectrum. In order to address this problem, other reagents, which give rise to colourless organic gels, can be used. An example of this is found on the patent literature, in which resorcinol is replaced by melamine [26].

The removal of one of the main reagents to obtain a transparent material requires modifications of certain parameters governing the process, such as time and temperature of synthesis, type of catalyst and concentration of the reagents. Thus, the process for preparing melamine formaldehyde (MF) aerogels differs from that used to prepare RF aerogels. Melamine (M) and formaldehyde (F) are first dissolved in an aqueous solution containing the basic catalyst. Unlike RF aerogels for which the R/F molar ratio is usually fixed at 0.5, MF aerogels are synthesized using a M/F molar ratio of 0.27 [26]. As mentioned above, an excess of formaldehyde favours cross-linkages between polymeric chains and a more reinforced structure is obtained. Because of the limited water solubility of crystalline melamine, the precursor solution should be heated at 70 °C for 15 min until it becomes clear. The solution is then cooled to room temperature and the final pH value is adjusted by adding an acidic catalyst. In order to obtain a transparent material, the pH value must be within the range of 1–2. In contrast to the conventional synthesis of RF aerogels, acidic conditions are necessary for MF aerogels to promote the condensation reaction which leads to the formation of gel. Furthermore, the pH is the most important variable for controlling the transparency of the material. The temperature and duration of the synthesis of MF aerogels depend on the percentage of solids in the precursor solutions. Solutions with a high percentage of solids ($\geq 20\%$wt.) become gel in 48 h at room temperature, while solutions with a low percentage ($\leq 7\%$wt.) need to be heated up to 90 °C for 4 weeks [26]. Then the samples are cured at 90 °C for 5 days.

Another advantage of this process is that the acid-wash stages can be removed, since it is the excess of formaldehyde that provides additional crosslinkages. Finally, the wet MF aerogels are directly subjected to solvent exchange, drying and carbonization processes. The materials thus obtained exhibit similar properties to those of RF aerogels, but in addition they are transparent and, hence, useful in optical applications.

2.4.1.2 Phenolic-Furfural Aerogels

The growing interest in using aerogels in a wide variety of applications, either in organic or carbon form, has led to further developments with these materials with the focus on reducing costs and improving their properties. To achieve this replacement with less expensive reagents may be a reasonable alternative. Amongst all the possible reagents available, phenolic-furfural mixtures have been used in patents as precursors for the synthesis of organic aerogels with promising results [16, 27].

Phenolic-furfural aerogels are prepared by mixing phenolic novolak resins and furfural in equal proportions. In this synthesis process, propanol and phosphoric acid are used as solvent and catalyst, respectively [27]. Phenolic novolak is a low molecular weight polymer derived from the acid-catalyzed reaction of a molar excess of phenol with formaldehyde. In order to promote the sol–gel reaction, furfural must be added. In fact, the polymerization reaction occurs between aromatic rings of phenolic resin and the aldehyde group on furfural. The sol–gel reaction, curing and drying are performed following the method used for RF aerogels. However, these materials are particularly more efficient than conventional aerogels, as the use of propanol as solvent allows direct exchange with CO_2 prior to the supercritical drying step. Hence, there is no need for further solvent exchanges [16].

2.4.1.3 Phenol-Formaldehyde Aerogels

Even through phenolic-furfural mixtures lead to well-developed porous structures, furfural is more expensive than formaldehyde, which counteracts the costs saved by replacing resorcinol with phenol [17, 25]. Furthermore, the handling of furfural is more problematic than formaldehyde and its use is not desirable in industrial scale processes [17]. It is more profitable to obtain porous materials based on phenol formaldehyde (PF) precursor solutions. PF organic gels can be synthesized by the sol–gel reaction between phenol and formaldehyde using n-propanol as solvent and hydrochloric acid or sodium hydroxide as catalyst [17]. By means of this formulation, the sol–gel reaction, curing and drying stages can be performed following the method used for RF aerogels.

Resorcinol can also be replaced by phenol to give rise to organic xerogels, which are easier to synthesize than carbon aerogels. In this case, water and sodium carbonate are used as solvent and catalyst, respectively. However, it has been demonstrated that additives, such as cationic surfactants, are also needed in order to avoid shrinkage of the polymeric structure [25]. Moreover, it should be noted that only up to 70% of resorcinol can be replaced by phenol. Above 70%, the precursor solution precipitates and hence, sol–gel reaction does not occur properly [12]. Furthermore, the greater the amount of phenol used to replace resorcinol, the greater the amount of additive needed to obtain a well-developed porous structure [25].

2.4.2 Introduction of Doping Agents and Additives

Regardless of which precursors are used to synthesize carbon gels, other treatments can be employed to modify one or more properties and to fit them more accurately to the requirements of a specific application. For example, redox reactions, charge transfer, the wetting process, blockage of reactive sites, etc., are used in various applications. Hence, not only the porous properties but also chemical processes play an important role due to interactions between the fluids in the applications and the surface of the carbon gels. Doping carbon gels with different species or introducing additives into the precursor solutions are common processes for modifying the porosity and the chemical nature of carbon gels. Some of these processes have been patented due to the advantages they offer compared to the conventional synthesis process. Enlargement of the surface area, modification of the functional groups at the surface or enhancement of electrical conductivity are some of ways in which the gels can be modified [14, 31].

As inferred from several patented methods, doping agents can be incorporated into the carbon structures at various stages of the process as long as the doping agent is compatible with the process: (i) during the polymerization stage, (ii) in the polymer wet, dried or carbonized gel or (iii) during carbonization process.

Doping agents can be added to the precursor solution prior to gelation, followed by the conventional synthesis process [8, 14, 32, 33]. Other methods may involve previous ultrasonic water bath heat treatment in order to promote the polymerization reaction by high-intensity sonication until gelation is achieved. Sol–gel reaction and curing can be then continued as in the conventional method [16]. However, this technique exhibits some drawbacks as the doping agent may remain encapsulated inside the porous structure, particularly into closed pores, becoming inaccessible [34].

Another technique involves the impregnation of the wet, dried or carbonized gel in a solution containing the doping agent [35]. Several methods can be applied such as impregnation in a solution containing an excess of the doping agent, wet impregnation incipient [8], chemical vapour infiltration [32], ion adsorption, ion exchange or physical mixing [36]. Among all, impregnation is the most commonly used. The degree of doping depends on the method selected, the doping agent used and the structure of the carbon gel. In these techniques, it is necessary that the doping agent is dispersed in a solution that can then diffuse into the pores of the carbon. Doped organic gels can then be carbonized, leading to a coating of dopant agent on the inner surface of carbon gels, i.e. inside the pores [8]. However, the tortuosity in low permeable carbon gels may present significant experimental difficulties for incorporating into the structure of carbon gels certain doping agents, such as metal nanoparticles, particularly when have been synthesized as monoliths [33]. In these cases, it may be more convenient to tailor macroporous carbon gels in order to facilitate the incorporation of the doping agents.

Furthermore, the introduction of doping agents inside the porous structure, might lead to a decrease in surface area and total pore volume. This effect is enhanced

Table 2.1 Improvements in the properties and applicability of carbon gels by the incorporation of a doping agent and additives

Doping agent	Properties modified	Application	References
Carbon forms: Graphene, fullerenes, carbon nanotubes and fibres	Electric and mechanical properties Internal surface area Enhances metal doping	Energy storage Catalysis Adsorption	[16, 31, 32, 37–41]
Inorganic materials: Silica gels and ceramic	Porous and mechanical properties Thermal conductivity	Thermal insulation	[14, 42]
Metals: Ni, Pb, Pt, Rh, Ti, Cu, Ag, etc.	Electric properties Chemical properties (reactivity)	Energy storage Catalysis	[8, 33, 34, 36, 43–45]
Functional groups: O, N, S, P, B, F, etc.	Chemical properties (hydrophilicity, reactivity, etc.) Enhances metal doping	Energy storage Catalysis Adsorption	[6, 8, 46]
Additives: Surfactants (anionic, cationic or nonionic)	Porous and mechanical properties		[7, 11, 14, 25]

when the impregnation is carried out on dry samples, prior to carbonization, due to the subsequent drying of the solvent containing the doping agent. The organic gel is not able to counteract the capillary forces produced during this second drying process and shrinkage occurs [8]. The decrease in pore volume can also be produced due to the incorporated doping agent occupying pores within the pore structure.

Impregnation can be performed either on previously milled material or to non-milled material. Depending on the agent used for doping it will be better use one type of material or another. For example, it has been reported that a greater degree of manganese can be incorporated into a milled carbon gel than into non-milled samples, while the reverse case was observed with nickel [8].

In other words, the effect of doping depends to a large extent on the type of doping agent used. In the patent literature numerous doping agents used to improve the properties of carbon gels can be found (Table 2.1). The most relevant are described below.

2.4.2.1 Carbon Gels Doped with Other Forms of Carbon

Forms of carbon are generally added to precursor solutions prior to the synthesis of organic gels. Graphene, fullerene, carbon nanotubes, or carbon fibres are the most commonly used carbon dopants.

Doping with graphene can be carried out with non-functionalized graphene such as graphene oxide (GO) or other functionalized graphenes such as carboxylated graphene. However, graphene oxide is more common as the hydrophilic groups on its surface make it easy for it to be dispersed in water [39]. Indeed, GO nanosheets are stable in an aqueous solution as a result of ionization of the carboxylic acid and phenolic hydroxyl groups that are known to be present on the GO sheets. The most common method to produce GO is by the Hummers method in which is suspended in a solution by ultrasonication. This solution can be used as the solvent in the synthesis of carbon gels [39]. Doping carbon gels with graphene allow more control over the density, shape, electric conductivity and internal surface, so that the resulting material displays improved electrical and mechanical properties [38]. This not only avoids shrinkage during drying but also leads to materials that are able to withstand harsher industrial conditions [39]. Furthermore, graphene can improve the dispersion of other doping agents, such a cationic metal, resulting in a material that is especially useful in catalytic applications [40]. Likewise, it has been reported that these materials have excellent adsorption properties. In fact, oil absorption on a graphene-doped carbon gel can reach 21–70 times its own mass [38]. Accordingly, carbon gels containing graphene, open up a whole new interesting range of possibilities for their application.

Fullerenes have great potential for use in energy storage devices due to their ability for become charged both on and into the ball. However, the balls are insulators and charges are not readily transported inside pure-fullerene solids. Hence, the idea of using carbon gels as a support for fullerenes has attracted the attention of researchers and a synthesis process has already been patented [32]. The porous structure of carbon gels enables fullerene molecules to be deposited on the surfaces of the pores, leading to an increase in surface area. These carbon gels are low-density materials and excellent mechanical and conductive supports. As a result, lightweight composites with a high electrical conductivity, ideal for energy-storage applications, are obtained [32].

Carbon nanotubes are also commonly used doping agents in the synthesis of carbon gel due to their high strength and resistance to corrosion [31]. These features along with their large specific surface area, pore structure that is suitable for mass transport and good thermal conductivity properties can lead to composite materials with improved properties. The improvements achieved by the addition of carbon nanotubes allow carbon gels to be used in several applications with a better performance. For example, carbon gels doped with nanotubes have been patented as catalyst supports in fuel cells. The incorporation of nanotubes increases the oxygen reduction reaction [31]. Importantly, the catalyst material obtained is less expensive than commercial materials, which can significantly reduce costs and help to promote the commercial application of fuel cells. Carbon gels doped with nanotubes have also been patented as oil-absorbing materials. A small amount of carbon nanotubes leads to larger pores and, hence, the absorption rate is increased from 27 to more than 50 g/g (oil/doped carbon gel) [37].

Carbon fibres can also be added to the precursor solutions [16, 41]. This doping agent provides mechanical strength and electric conductivity to carbon gels, resulting in interesting materials for use as a functional material in aircraft and vehicles [41] or as electrode material in electrochemical devices [16].

2.4.2.2 Carbon Gels Doped with Inorganic Materials

Inorganic materials such as silica gels [14] have also been reported to be good doping agents. The addition of silica gel leads to really interesting materials for use as thermal insulators. Thermal conductivity decreases by as much as 30% when 50%wt. of silica gel is added [14]. Furthermore, the addition of reinforcing inorganic aerogel particles significantly mitigate shrinkage of organic aerogels that inevitably occurs during sol–gel processing and drying stages [14].

Carbon gels can also be produced by premixing the precursor solution with ceramic materials [42]. Ceramic materials offer internal support to carbon gels, which makes them suitable for use as electrode material [42]. Examples of effective ceramic materials include silica, alumina-silicates or glassy materials based on borate phosphates or silicates with alkaline earth or transition metals [42].

2.4.2.3 Carbon Gels Doped with Metals

Nowadays, several patents can be found on doping carbon gels with metals due to the great benefits this process offers in many applications. A feasible way to incorporate metals is by impregnation, in which carbon gels are exposed to a colloidal suspension containing the metal. Suitable metals include nickel, copper, platinum, palladium, silver, noble metals, etc. which can be added in elemental form, oxidized form such as oxides, salts, etc., or as part of a chemical compound or alloy [8, 36, 43, 44, 47]. Then, a thermal treatment is applied to reduce the doping agents to their metal state [45]. The maximum theoretical metal content that can be introduced into the precursor solution depends on the amount of water used in the synthesis and the solubility of the metal or metal compound [8, 33]. A metal content, expressed in terms of percentage weight of the total weight of carbon-doped gel, of up to 25% can be obtained depending on the metal precursor used and the post-thermal treatment applied [33]. The phase and crystal structure of the metal or oxide can be adjusted by controlling the temperature and time of the heat treatment [8]. Milling conditions to which the carbon gels are subjected are also relevant, as an excessive degree of milling, could result in a material with the metal filling the mesopores and not in the micropores [8]. Furthermore, a successful dispersion of the metal nanoparticles would be facilitated by the appropriate arrangement of the chemical anchoring sites inside the polymeric structure [34]. For example, the incorporation of functional groups [44] or the introduction of surfactant into the precursor solution could increase the number of anchoring sites, thereby facilitating the dispersion of metals [34].

Metals enhance electrical properties when incorporated into a carbon material, but doping with metals is commonly used to incorporate catalysts into the structure of carbon gels in order to enhance certain catalytic reactions. The catalyst can be deposited by co-precipitation, the urea method, induced hydrolysis reconstruction, the sol–gel technique, hydrothermal, microwave, ultrasound treatments and anion exchange reactions [36]. Regardless of the technique of deposition used, it is essential to ensure an effective dispersion and tethering of the metal through the pores in order to significantly increase the catalytically active electrode surface and to avoid lixiviation of the metal during catalysis. This can be achieved by using a properly porous-tailored carbon gel. Microporosity on carbon gels favours the tethering and dispersion of transition metals [33], while meso–macroporosity facilitates the transport of molecules (fuels, oxidants reactants, etc.) to the active catalytic sites [44].

2.4.2.4 Functionalization

The main objective of functionalization is to introduce heteroatoms onto the surface of carbon gels. The most common functionalization methods are the incorporation of oxygen and nitrogen groups. However, recently the possibility of incorporating other heteroatoms onto the surface of the gels, such as sulphur, phosphorus, fluorine or boron, excited a great deal of scientific interest. Functionalization enhances the performance of carbon gels in certain applications, such as in some catalytic processes or energy storage systems. In the latter, interest has been especially intense due to the improvement in the interaction between the electrolytes and the surface of the carbon gels [6, 8].

Functionalization can be performed on the outer surface of carbon gels or the inner pore surface. There is a variety of techniques that can be applied independently or in conjunction, depending on the type of functionalization desired. Surface functionality can be modified by replacing the main reagents by others containing boron [48] or nitrogen [49] or by introducing additives [50]. Alternatively, the inner chemical surface of the pores can be adjusted by exposing the carbon gels to various gases or solutions at certain temperatures [8]. Functional groups can thus be incorporated into the carbon structure as substitutional atoms, during the polymerization or pyrolysis process or by means of a post-treatment process.

The term functionalization is known as oxidation when oxygenated functional groups are introduced [47]. The oxidation process can be performed with an oxidizing gas (ozone, oxygen, nitrous oxide, nitric oxide, carbon dioxide, etc.) or with oxidants aqueous solutions (sodium hypochlorite, hydrogen peroxide, ammonium peroxydisulfate, nitric acid, etc.). The presence of oxygen-based functional groups can affect the wettability and chemical and electrochemical reactivity of carbon gels, which is of great importance for establishing the electrochemical double layer generated in supercapacitors [46].

The process of incorporating nitrogen groups can be accomplished with various compounds such as nitric monoxide, amines or nitriles [15]. The degree of functionalization and the nature of the functional groups depend on the reactivity of the oxidant or dopant, the nature of the carbonaceous material and the temperature and treatment time. The increase in the number of functional groups favours the subsequent anchoring of metallic particles, property of great interest in catalyst supports [15]. Moreover, hydrophilicity, which is a property necessary when using the carbon gel as an adsorbent, can also be improved.

The incorporation of sulphur groups can be performed by post-synthesis treatments such as impregnation. The sulphonation patented process consists of bringing the carbon gel into contact with a concentrated solution of sulphuric acid [46]. After impregnation, a heat treatment is performed in order to decompose the functionalizing agent to give rise to sulphur trioxide, which is then released through the pores of the polymeric structure. Ultrasounds can be also applied before any temperature increase in order to accelerate the filling of the pores [46]. The introduction into the precursor solution of heterocyclic sulphur such as thiophenecarboxaldehyde can be also used [43]. Its aldehyde group also contributes to the formation of the structure by allowing thiophene-like functionalities to be retained inside mesoporosity. This is quite an interesting technique if further doping with Pt is performed, as the incorporated functionalities can bind to Pt nanoparticles from colloidal suspensions [44]. Another process involves the introduction of sulphur using gaseous agents such as H_2S or SO_2 under different heating conditions. However, this process has a major industrial disadvantage as the gases produced are toxic and its implementation is complex [46].

Phosphor-doped carbons are of interest for use as electrode materials in lithium-ion batteries [36]. Phosphorus impregnation can be performed by bringing the carbon gel into contact with a H_3PO_4 solution prior to carbonization. By this technique, the electrochemical capacity of the batteries can be enhanced due to the greater intercalation of lithium into the carbon structure [51]. Phosphoric acid can also be used as catalyst in the preparation of the precursor solution to obtain carbon gels doped with phosphor [36].

2.4.2.5 Introduction of Additives

Preservation of the pore structure of the gel during removal of the solvent is important. During drying, the gel can shrink substantially, leading to an increase in density of the final material or even to non-porous materials. In order to minimize this effect, some patents suggest the incorporation of reinforcing additives into the precursor solution so as to moderate shrinkage and densification [7, 11, 14, 25]. However, it should be noted that additives not only avoid shrinkage but can also alter the polymerization reaction and in turn the final properties. Indeed, additives can also precipitate at a certain pH value or even modify the initial pH value, which may lead to a poorer control over the final properties [11, 25]. Furthermore, depending on the processing conditions, a certain amount of additive may remain after drying. This can be a drawback for some applications in which organic gels are needed and other variables such as the concentration of the main reagents and/or the synthesis conditions may need to be adjusted to avoid shrinkage [7], or alcohol washing stages may need to be applied to remove the surfactant [25]. In the case of carbon gels, this inconvenience disappears as additives can be removed during carbonization process [7, 8]. Hence, it is essential to select the appropriate nature and concentration of the additive used.

2.4.3 Thermal Treatments

Thermal treatments such as carbonization or activation are well-known techniques, used in multiple processes in order to obtain carbonaceous materials from different precursors. Due to pre-gel knowledge about these processes, the number of patents regarding standard thermal treatments applied in the synthesis of carbon gels is quite limited. However, these processes are of great importance and deserve to be mentioned in the present chapter.

2.4.3.1 Carbonization

Once dried organic gels have been obtained, carbonization must be applied in order to produce the final carbon gel. This process consists in bringing the organic gel into contact with an inert gas in a tube furnace at high temperature. Nitrogen, helium or argon are usually used as carrier gas and constant and moderate flows are selected according to the amount of sample and the size of the tube furnace. Temperatures between 600 and 1000 °C are generally used with slow heating ramps to prevent excessive shrinkage of the structure as a result of the removal of the gases produced inside the material [8, 11].

The main aims of carbonization processes are first to stabilize the material, and second to increase the micropore volume and surface area and also to widen the narrow pores formed during the process of synthesis. During this process, non-cross-linked organic chains, labile oxygen and hydrogen surface groups are removed, due to breakages of the C-O and C-H bonds at low and high temperatures respectively, which results in thermally stable nanostructured materials made up mainly of carbon. Furthermore, the elimination of volatiles and labile matter leads to the formation of microporosity, mainly in the primary particles of the polymer, while the nanostructure developed during the synthesis of the organic gel usually remains intact during the carbonization step [4, 25]. Thus, microporosity can be controlled during the carbonization step and meso–macroporosity can be controlled during the synthesis of organic gels, which is a great advantage for the design and control of the porous properties of carbon gels [8]. This type of nanostructure endows the materials with a series of electrical, thermal and mechanical properties that are very useful in applications related to catalysts and energy storage.

Others gases such as fume of ammonia can be used during carbonization. In this case, in addition to the development of the microporosity, nitrogen functionalities are incorporated into the structure.

2.4.3.2 Activation

The carbonization process converts not only organic gels into carbon gels but also increases the surface area. However, certain applications, such as some energy storage systems require very high areas and it is then necessary to apply other heat treatments. Activation processes can increase the surface area to 2000–3000 m^2/g. Carbon gels can be also subjected to activation processes after or during the carbonization step. During activation, the volume of narrow pores created during the synthesis of the organic gel and/or the carbonization process is also increased. Therefore, it is very common to activate carbon xerogels for use in applications where besides narrow mesopores a high volume of micropores are also needed. It is generally accepted that there are two main types of activation: chemical and physical.

Chemical activation consists in bringing the carbon or organic gel into contact with an activating agent (phosphoric acid, zinc chloride, alkali carbonates, alkali hydroxides, etc.) at temperatures between 400 and 800 °C [11, 46]. After this heat treatment, the material must be washed to remove residual chemical agents and possible inorganic products generated in the reaction. This step is one of the main disadvantages of this type of activation, since in addition to lengthening the process, sometimes it is extremely difficult to completely remove all the traces of the residual activating agent.

Physical activation is the most common method and consists in bringing the material into contact with a gas (water vapour or carbon dioxide) at temperatures between 700 and 1000 °C [6, 8, 46]. Activation time and temperature have a big impact on the resulting activated carbon material as well as on the manufacturing costs. Increasing the temperature and time leads to higher activation percentages but also increases the cost by reducing the overall yield [8]. Carbon gels obtained by this method exhibit different porous properties to those obtained by chemical activation. Therefore, the final application of the activated carbon gels will determine which activation process will be more suitable.

2.5 Applications

Carbon gels have generated widespread interest over the past decades due to the possibility of designing their properties to fit the requirements of specific applications. The right combination of properties makes organic and carbon gels suitable for use in various applications, many of which have been patented. Some of the most outstanding patents are described below.

2.5.1 Electronic and Energy Storage Applications

Interest in electronic applications and energy storage has increased considerably in recent years because of the importance these devices have acquired around the world. In most of these applications carbon gels are used as electrode material. However, each of these devices operates differently, and hence carbon gels with different properties are required for each application. In general, a high affinity between the carbon surface and certain particles that can be easily wetted by electrolyte is required for ion conduction and electrochemical reactions, while particles that repel electrolyte are convenient for gas diffusion channels [8, 52]. Furthermore, pore size also plays an important role, as carbon materials with small mesopores (less than 10 nm) may be useful to enhance the transport of ions and maximize power [8, 53]. In general, an optimized blend of micropores and mesoporores within the carbon structure contributes to the enhancement of electrochemical performance in energy storage devices.

The use of carbon gels with a high density that are electrically conductive, dimensionally stable and machinable as electrode material in supercapacitors can increase specific capacitance considerably compared to that of conventional capacitors [6, 53]. This is because it is possible to design the properties of carbon gels. Capacitances increase with pore volume and surface area. Small mesopores and large micropores seem to be the ideal combination of properties as these pores are large enough for electrolyte ions to diffuse through and yet small enough to ensure that the material has a large surface area. However, should be noted that the surface area not only needs to be high (preferably of the order of 1500 m^2/g) but also accessible, inasmuch that micropores that are too small may not be accessible to large electrolyte ions, especially cations [52, 53]. Therefore, an optimized structure will have large pores to act as a transmission and electrolyte transport pathway and fully accessible micropores for developing the highest specific capacitance.

Carbon gels with optimized properties for use as electrode material in different type of batteries are also expected to address the deficiencies provided by current commercial materials. In lithium-ion batteries, pore size distribution is important for both the storage capacity of the material and the kinetics and power capability of the system [36]. On the one hand, micropores with an average pore size smaller than 1 nm may create additional storage sites as well as lithium-ion diffusion paths, while surface areas smaller than 50 m^2/g have shown even better efficiencies. On the other hand, mesoporosity provides pathways for electrolyte, the conduction of ions and electrons. Furthermore, mesopores are especially suited for doping, which has been demonstrated to improve the electrochemical capacitance of batteries [36, 51]. The use of properly designed carbon gels leads to efficiencies in the first cycle of more than 80%.

Metal–air batteries have a much higher theoretical specific energy than most available primary and rechargeable batteries. However, their energy density is limited by the deposition of lithium peroxide (a product of the air electrode reaction) within the pores of the air electrode, which interrupts the low of current thereby limiting the usefulness of the device. Accordingly, recent research has focused on increasing the energy density of lithium–air batteries by using carbon gel with an appropriate tailored porous structure. The mesoporosity of carbon gels accommodates a large volume of reaction products per gram of carbon and provides a catalytic surface for the Li/O_2 reaction. In addition, the pore structure is able to accommodate various bifunctional catalysts, while maintaining the desired mesoporosity. Thus, electrodes prepared from carbon gels comprise a high specific energy and resist the clogging and inactivation of the pores. Accordingly, it is reported that metal–air batteries have a greater energy density and active life when carbon gels are used, compared to devices prepared with other already commercialized carbon materials [8].

Fuel cells are direct energy conversion devices, which are being developed for numerous uses including electric vehicles, space power, remote power source, etc. [54]. Fuel cells typically use carbon composite electrodes, impregnated with platinum, rhodium or other noble metals. However, these electrodes suffer from high polarization resistance, resulting in a limited power capability [54]. The problem associated with high polarization resistance is largely due to the rate-limiting catalysis of the oxygen reduction reaction, which is proportional to the surface area of the electrodes. It is clear therefore that there is a need to provide electrodes with a high surface area so that they can be synthesized at an industrial scale. For this purpose, carbon gels are seen as potential material to prepare electrodes for fuel cells, since such materials not only have high surface area but an excellent electrical conductivity and controllable porosity [54]. Besides, carbon gel can be synthesized in several forms and from different precursor solutions, which could lead to the preparation of carbon-doped. Both features are of great relevance. On the one hand, carbon gel synthesized in the form of monoliths could be used to great advantage in fuel cell electrodes due to their higher electronic connectivity compared to powder, which reduces contact resistance. Furthermore, monolithic material also simplifies the fabrication of fuel cell assemblies since no binder is needed to prepare the electrodes [44]. On the other hand, as already mentioned throughout this chapter, carbon gel can be doped with nanotubes [31] or functionalized with sulphur to improve its catalytic activity in fuel cells [44].

2.5.2 Catalyst Support

Carbon gels are also used as catalysts in various catalytic support processes. For this application it is essential to dope the carbon gel with metals (active phase), as previously described. In general, the carbonaceous material must have a high mesoporosity, preferably with wide pores to allow the incorporation of the metal and the adsorption–desorption of reactants and products. Surface area must also be raised to allow a better dispersion of the active phase. Furthermore, the modification of the surface chemistry, especially in relation to oxygen and nitrogen groups, could enhance both catalytic activity and the durability of the catalysts.

2.5.3 Gas Storage

Great progress has been made in the field of hydrogen fuel generated from renewable power. However, there is still a need to improve the materials used in these devices. Carbon gels have great potential in this application and, for this reason, have been patented as hydrogen storage materials. It has been shown that microporosity and doping carbon gels with metals such as titanium, manganese, nickel or cerium enhances hydrogen storage [43, 45, 55].

2.5.4 Coatings

The use of organic and carbon gels as coating material has recently aroused a great deal of interest due to the stabilization and inertness of these materials under solidification conditions. Several patents can be found for this application, particularly in relation to thermal and acoustic insulating systems [56, 57]. The processes involve the application of the precursor solution to a surface of an object by spraying, screening, trowelling, slot die coating, cast coating, extrusion or a combination thereof to form the coating. These coating techniques are already applied at an industrial scale and allow several working cycles so that the thickness of the layer can be adapted to the requirements of the application of the coated surface, such as thermal and acoustic insulator systems or to meet the contours of aircraft and vehicles shapes [56, 57]. Furthermore, precursor solutions can be doped with fibres or inorganic compounds to improve the flexibility and mechanical properties of the coatings [56, 57].

Different types of coating can be applied to the surface depending on the final application. For example metal wires used in solid phase microextraction systems have been coated with a precursor solution of carbon gels. The devices obtained are robust, stable in fields of high radiation, and highly effective for collecting gas and liquid samples while maintaining superior mechanical and thermal stability during their routine use [58]. Furthermore, the time required for the extraction can be improved since a highly porous material allows a quicker uptake of analyte. Besides these applications, another interesting example is worth mentioning, i.e. organic gels which have been patented as molding materials [59]. The patented process consists in wetting a wax mold with the precursor solution and then converting the sol to a gel at a temperature below the softening point of the wax. The coated wax mold is then submitted to thermal treatment at a temperature above the pouring point of the wax in order to remove it. By this process, lightweight quasi–adiabatic molds with highly porous walls can be easily prepared.

Precursor solutions from which carbon gels are obtained have been also patented as impregnators for membranes such as cellulose [60, 61]. The impregnated membranes are then submitted to heat for the gelation and curing, then subjected to solvent exchange and drying after which they are finally carbonized.

A mechanically stable hydrophobic layer of carbon aerogel-cellulose membrane is obtained useful for use as active electrode material in fuel cells.

2.5.5 *Other Applications*

Apart from the patents already mentioned, there are other interesting cases to be highlighted. For example carbon gels have been patented as inner cushion materials for helmets to absorb the impact energy [62]. In general the porous structure of carbon gels requires the application of a large amount of energy to collapse. So that, upon impact the energy is used to collapse the structure and as a result the air inside the pores diffuse so quickly that they generate turbulences on a microscopic scale, converting the kinetic energy into heated air. Carbon gels have also been patented as filler filters for use in smokable products such as cigarettes [63]. Although this patent is quite striking, the incorporation of carbon gels into cigarette filters does not increase their capacity to absorb compounds such as nicotine, water or carbon monoxide. So this patent, although curious, does not seem to be of much commercial interest.

2.6 Conclusions

The aim of this chapter has been to provide the reader a general view of the current state of the art in the field of carbon gels production and uses. These synthetic carbonaceous materials have clearly great potential, due to the possibility to modify and design their properties according to each application. However, the number of patents in relation to the amount of research documents is still very low.

Most of the current patents deal with the improvement of some of their properties or their use in one of the numerous possible applications, but more effort to produce them at industrial scale in a cost effective process is needed. Very few of the current options of the synthesis process can be transferred effectively to the industry. Nevertheless, carbon gels have numerous potential applications with superior performance in comparison to the current materials used. Consequently, the synthesis option that achieves carbon gel production at industrial scale and at competitive prices will reach many different markets.

References

1. Calvo EG, Arenillas A, Menéndez JÁ (2011) Designing nanostructured carbon xerogels. Intech Open Access Publisher
2. Rey-Raap N, Arenillas A, Menéndez JA (2015) Formaldehyde in the synthesis of resorcinol-formaldehyde carbon gels. In: Patton A (ed) Formaldehyde: synthesis, applications and potential health effects. Nova Science

3. Pekala RW (1989) Low density, resorcinol-formaldehyde aerogels. US patent 4873218
4. Droege MW (1999) Low density open cell organic foams, low density open cell carbon foams, and methods for preparing same. US patent 5945084
5. Rey-Raap N, Menéndez JA, Arenillas A (2014) RF xerogels with tailored porosity over the entire nanoscale. Microporous Mesoporous Mater 195:266–275
6. Kaschmitter JL, Mayer ST, Pekala RW (1993) Supercapacitors based on carbon foams. US patent 5260855
7. Albert DF, Andrews GR, Bruno JW (2006) Organic, open cell foam materials, their carbonized derivatives, and methods for producing same. US patent 7005181
8. Feaver AM, Costantino HR, Varjian RD (2011) Mesoporous carbon materials comprising bifunctional catalysts. US patent 20110223494
9. Costantino HR, Feaver A, Scott WD (2013) Manufacturing methods for the production of carbon materials. US patent 10140094572 A1
10. Feaver A, Cao G (2010) Activated carbon cryogels and related methods. US patent 7723262
11. Geramita K, Kron B, Costantino HR, Feaver AM, Sakshaug AJ, Thompkins LA, Chang ATY (2013) Preparation of polymeric resins and carbon materials. US patent 20130280601
12. Rojas-Cervantes ML (2014) Some strategies to lower the production cost of carbon gels. J Mater Sci 50(3):1017–1040
13. Mayer ST, Kaschmitter JL, Pekala RW (1995) Method of low pressure and/or evaporative drying of aerogel. US patent 5420168
14. Lee JK (2007) Organic aerogels reinforced with inorganic aerogel fillers. US patent 20070259979
15. Setoyama N, Kajino T, Takagi H, Asaoka T, Fukushima Y (2007) Carbon gel composite material. US patent 20070196659
16. Cooper JF, Tillotson TM, Hrubesh LW (2005) Aerogel and xerogel composites for use as carbon anodes. WO patent 2005101553 A2
17. Scherdel C, Reichenauer G (2012) Microporous and mesoporous carbon xerogel having a characteristic mesopore size and precursors thereof and also a process for producing these and their use. WO patent 2010000778 A1
18. Awadallah AF, Al-Muhtaseb SA (2014) Electrochemical polymerization process for preparation of crosslinked gel. US patent 8877826
19. Tonanon N, Wareenin Y, Siyasukh A, Tanthapanichakoon W, Nishihara H, Mukai SR, Tamon H (2006) Preparation of resorcinol formaldehyde (RF) carbon gels: use of ultrasonic irradiation followed by microwave drying. J Non Cryst Solids 352:5683–5686
20. Tonanon N, Siyasukh A, Tanthapanichakoon W, Nishihara H, Mukai SR, Tamon H (2005) Mprovement of mesoporosity of carbon cryogels by ultrasonic irradiation. Carbon 43(3): 525–531
21. Zubizarreta Sáenz de Zaitegui L, Gómez Calvo E, Menéndez Díaz JÁ, Arenillas de la Puente A (2011) Procedimiento de obtención de xerogeles orgánicos de porosidad controlada. ES patent 2354782 A1
22. Calvo EG, Juarez-Perez EJ, Menendez JA, Arenillas A (2011) Fast microwave-assisted synthesis of tailored mesoporous carbon xerogels. J Colloid Interface Sci 357(2):541–547
23. Juárez-Pérez EJ, Calvo EG, Arenillas A, Menéndez JA (2010) Precise determination of the point of sol–gel transition in carbon gel synthesis using a microwave heating method. Carbon 48(11):3305–3308
24. Rey-Raap N, Menéndez JA, Arenillas A (2014) Simultaneous adjustment of the main chemical variables to fine-tune the porosity of carbon xerogels. Carbon 78:490–499
25. W. Bell (2002) Mesoporous carbons and polymers. US patent 20020065333 A1
26. Pekala RW (1992) Melamine-formaldehyde aerogels. US patent 5081163
27. Pekala RW (1996) Organic aerogels from the sol-gel polymerization of phenolic-furfural mixtures. US patent 5556892
28. Aegerter MA, Leventis N, Koebel MM (2011) Aerogels handbook. Springer

29. Fairén-Jiménez D, Carrasco-Marín F, Moreno-Castilla C (2006) Porosity and surface area of monolithic carbon aerogels prepared using alkaline carbonates and organic acids as polymerization catalysts. Carbon 44(11):2301–2307
30. Rey-Raap N, Arenillas A, Menéndez JA (2016) A visual validation of the combined effect of pH and dilution on the porosity of carbon xerogels. Microporous Mesoporous Mater 223:89–93
31. Carbon nanotube doped carbon gel catalyst for fuel cell and its application (2013). CN patent 102856562
32. Lamb LD, Huffman DR (1997) Aerogel/fullerene hybrid materials for energy storage applications. US patent 5698140
33. Pérez CAF, Moreno CC, Carrasco MF, Maldonado HF, Morales TS, Kapteijn F, Ros CH (2013) Doped carbon material for the electrocatalytic conversion of CO_2 into hydrocarbons, uses of the material and conversion method using said material. WO patent 2013004882 A9
34. Maldonado HFJ, Pérez CAF, Jirglová H (2013) Method for obtaining doped carbon gels, gels thus obtained and use thereof as catalysts. WO patent 2012164128 A3
35. Job N, Lambert SD, Zubiaur A, Cao C, Pirard JP (2015) Design of Pt/carbon xerogel catalysts for PEM fuel cells. Catalysts 5(1):40–57
36. Sakshaug A, Kron BE, Thompkins LA, Geramita K, Mcadie A, Costantino HR, Feaver AM (2014) High capacity hard carbon materials comprising efficiency enhancers. WO patent 2014201275 A2
37. Oil-absorbing aerogel and preparation method thereof (2014). CN patent 103752277
38. Graphene oxide reinforced carbon aerogel material, and preparation method and application thereof (2015). CN patent 103274384
39. Gallastegui AG, Shaffer M, Alyoubi AO, Basahel S (2013) Graphene and graphene oxide aerogels/xerogels for CO_2 capture. WO patent 2013132259
40. Pauzauskie PJ, Worsley MA, Baumann TF, Satcher JH, Biener J (2012) Graphene aerogels. US patent 20120034442
41. Ratke LP (2004) Composite material used as a functional material in the construction of vehicles and aircraft, consists of fiber-reinforced plastic and/or carbon aerogels. DE patent 10300979 A1
42. Cherepy NJ, Jankowski AF, Tillotson TM, Fiet K (2006) Carbon aerogel and xerogel fuels for fuel cells and batteries. WO patent 2006025993 A2
43. Feaver A, Cao G (2010) Carbon-based foam nanocomposite hydrogen storage material. US patent 7816413
44. Long JW, Rolison DR, Baker W (2008) Sulfur-functionalized carbon nanoarchitectures as porous, high surface area supports for precious metal catalysts. US patent 7442747
45. Struthers RC (1994) Metal alloy laded carbon aerogel hydrogen hydride battery. US patent 5366828
46. Ayme-Perrot D, Dieudonné M, Sonntag P, Pasquier AC (2014) Sulphur-modified monolithic porous carbon-based material, process for the preparation thereof and uses thereof in the storage and release of energy. US patent 20120156567
47. Mager N, Meyer N, Léonard AF, Job N, Devillers M, Hermans S (2014) Functionalization of carbon xerogels for the preparation of palladium supported catalysts applied in sugar transformations. Appl Catal B 148–149:424–435
48. Zapata-Benabihe Z, Moreno-Castilla C, Carrasco-Marín F (2014) Influence of the boron precursor and drying method on surface properties and electrochemical behavior of boron-doped carbon gels. Langmuir 30(6):1716–1722
49. Kiciński W, Norek M, Jankiewicz BJ (2014) Heterogeneous carbon gels: n-doped carbon xerogels from resorcinol and n-containing heterocyclic aldehydes. Langmuir 30(47): 14276–14285
50. Rey-Raap N, Szczurek A, Fierro V, Celzard A, Menéndez JA, Arenillas A (2016) Advances in tailoring the porosity of tannin-based carbon xerogels. Ind Crops Prod 82:100–106
51. Mayer ST, Pekala RW, Morrison RL, Kaschmitter JL (1994) Doping of carbon foams for use in energy storage devices. US patent 5358802

52. Nair VMP, Margolese D, Anz SJ, Wang S (2009) Nanoporous electrodes and related devices and methods. US patent 20090303660
53. Feaver AM, Cao G (2010) Electric double layer capacitance device. US patent 7835136
54. Mayer ST, Kaschmitter JL, Pekala RW (1997) Carbon aerogel electrodes for direct energy conversion. US patent 5601938
55. Schuth F, Bogdanovic B, Akira T (2006) Materials encapsulated in porous matrices for the reversible storage of hydrogen. US patent 20060264324
56. Coronado PR, Poco JF (2000) Flexible aerogel composite for mechanical stability and process of fabrication. US patent 6087407
57. Droege M, Downey S, Hrubesh L (2012) A coating composition for thermal protection on substrates, processes for manufacturing, and methods of applying same. EP patent 2440400 A1
58. Miller FS, Andresen BD (2005) Solid phase microextraction device using aerogel. US patent 6905031
59. Ratke L, Fricke J (2003) Precision casting and dead-mold casting in plastic/carbon aerogels. US patent 6599953
60. Petricevic R, Fricke J, Leuschner R, Lipinski M (2003) Gas diffusion electrode and its production. US patent 6503655
61. Gebhardt U, Leuschner R, Lipinski M, Waidhas M (2003) Method of discharging reaction water in PEM fuel cells and fuel cell for carrying out the method. US patent 6576358
62. Clavell RC (2004) Inner cushions for helmets. US patent 6704943
63. Branton P, Schuth F, Schwickardi M (2013) Smoke filtration. EP patent 2475272 B1

Chapter 3
Robust Mesoporous Polymers Derived from Cross-Linked Block Polymer Precursors

Myungeun Seo

3.1 Introduction

Mesoporous materials have a porous morphology comprised of open pores ranging in diameter from 2–50 nm. The porous surfaces are only accessible to substances smaller than the pore size, which makes such materials useful in a wide range of applications, including filtration, separation, storage, and heterogeneous catalysis. Particularly compared with microporous materials, where the pore diameter is <2 nm, mesoporous materials are advantageous in applications that involve molecules with larger size, such as proteins. For example, mesoporous anodized aluminum oxide (AAO) membranes are used to estimate the number of cells in a solution by counting the cells remaining on the surface of the membrane after filtration [1].

Among mesoporous materials, MCM-41 is probably one of the most famous [2]. MCM-41 was synthesized by using cylindrical aggregates of alkyltrimethylammonium surfactants formed in water as templates. A sol-gel reaction of a silica precursor occurred on the surface of the surfactant aggregates, and the surfactant was subsequently removed by calcination. As a result, well-defined and hexagonally packed cylindrical pores with diameters of 2—10 nm were produced by adjusting the alkyl chain length of the surfactant and using an additive [3]. Mesoporous silica with different pore structures such as double gyroid and perforated lamellae could be also obtained by varying the composition of the synthetic mixture [4]. The synthesis of the MCM family stimulated research on inorganic mesoporous materials leading to huge advances in the field since then (for example, see [5–7]).

Compared with inorganic mesoporous materials, organic mesoporous materials have even more attractive features. Organic materials, and polymers in particular, are typically lighter in weight, can be processed into various shapes, such as films

M. Seo (✉)
Graduate School of Nanoscience and Technology, Korea Advanced Institute of Science and Technology (KAIST), Daejeon 34141, Republic of Korea
e-mail: seomyungeun@kaist.ac.kr

P. Bettotti (ed.), *Submicron Porous Materials*,
DOI 10.1007/978-3-319-53035-2_3

and fibers on an industrial scale, and can be functionalized much more easily. For example, mesoporous polymer membranes for ultrafiltration of water have been typically obtained by carefully controlling the phase inversion process: the phase separation in the polymer solution occurs upon contact with a nonsolvent, or a change in temperature [8]. Asymmetrically porous membranes have been produced, where mesopores with rather broad size distributions, located on the surface, act as a discrimination layer, while a sponge-like macropore structure is generated beneath that layer providing high flux. But while porous polymers of this type are extremely useful for water treatment, this macrophase separation-based fabrication process does not allow the production of well-defined pore structures with precise control over pore size and pore size distribution as can be achieved in inorganic mesoporous materials such as MCM-41.

The microphase separation of block polymers is a unique route to producing well-defined nanostructures, because it allows extraordinary control over feature size and morphology. This chapter aims to explore how well-defined mesoporous polymers comparable to inorganic mesoporous materials can be synthesized from microphase-separated block polymer precursors. Since the pioneering work reported by Lee et al. [9, 10], tremendous efforts have been devoted to this field realizing the promising potential of the materials as advanced separation membranes. The concept of transforming the microphase-separated morphology of a block polymer precursor into a voided nanostructure with topological contrast is also very closely related to block copolymer lithography, which is considered to be one of the next-generation lithographic techniques [11].

A number of excellent reviews are available highlighting various aspects of mesoporous polymers derived from block polymer precursors (for example, see [12–15]). The present contribution will specifically focus on methods of synthesizing robust mesoporous polymers with enhanced pore stability by cross-linking the block polymer precursors. To this end, the chapter is organized as follows. First, the key conditions of the microphase-separated block polymer precursors required to successfully produce corresponding mesoporous polymers will be presented, and essential techniques for analyzing the block polymer microphase separation and pore characteristics will be introduced. Then cross-linking of the block polymer precursors will be discussed as a means of increasing the pore stability of the resulting mesoporous polymers. Synthetic routes to various cross-linked block polymer precursors will be reviewed in detail highlighting recent examples from the literature.

3.1.1 How to Derive Mesoporous Polymers from Block Polymer Precursors

Figure 4.1a schematically shows the key steps involved in obtaining mesoporous polymers from block polymer precursors, utilizing a diblock copolymer as the precursor. A block polymer is a class of macromolecules consisting of chemically distinct blocks connected via covalent bonds. A diblock copolymer is composed of

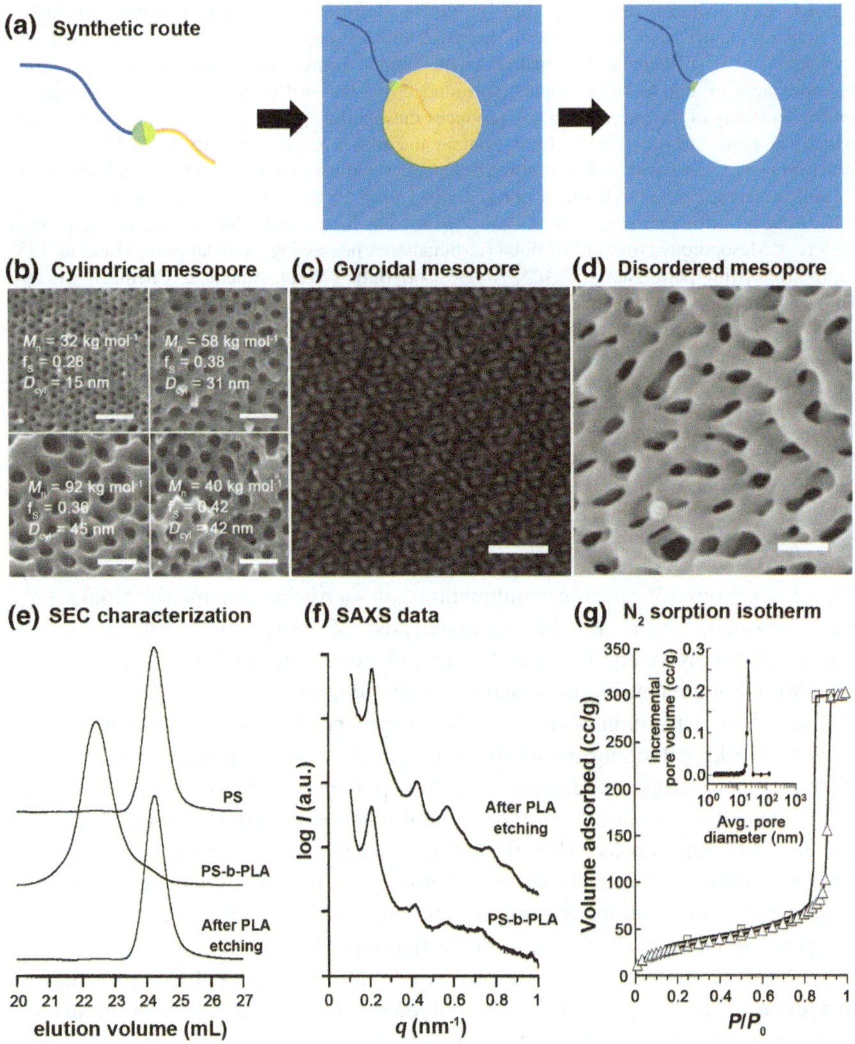

two different A and B blocks joined together at a covalent junction, and this arrangement is denoted as A-b-B. To produce a mesoporous polymer, a suitable block polymer precursor should consist of a sacrificial block which forms a distinct microdomain as a result of microphase separation, and which can be selectively etched out of the precursor to form well-defined mesopores. The remaining polymer block will form a matrix that supports the pore structure, and the pore surface will be decorated with functional groups originating from the matrix chain end.

To achieve a porous polymer which retains the microphase-separated morphology, the sacrificial block serves as a porogen and must be removed without damaging the matrix or destroying the microphase-separated morphology. Suitable

Fig. 4.1 Key features of mesoporous polymers derived from block polymer precursors. **a** Schematic depiction of the synthesis route. A diblock copolymer precursor composed of a sacrificial block (*yellow*) and a matrix block (*blue*) forms hexagonally packed cylindrical microdomains of the sacrificial block surrounded by the matrix via microphase separation. Selective etching of the sacrificial block produces mesoporous polymers with hexagonally packed cylindrical pores, whose pore surfaces contain the matrix chain ends. (**b–d**) SEM images of mesoporous polymers with different pore structures. Scale bars represent 100 nm. **b** Mesoporous PS with differently sized cylindrical pores derived from PS-b-PLA with different molar masses [21]. Adapted with permission of *J. Am. Chem. Soc.* Copyright (2002) American Chemical Society. **c** Mesoporous cross-linked poly(1,2-butadiene) possessing gyroidal pores (Li et al. [23]. Reproduced with permission of *ACS Nano*. Copyright (2011) American Chemical Society. **d** Mesoporous polyethylene possessing 3D continuous disordered pores [25]. Reproduced with permission of *Macromolecules*. Copyright (2006) American Chemical Society. **e–g** Characterization of mesoporous PS possessing cylindrical pores derived from PS-b-PLA. **e** SEC traces of PS, PS-b-PLA, and PS-b-PLA after PLA etching [16]. Adapted with permission of *J. Am. Chem. Soc.* Copyright (2001) American Chemical Society. **f** SAXS data of PS-b-PLA and mesoporous PS showing C morphology [16]. Adapted with permission of *J. Am. Chem. Soc.* Copyright (2001) American Chemical Society. **g** Nitrogen sorption isotherm of mesoporous PS. Inset shows pore size distribution obtained by Barrett-Joyner-Halenda analysis [21]. Adapted with permission of *J. Am. Chem. Soc.* Copyright (2002) American Chemical Society

etching reactions for specific combinations of sacrificial and matrix blocks have been reported, such as the methanolysis of polylactide (PLA) out of polylactide-b-polystyrene (PLA-b-PS) diblock copolymer under a basic condition [16]. More comprehensive information can be found elsewhere [17].

Microphase separation occurs in block polymers when the segregation strength between blocks belonging to the block polymer is large enough (for more information on the microphase separation of block polymers, see [18, 19]). The polymer chain is stretched outward from the covalent junction to avoid unfavorable contact between different blocks. However, this arrangement is compromised by the entropic contribution, which seeks to restore the stretched chain to its original random coil conformation, and this competition ultimately results in the formation of a periodic nanostructure with minimal free energy.

Segregation strength is represented as a multiplication of the Flory–Huggins interaction parameter χ and degree of polymerization N. Chemically more distinct blocks typically show higher χ. For diblock copolymers, it is known that domain spacing (the distance between microdomains) d scales with N following $d \sim N^{2/3}$ when $\chi N > 100$ (strong segregation regime) [20]. Thus, the size of sacrificial microdomains and pores derived by etching the sacrificial block can be precisely controlled by varying N, which is directly correlated to the molar mass of the block polymer if the volume composition is fixed. The typical size of microdomains that can be obtained by the microphase separation of block polymers is 10–50 nm, and so the microphase separation of block polymers offers a robust route to synthesizing mesoporous polymers with exquisite control of pore size comparable to that of mesoporous silica. The scanning electron microscopy (SEM) images shown in Fig. 4.1b demonstrate the various pore sizes that can be obtained in the mesopore regime simply by varying the molar mass of the block polymer precursors [21].

To avoid pore collapse after the removal of the sacrificial block, the matrix should be in a solid state and three-dimensionally (3D) continuous. To generate open pores accessible to external substances the sacrificial microdomain should also be 1D continuous, at least. For applications where the transport of substances through the pores is important, a 3D continuous percolating pore structure is more beneficial since that structure is more resistant to pore clogging, and the methods of aligning 1D continuous microdomains is often tedious, costly, and not straightforward [22]. Among the classic morphologies found in ordered diblock copolymers, hexagonally packed cylinder and gyroid phases can be used to generate mesoporous polymers. While porous polymers with independently interpenetrating 3D porous spaces and matrix, which are advantageous for transport applications, can be derived from bicontinuous gyroid precursors, obtaining precursors in the gyroid phase has been challenging, because they have a narrow composition window. An SEM image of a mesoporous polymer possessing gyroidal pores is shown in Fig. 4.1c [23].

The majority of mesoporous polymers are derived from a cylindrical phase, as shown in Fig. 4.1b, where the sacrificial block with ca. 30% of volume fraction forms 1D continuous cylinders and the rest forms a 3D continuous matrix. Alignment of the cylindrical pores has been achieved by processing the block polymer precursors, for example, using channel die alignment in case of bulk materials [21] and solvent annealing for thin films [24]. It should be also noted that a disordered bicontinuous morphology obtained by arresting microphase separation process via crystallization [25] or cross-linking [26] has been also successfully utilized to produce 3D continuous mesoporous polymers, as shown in Fig. 4.1d.

Since this approach involves the synthesis of a block polymer precursor, nanostructure formation via microphase separation, and conversion of the precursor into mesoporous polymer, various analytical techniques are required to properly characterize the molecular information of the precursor, its microphase separation behavior, and the pore characteristics of the mesoporous polymer. Size exclusion chromatography (SEC) and ^{1}H nuclear magnetic resonance (NMR) spectroscopy are used to determine molar mass and the composition of the precursor. To determine whether the precursor has formed the desired nanostructures via microphase separation, the microphase separation behavior of the precursor is typically studied by small angle X-ray/neutron scattering (SAXS/SANS) and transmission electron microscopy (TEM). Selective etching of the sacrificial block can be monitored by several techniques, and the use of SEC and ^{1}H NMR spectroscopy can directly probe whether the sacrificial block has been removed from the precursor without degradation of the matrix block. The pore structure can be examined in comparison with the microphase-separated morphology of the precursor by scattering and SEM, to confirm that the microphase-separated morphology was successfully transformed into a mesopore structure. Pore characteristics can be further analyzed by conducting nitrogen sorption isotherm experiments, which provide quantitative information about specific surface area, average pore size, pore size distribution, and porosity. Exemplary data following the synthesis of mesoporous PS derived from PS-b-PLA are shown in Fig. 4.1e–g, including SEC traces, SAXS patterns, and a nitrogen sorption isotherm [16, 21].

3.1.2 Pore Stability of Mesoporous Polymers

Because etching of the sacrificial block to form mesopores creates a large surface area, thus exposing the matrix to the air, a mesoporous polymer tends to release excess surface energy as heat by collapsing the pore structure and eliminating the surface area. In the case of mesoporous silica, the pore structure is highly stable up to 900 °C due to a strong covalent Si–O network structure. However, the thermal stability of mesoporous polymers is intrinsically lower than that of inorganic materials because the matrix polymer becomes fluid above the glass transition temperature (T_g) and fills in the porous space, leading to pore collapse. This is an important issue since the operating temperature range of mesoporous polymers in applications will be limited by their thermal stability. Also, pore stability is one of the factors which limits efforts to produce smaller pore sizes because the specific surface area and the excess surface energy increases as the pore size decreases, making the mesoporous polymer even more unstable.

Zalusky et al. [21] investigated the thermal stability of mesoporous PS derived from PS-b-PLA. From a 2D SAXS experiment of mesoporous PS, where the cylindrical pores were aligned vertically, they observed that the scattering intensity along the meridional direction diminished as temperature was increased up to 98 °C, which is close to the T_g of PS (Fig. 4.2a). Differential scanning calorimetry (DSC) of an identical specimen showed a large exothermic transition around the temperature during the first heating cycle (Fig. 4.2b). In a second heating scan, the specimen showed an endothermic transition corresponding to the glass transition of PS.

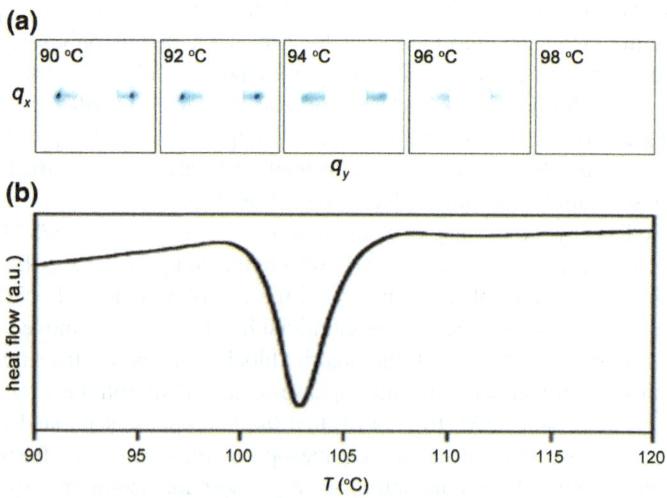

Fig. 4.2 Pore collapse of mesoporous PS with vertically aligned cylindrical pores derived from PS-b-PLA precursor [21]. Adapted with permission of *J. Am. Chem. Soc.* Copyright (2002) American Chemical Society. **a** Snapshots of 2D SAXS pattern with increasing temperature. **b** DSC thermogram obtained during the first heating cycle

The results strongly support that the mesopore structure collapsed around the glass transition temperature of the PS matrix, releasing the excess surface energy as heat.

In addition to thermal stability, chemical, and mechanical stabilities of the pore structure should be also considered in applications of porous polymers. As a non-cross-linked porous polymer can still be readily dissolved in a good solvent and lose its pore structure, resistance to chemicals including solvents can be an important issue. Mechanical stability is also important, especially in filtration and separation applications, since the polymer needs to be able to withstand significant hydraulic pressure for a long period of time without losing its mechanical integrity.

3.1.3 Robust Mesoporous Polymers Derived from Cross-Linked Block Polymer Precursors

Cross-linking of the matrix block in the block polymer precursor prior to removal of the sacrificial block is an effective method of increasing the stability of porous polymers. Cross-linking is a process which generates covalent bonds between linear polymer chains to form a 3D polymer network in which all the polymer chains are essentially connected. The cross-linked polymer has a higher T_g as its cross-linking density increases. It is not soluble in solvents (although it may swell), and it is mechanically more robust due to the existence of the percolating covalent network. If the matrix block in the microphase-separated block polymer precursor can be selectively cross-linked and then the sacrificial block can be etched, the resulting cross-linked mesoporous polymer will exhibit enhanced thermal, chemical, and mechanical stabilities compared to the non-cross-linked analog. The schematic depiction shown in Fig. 4.3 illustrates a synthetic route to a robust mesoporous polymer from a cross-linked block polymer precursor, in contrast to the process shown in Fig. 4.1.

If the matrix polymer is rubbery, where the polymer T_g is below room temperature and so the chain is mobile at room temperature, as in polyisoprene (PI), cross-linking is absolutely required to achieve porous polymers that are at least

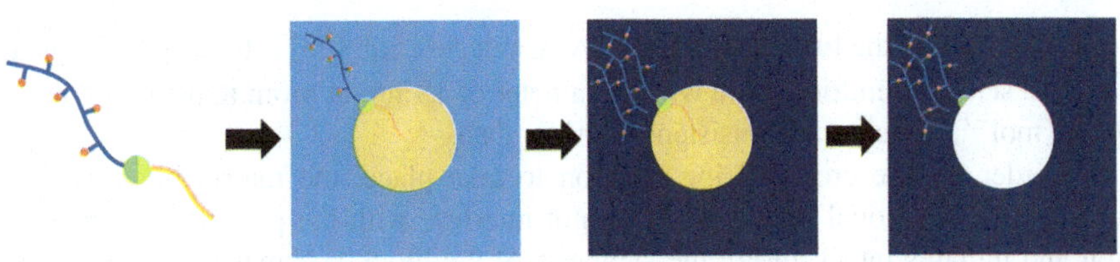

Fig. 4.3 Synthetic route to robust mesoporous polymer from cross-linked block polymer precursor. A reactive block polymer precursor consisting of a sacrificial block (*yellow*), a matrix block (*blue*), and cross-linkable moieties (*red*) on the matrix block. Self-assembling the polymer to form a target microphase-separated morphology and cross-linking the matrix results in a cross-linked block polymer precursor. The cross-linked matrix is depicted as dark blue, and a part of the cross-linked network is illustrated. Removal of the sacrificial block produces cross-linked mesoporous polymers

stable at room temperature. The cross-linking of a rubbery polymer produces an elastomer, and the modulus of the elastomer increases as a function of the cross-linking density. Consequently, a properly cross-linked matrix polymer may exhibit a modulus sufficient to endure the Laplace pressure generated at the pore surface and avoid pore collapse.

Pore stability as a function of the cross-linking density of an elastomer was theoretically discussed by Muralidharan and Hui [27]. The discussion was motivated by the work of Cavicchi et al. [28], in which mesoporous PI was produced from a polyisoprene-b-polydimethylsiloxane (PI-b-PDMS) precursor by cross-linking of PI via radical reaction, and the subsequent etching of PDMS in the presence of tetrabutylammonium fluoride. The authors observed that mesoporous PI with 13 nm diameter cylindrical pores was successfully derived when the molar mass between cross-links ($\overline{M_c}$, i.e., the inverse of the cross-linking density) was 500 g mol^{-1}. An identical PI-b-PDMS with $\overline{M_c} = 11000$ g mol^{-1} was nonporous after PDMS etching presumably because the Laplace pressure overwhelmed the modulus of the PI at low cross-linking density, leading to pore collapse.

According to the theoretical discussion, a stability criterion can be determined from the modulus E, Poisson's ratio ν, and the surface tension γ using Lame's solution. For an infinite cylindrical pore having the radius r in a linear elastic solid (i.e., incompressible so $\nu = 0.5$), the pore is expected to be stable if the following equation is satisfied:

$$\frac{2Er}{3\gamma} > 1$$

The criterion predicts that pore stability increases as the modulus of the matrix and the pore size increase, but decreases as the surface tension decreases. In the case of an elastomeric material formed by the cross-linking of rubbery polymers, the modulus E can be estimated as follows, where ρ is density, R is a gas constant, and T is temperature.

$$E \approx \frac{3\rho RT}{\overline{M_c}}$$

By combining the two equations above, Cavicchi et al. [28] estimated the critical $\overline{M_c}$ for a stable cylindrical pore with a diameter of 13 nm at room temperature to be 900 g mol^{-1}, which was consistent with the data.

In order for the cross-linking reaction to take place, the matrix block should consist of a functional group that does not interfere with the polymerization reaction and remains intact during the synthesis of the block polymer precursor. Also, the functional group should readily undergo a cross-linking reaction within the matrix microdomain without destroying the target microphase-separated morphology, so the nanostructure can be transformed into a mesopore structure.

Popular cross-linking schemes reported in the literature are summarized in Table 4.1. Approaches to cross-linking of the matrix block can be classified into four categories, depending on the position of the functional group. The first

Table 4.1 Cross-linking of the matrix block in the block polymer precursor

Matrix block	Functional group	Cross-linking method	References
Cross-linking reactions utilizing functional groups in the matrix block			
poly(4-vinylphenyl)dimethyl-2-propoxysilane	siloxyl group (protected silanol)	acid-catalyzed hydrolysis and condensation	[9, 10]
poly(pentamethyldisilylstyrene)	silyl group	formation of silicon oxycarbide by UV/ozone treatment	[29]
poly(2-cinnamoylethyl methacrylate)	cinnamoyl group	dimerization by UV irradiation	[30–32]
polystyrene	aromatic ring	UV irradiation on polymer thin film	[33, 64, 65]
polystyrene and 1,4-polybutadiene	double bonds	γ-irradiation	[34]
polyisoprene	allylic hydrogen	radical reaction (combination)	[28, 35]
1,2-polybutadiene	pendent double bond	radical reaction (chain reaction)	[23, 37, 38]
polystyrene	aromatic ring	ozone treatment on polymer thin film	[40]
poly(2-vinylpyridine)	pyridine ring	quaternization with 1,4-diiodobutane	[41]
Cross-linking reactions utilizing functional groups introduced by copolymerization			
poly(styrene-co-4-vinylbenzocyclobutene)	benzocyclobutene group	dimerization at >200 °C	[42, 43]
poly(styrene-co-*p*-norbornenylethylstyrene)	norbornene group	ROMP with dicyclopentadiene in the presence of Grubbs catalyst	[46–48]
poly(cyclooctene-s-5-norbornene-2-methylene methacrylate)	methacrylic double bond	radical copolymerization with vinyl monomers	[45]
poly(styrene-co-2-hydroxyethyl methacrylate) and poly (styrene-co-2-hydroxyethyl acrylate)	hydroxyl group	urethane formation with methylene diphenyl diisocyanate	[26]
poly(styrene-co-4-vinylbenzoyl chloride)	acid chloride group	amide formation with *m*-phenylenediamine	[49, 51]

(continued)

Table 4.1 (continued)

Matrix block	Functional group	Cross-linking method	References
In situ cross-linking by copolymerization of a cross-linker with the matrix monomer			
poly(styrene-co- divinylbenzene)	styrenic double bond	in situ cross-linking by copolymerization	[52–54]
poly(vinylbenzyl chloride-co-divinylbenzene)	styrenic double bond	in situ cross-linking by copolymerization	[55]
poly(isobornyl acrylate-co-ethylene glycol diacrylate)	acrylic double bond	in situ cross-linking by copolymerization	[57]
Thermosetting polymerization in the presence of block polymer precursors			
poly(ethylene oxide)	none	condensation of phenol and formaldehyde	[58, 62]
poly(ethylene oxide)	none	thermal imidization with resol condensation	[63]

approach utilizes functional groups of the matrix polymer inherited from the monomer structure. The second approach introduces more specific functional groups, with control over the density of the functional groups along the chain, by copolymerizing functional comonomers with the matrix monomer. The first and the second approaches can be considered post-polymerization modification routes, while the third approach differs, as it involves in situ cross-linking during polymerization, by copolymerizing a cross-linker with the matrix monomer. The fourth approach does not require a particular functional group in the matrix block; instead, condensation polymerization is conducted in the presence of the block polymer precursor to produce a thermoset, where the matrix block is mixed with the thermoset but the sacrificial block is segregated to form individual microdomains.

3.1.3.1 Cross-Linking Reaction Utilizing Functional Groups in the Matrix Block

In order to maximize the cross-linking density of the matrix block, it is desirable to introduce cross-linkable moieties in every repeating unit. Diene monomers such as butadiene and isoprene intrinsically leave a double bond in the repeating unit when they are anionically polymerized, which can be further utilized for cross-linking. Functional monomers such as 2-vinylpyridine containing a pyridine ring can be also post-functionalized after polymerization. More specific cross-linkable moieties such as siloxyl groups and cinnamoyl groups can be introduced as pendent groups of monomers via organic synthesis.

Because the microphase-separated morphology of the block polymer precursor should be preserved during cross-linking, the cross-linking reaction is typically conducted on the precursor in the solid state (which is already microphase-separated) or in a highly concentrated solution of the precursor. FTIR or UV spectroscopy is commonly used for monitoring the progress of the cross-linking reaction since ^{1}H NMR and SEC analyses are not applicable for insoluble cross-linked precursors (and the corresponding mesoporous polymers). Cross-linking density is often estimated by swelling experiments.

Condensation reaction

The contribution by Lee et al. [9, 10] not only demonstrated for the first time that well-defined mesoporous polymers can be obtained from block copolymer precursors, but also described all the key components of this approach, including formation of a microphase-separated precursor, cross-linking of the matrix block, and selective removal of the sacrificial polymer block to generate voids. They synthesized poly(4-vinylphenyl)dimethyl-2-propoxysilane-b-polyisoprene-b-poly(4-vinylphenyl)dimethyl-2-propoxysilane as the triblock copolymer precursor, to form a mesoporous polymer by anionic polymerization. The matrix poly(4-vinylphenyl)dimethyl-2-propoxysilane was cross-linked by immersing the precursor film (20 μm) in an aqueous solution of HCl at room temperature for 2 days, which hydrolyzed the isopropoxysilyl group and subsequently condensed the resulting silanol. The reaction was monitored by FTIR, which indicated the disappearance of a vibrational band corresponding to Si–O–C stretching, and the emergence of a new band corresponding to Si–O–Si stretching, which corroborated the quantitative conversion of isopropoxysilyl groups into siloxane linkages. Then the cross-linked film was subjected to ozonolysis in dichloromethane at −40 °C to selectively remove the PI middle block. Although the authors did not investigate differences in pore stability between the non-cross-linked and cross-linked porous polymer films, it appears that the cross-linking step was critical to the formation of the pore structure, as the non-cross-linked precursor could be dissolved in dichloromethane resulting in a nonporous product after ozonolysis.

Radiation

Chan et al. [29] utilized the different oxidation behaviors between a hydrocarbon polymer and a silicon-containing polymer to prepare a mesoporous ceramic thin film with gyroid morphology. A thin film of PI-b-poly(pentamethyldisilylstyrene)-PI triblock copolymer precursor synthesized via anionic polymerization was subjected to ozonolysis and UV radiation. While the PI was degraded by the ozonolysis and readily removed, the poly(pentamethyldisilylstyrene) was converted into silicon oxycarbide producing mesoporous silicon oxycarbide.

UV was also utilized to cross-link poly(t-butyl acrylate)-b-poly(2-cinnamoylethyl methacrylate) thin film; the cinnamoyl moiety underwent 2 + 2 dimerization under irradiation with UV [30–32]. For 25–50 nm thick film possessing hexagonally packed cylinders of poly(t-butyl acrylate) (PtBA), it was estimated by UV spectroscopy that ca. 35% of the cinnamoyl groups were consumed by 8 s of UV radiation. PtBA was converted into poly(acrylic acid) by soaking the cross-linked precursor in a dichloromethane solution of trimethylsilyl iodide for 2 weeks, followed by subsequent hydrolysis in a water/methanol mixture. Removal of the bulky t-butyl group from the PtBA produced cylindrical mesopores coated with poly (acrylic acid). Figure 4.4 shows TEM images of the non-cross-linked precursor and the cross-linked mesoporous polymer, showing that the hexagonally packed cylindrical morphology was preserved during the cross-linking and etching processes [32].

Using a 2 μm-thick cross-linked film with cylindrical mesopores having diameters of ca. 17 nm, the authors showed that water could permeate through the film, and that the transport was sensitive to pH and the existence of divalent cations due to the poly(acrylic acid) in the pores [31]. The porous polymer membrane could be also used as a host for inorganic nanoparticles [32].

While the mechanism is not clear, PS can be also cross-linked upon exposure to deep UV in a thin film state [33]. A combination of PS and poly(methyl methacrylate) (PMMA) in the block architecture makes a very interesting pair, since PMMA is a well known negative photoresist because of its photodegradability under UV exposure. Thurn-Albrecht et al. [33] showed that UV exposure of a thin film of PS-b-PMMA followed by rinsing with acetic acid could transform cylindrical microdomains of PMMA perpendicularly oriented to the substrate, into cylindrical mesopores while simultaneously cross-linking the PS matrix. This kind of topographic pattern can be utilized as a template for nanomaterial

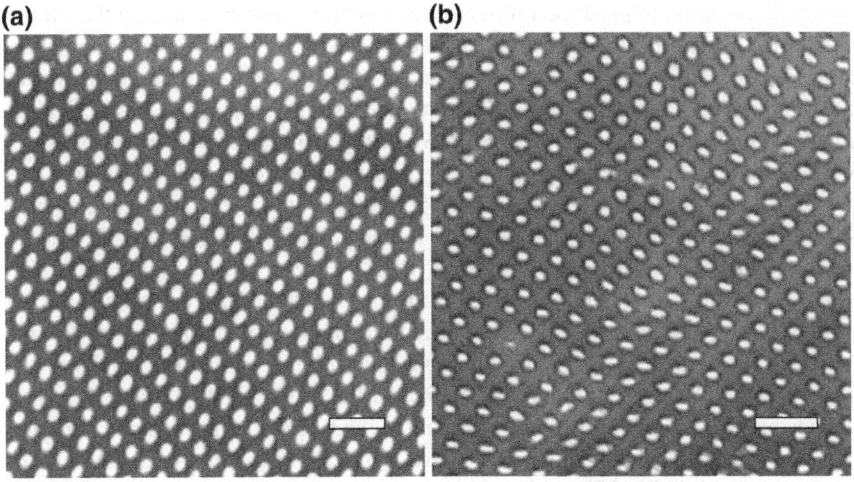

Fig. 4.4 TEM images of PtBA-b-poly(2-cinnamoylethyl methacrylate) thin film after thermal annealing (**a**), and the corresponding cross-linked mesoporous film obtained by UV irradiation and the subsequent removal of the t-butyl groups (**b**) [32]. Scale bars represent 100 nm. The observed distorted cylinders were probably produced by shear-induced deformation during microtomy. The image **a** was obtained after staining with OsO_4, while the image **b** was obtained without staining, exploiting the natural contrast between the cross-linked matrix and void. Adapted with permission of *Chem. Mater.* Copyright (1999) American Chemical Society

fabrication, or can be transferred to an underlying substrate using etching processes. This emerging technology is commonly referred to as block copolymer lithography [11]. It should be noted that UV irradiation has been also utilized with other PS-containing block polymers such as PS-b-PLA and PS-b-poly(2-vinylpyridine) (PS-b-P2VP) thin films, to cross-link the PS matrix.

Sekine et al. [34] demonstrated that γ-ray radiation can cross-link a 1,4-polybutadiene (1,4-PB) matrix by consuming the double bonds located on the main chain, and simultaneously degrade cylindrical PMMA microdomains in the PB-b-PMMA thick film (ca. 700 μm) to produce cross-linked mesoporous PB. PS-b-PMMA was also cross-linked by γ-ray radiation, but the cross-linking density was lower and the PMMA etching was less favored as compared with PB-b-PMMA presumably due to the protective effect of the aromatic rings in PS against γ-ray radiation.

Radical reaction

PI can be cross-linked via radical pathway to produce an elastomeric material, as discussed in the example reported by Cavicchi et al. [28]. It is believed that free radicals abstract allylic hydrogens in the PI backbone and the radicals generated on the allylic position of different PI chains undergo a combination reaction to produce a cross-link [35]. Dicumyl peroxide was used as a free radical source for the cross-linking of PI-b-PDMS. The reaction was conducted either by dispersing dicumyl peroxide on the surface of the polymer film and heating to 105–115 °C [35] or by preparing a solution of PI-b-PDMS and dicumyl peroxide in dichloromethane, casting the solution to produce a film after evaporation, and then heating the film to 150 °C [28]. Treating the cross-linked precursor with HF or tetrabutylammonium fluoride successfully produced mesoporous PI when the cross-linking density was sufficient to withstand the Laplace pressure. Interestingly, a pore-collapsed sample, and therefore featureless under SAXS showed a regeneration of scattering intensities corresponding to the parent microphase-separated morphology when the sample was swollen with toluene [36]. The reversible porosity was attributed to the fact that there is a lower interfacial tension between the toluene and PI than between air and PI, which would reduce the Laplace pressure [28].

1,2-PB containing a pendent double bond per repeating unit can be selectively synthesized by optimizing the polymerizing condition of butadiene in anionic polymerization, and it is more attractive than PI in terms of efficiency of cross-linking. The pendent double bonds in 1,2-PB can be readily attacked by free radicals and the reaction proceeds following a chain reaction mechanism, allowing one to effectively cross-link the PB matrix with significantly lower amounts of radical source than PI [37].

1,2-PB-b-PDMS has been extensively studied as a PB-containing block polymer precursor. In addition to cross-linked mesoporous polymers containing cylindrical mesopores [38], gyroid mesoporous membranes made of cross-linked PB have also been synthesized and tested as ultrafiltration membranes taking advantage of the 3D continuous pore structure (Li et al. [23, 39]).

It was also found that by varying the temperature of the cross-linking reaction, different phases can be arrested, to produce mesopores based on metastable morphologies [37]. For example, PB-b-PDMS with a molar mass of 14.2 kg mol^{-1} and 39% of PDMS volume fraction exhibited lamellar morphology at room temperature. With increasing temperature, the polymer showed an order-to-order transition from lamellae to tentatively hexagonally perforated lamellae (HPL), and then became gyroid, at 80 and 120 °C, respectively. It should be noted that the HPL phase is metastable. While cross-linking the precursor at 140 °C produced the cross-linked precursor with gyroid morphology, two-step cross-linking at 85 °C and then 140 °C successfully arrested the HPL morphology. This example demonstrates that a cross-linking methodology can be used to kinetically "lock-in" metastable morphologies, and even more enrich a pool of accessible nanostructures via block polymer self-assembly (Fig. 4.5).

Other external agents

Gaseous or readily vaporizable external agents have been utilized in the cross-linking reactions of matrix blocks, by exploiting the high diffusivity of the agents in the solid matrix. Ozone treatment was employed to cross-link the PS matrix block in PS-b-PMMA thin film containing PMMA homopolymer [40]. Cylindrical mesopores were generated by leaching the PMMA homopolymer out of the film. 1,4-diiodobutane vapor was used to cross-link a P2VP block in a thick film (ca. 1 mm thick) containing a mixture of PI-b-P2VP-b-PI with PI homopolymer [41]. The 1,4-diiodobutane links two pyridine rings in P2VP by forming alkyl pyridinium iodide salt (i.e., quaternization). By varying the volume fraction of PI, the blend formed different morphologies, including hexagonally packed cylinders,

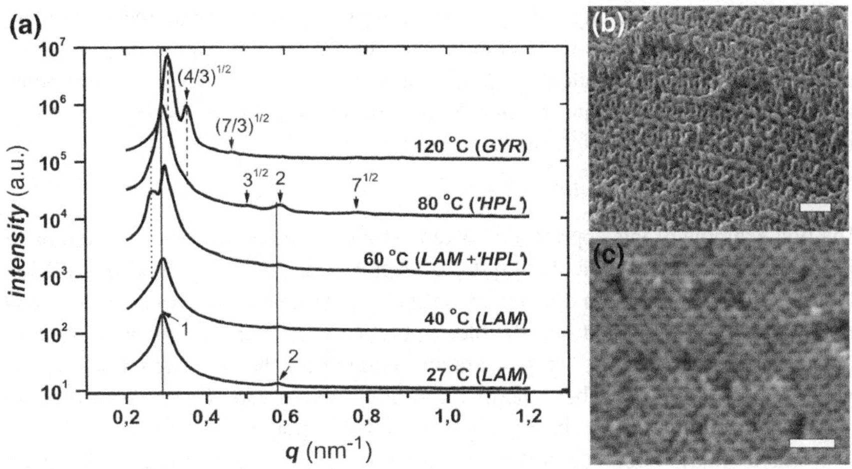

Fig. 4.5 Cross-linked mesoporous PB derived from PB-b-PDMS with a molar mass of 14.2 kg mol^{-1} and 39% of PDMS volume fraction [37]. **a** Temperature-dependent SAXS data of PB-b-PDMS. **b** SEM image of cross-linked mesoporous PB obtained by arresting the gyroid phase of the precursor via cross-linking at 140 °C. **c** SEM image of cross-linked mesoporous obtained by arresting the HPL metastable phase via two-step cross-linking at 85 °C and then 140 °C. Scale bars represent 100 nm. Adapted with permission of *Polymer.* Copyright (2011) Elsevier

gyroids, and lamellae. The gyroid morphology consisting of interpenetrating P2VP and PI microdomains was fixed by reaction with 1,4-diiodobutane, and then subjected to ozonolysis to remove the PI, which resulted in cross-linked mesoporous PI with gyroidal mesopores.

3.1.3.2 Cross-Linking Reaction Utilizing Functional Groups Introduced by Copolymerization

Incorporating cross-linkable functional groups into a matrix block by copolymerizing a functional monomer with a matrix monomer can conceptually decouple the degree of polymerization of the matrix block and its cross-linking density. These two variables can be independently controlled by varying the number of repeating units in the matrix block that are composed of the functional monomer, and those that are composed of the matrix monomer. The matrix monomer is supposed to play a primary role in microphase separation by providing sufficient incompatibility with the sacrificial block, so incorporation of the functional monomer may not significantly influence the microphase separation behavior of the block polymer precursor. A reactive block polymer precursor can be designed to employ a suitable cross-linking reaction, by devising a scheme for copolymerization of the matrix monomer with a functional monomer that possesses a reactive group, based on known block polymer precursors.

Since polymerization of the matrix monomer is typically well-controlled and produces a well-defined polymer with controlled molar mass and narrow dispersity, even monomers possessing highly reactive groups can be copolymerized up to some extent of incorporation, as long as the copolymerization proceeds in a controlled fashion. Controlled radical polymerizations such as reversible addition-fragmentation chain transfer (RAFT) polymerization would be desirable in this regard, owing to its high functionality tolerance.

Dimerization of benzocyclobutene

The first example of this approach involved benzocyclobutene as the reactive moiety for the cross-linking reaction [42, 43]. Benzocyclobutene effectively dimerizes at 180–250 °C without the addition of external agents [44]. As a result, the benzocyclobutene moiety can be incorporated into the block polymer precursors via copolymerization and remains intact, because the temperature required for the polymerization steps is typically below 150 °C. Then, thermal annealing above 180 °C can "turn on" the cross-linking reaction to produce a cross-linked block polymer precursor.

4-Vinylbenzocyclobutene, a styrenic monomer containing the benzocyclobutene moiety as a pendent group was synthesized and copolymerized with styrene via the RAFT process to form poly(styrene-co-4-vinylbenzocyclobutene) as the matrix block [42]. PMMA and PLA were investigated as sacrificial blocks, respectively [42, 43]. Cross-linking of the precursors was performed by thermal annealing of thin films at high temperatures, which also induced improved lateral ordering of the microdomains. A swelling experiment suggested that the cross-linking density may be lower than the highest possible value, assuming 100% dimerization, presumably due to limited diffusion of the benzocyclobutene moieties in the solid state. The cross-linked mesoporous films showed significantly enhanced dimensional stability against solvents and thermal treatments.

Addition polymerization

If a reactive block polymer precursor contains functional groups along the matrix block which can participate in polymerization, such as methacrylic double bonds for radical polymerization, and strained double bonds of norbornene group for ring opening metathesis polymerization (ROMP), cross-linking can be achieved by conducting addition polymerization of a new monomer in the presence of the reactive block polymer precursor. A multifunctional monomer (i.e., cross-linker) can be used to increase cross-linking density.

The resulting polymer will be composed of the block polymer precursor and the new thermoset polymer, where the matrix block and the thermoset polymer are covalently connected. The thermoset polymer should be properly chosen to ensure that it is miscible with the matrix block but incompatible with the sacrificial block, so that the microphase separation takes place between the sacrificial block and the remainder (i.e., matrix block and thermoset). For example, PMMA may not be suitable as the thermoset polymer if PLA is used as the sacrificial block because of their apparent miscibility [45]. Typically, a quite large amount of the new monomer is added (often with solvent) to the reactive block polymer precursor to make an initially homogeneous polymerization mixture. Incompatibility in the polymerization mixture becomes progressive as polymerization proceeds to grow the thermoset polymer, finally resulting in a phase-separated mixture. This process is called polymerization-induced phase separation (PIPS).

This concept was first demonstrated with a reactive block polymer precursor based on PLA-b-PS; the ROMP-active norbornene moiety was introduced to the matrix PS block by the copolymerization of styrene with *p*-norbornenylethylstyrene via the RAFT process [46]. A polymerization mixture was prepared consisting of the precursor, dicyclopentadiene (DCPD) as a cross-linker for ROMP, tetrahydrofuran (THF) as a solvent, and the second generation Grubbs catalyst. The double bond in the norbornene group as well as those in DCPD (i.e., the cross-linker) were spontaneously polymerized in the presence of the Grubbs catalyst to form a cross-linked gel within several minutes. After evaporation of the solvent, a transparent and bendable polymer film was obtained. The mesoporous polymer was then derived by PLA etching.

SAXS, TEM, and SEM were used to conduct structural investigation of the resulting thermoset and the corresponding porous polymer. The characterizations revealed a disordered bicontinuous structure with a length scale of several tens of nanometer, which was mostly developed by the PIPS process, was converted into a percolating mesopore structure (Fig. 4.6a, b). The 3D continuity of the porous space was evidenced by electrical conductivity measurements (after filling the pores with an ionic liquid) and gas permeation experiments. A mechanistic investigation indicated that, to prevent macrophase separation, a sufficient number of norbornene moieties needed to exist in the PS matrix, to participate in the ROMP process and generate a covalently joined network with poly(DCPD) [47]. Concentration of the block polymer precursor was also important for achieving the desired bicontinuous morphology. A transition to a morphology of encapsulated PLA microdomains was observed when the concentration was decreased below the overlap concentration. Control of pore size and surface area was demonstrated by tailoring molar mass of PLA and DCPD composition.

The PIPS process kinetically traps the bicontinuous morphology by cross-linking. A polymerization mixture was prepared to intentionally slow down the ROMP process, using a first generation Grubbs catalyst and triphenylphosphine as an inhibitor [48]. A waxy solid with a hexagonally packed cylindrical morphology could be obtained from a solution containing the block polymer precursor, DCPD (60 wt% relative to the precursor), the Grubbs catalyst and the inhibitor by evaporation of the solvent. The solid was slowly polymerized by heating to produce a cross-linked thermoset with ordered cylindrical microdomains, and was converted into a porous polymer with cylindrical mesopores. This indicates that morphological control of the cross-linked thermoset, from disordered bicontinuous to ordered morphologies, is possible by deliberately balancing the kinetic competition between the cross-linking of the matrix and the ordering of the microphase-separated structure. A study on the PIPS of vinyl monomers including styrene, divinylbenzene (DVB), MMA, and ethylene glycol dimethacrylate with PLA-b-poly (cyclooctene-co-5-norbornene-2-methylene methacrylate)-b-PLA also suggested that too early gelation may prevent the phase separation process, resulting in homogenous thermosets without nanostructures [45].

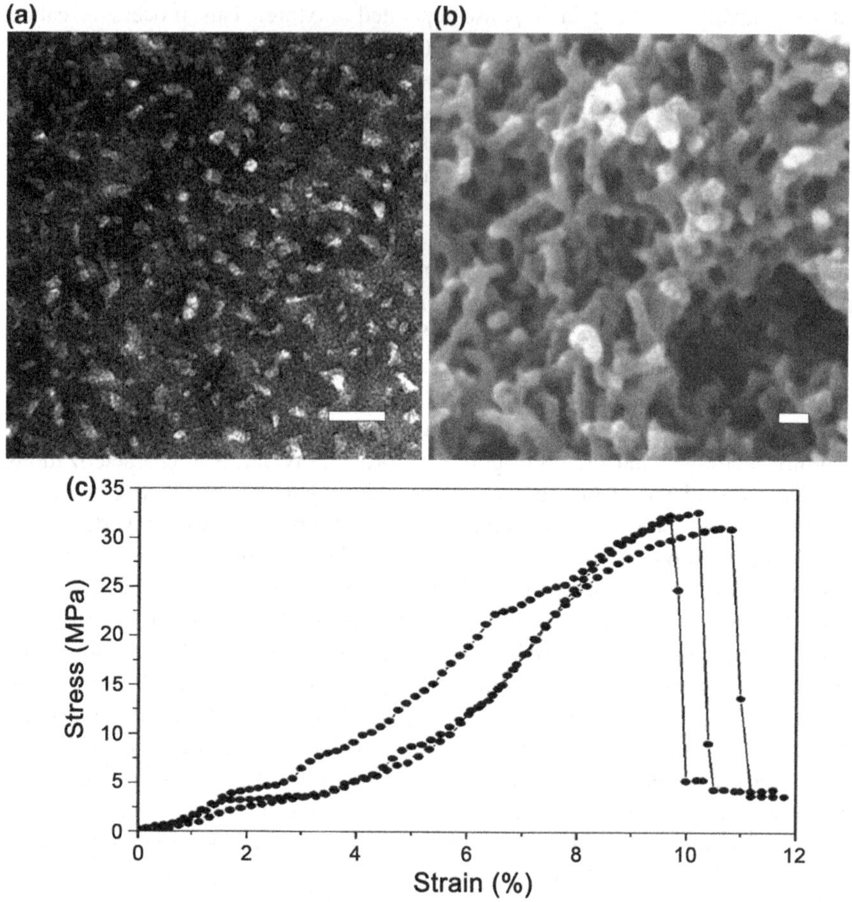

Fig. 4.6 Mesoporous poly(styrene-co-*p*-norbornenenylethylstyrene)/poly(DCPD) thermoset generated by PIPS process and subsequent PLA etching [46]. Scale bars represents 50 nm. **a** TEM image of an ultrathin section of the thermoset prior to PLA etching. The sample was stained with OsO_4. **b** SEM image of the mesoporous thermoset. **c** Stress-strain curve of the mesoporous thermoset (tested three times). Adapted with permission of *J. Am. Chem. Soc.* Copyright (2007) American Chemical Society

The mesoporous poly(styrene-co-*p*-norbornenenylethylstyrene)/poly(DCPD) thermoset showed remarkable mechanical properties. In contrast to non-cross-linked mesoporous PS which was very brittle and not suitable for tensile tests, the thermoset showed high ultimate elongation at break (ca. 10%), and tensile strength (32 MPa on average), as shown in Fig. 4.6c, comparable to pure poly(DCPD) (55 MPa) [46]. In addition, the thermoset retained its pore structure after annealing at 140 °C for 1 h (note that non-cross-linked mesoporous PS exhibited pore collapse at ca. 100 °C) or after swelling in THF and slow drying. All of these dramatically improved characteristics strongly indicate that cross-linking renders the pore structure much more stable against mechanical, thermal, and chemical stresses.

Reactions for step-growth polymerization

Cross-linking of the reactive block polymer precursor can also be achieved by utilizing organic reactions for step-growth polymerization, in which A and B functional groups react to form a covalent bond. The A functional group can be introduced into the matrix block by the copolymerization strategy, and reaction of the precursor with a molecule containing more than one B group produces a cross-linked block polymer precursor. It is important to choose functional groups that are highly reactive to each other so that the cross-linking reaction proceeds efficiently. The effect of the reaction on the microphase separation of the precursor should be also considered. The chemical structure of the B-containing monomer, the covalent bonds formed as a result of the reaction, and the resulting side products (if any) can affect the microphase separation behaviors of the precursor.

A hydroxyl group readily reacts with an isocyanate group to form a urethane linkage without any side products, and has been used for the industrial synthesis of polyurethane. Seo et al. [26] investigated the use of the alcohol/isocyanate reaction for the cross-linking of reactive block polymer precursors based on PLA-b-PS by introducing hydroxyl groups into the PS matrix block and reacting with methylene diphenyl diisocyanate (MDI). 2-Hydroxyethyl methacrylate and 2-hydroxyethyl acrylate were used as hydroxyl group-containing functional monomers and copolymerized with styrene via the RAFT polymerization process. A homogeneous solution of the precursor with MDI was prepared by using dichloromethane as a solvent, and a lightly cross-linked polymer film was obtained after evaporation of the solvent at room temperature.

Thermal annealing at 150 °C produced a fully cross-linked film as evidenced by FTIR. By keeping the incorporation ratio of hydroxyl groups in the PS matrix and varying the volume fraction of the PLA in the precursor, a variety of morphologies was obtained after cross-linking the matrix. In addition to lamellae and cylinders, the metastable HPL structure was also arrested by cross-linking. Figure 4.7 shows the evolution of the HPL structure as a function of annealing time at 150 °C, observed by SEM after PLA etching. The SEM, SAXS, and FTIR data suggested that incorporating MDI containing two aromatic rings into the PS matrix via urethane bond formation effectively increases the PS volume fraction in the precursor, inducing a phase transition from lamellar to more PS-rich phases. As the precursor is already lightly cross-linked prior to the thermal annealing, the phase transition appears to be limited within the lamellar structure, leading to the emergence of the HPL morphology. Acceleration of the cross-linking reaction by adding dibutyltin dilaurate arrested the disordered bicontinuous morphology consistent with previous studies.

Cross-linked mesoporous PS was successfully derived from precursors possessing disordered bicontinuous, HPL and hexagonally packed cylindrical morphologies, and showed higher T_g than non-cross-linked mesoporous PS, suggesting improved thermal stability could be achieved by cross-linking.

Seo and Hillmyer [49] also investigated cross-linking of the block polymer precursor via the reaction of acid chloride and amine. The amide-forming reaction

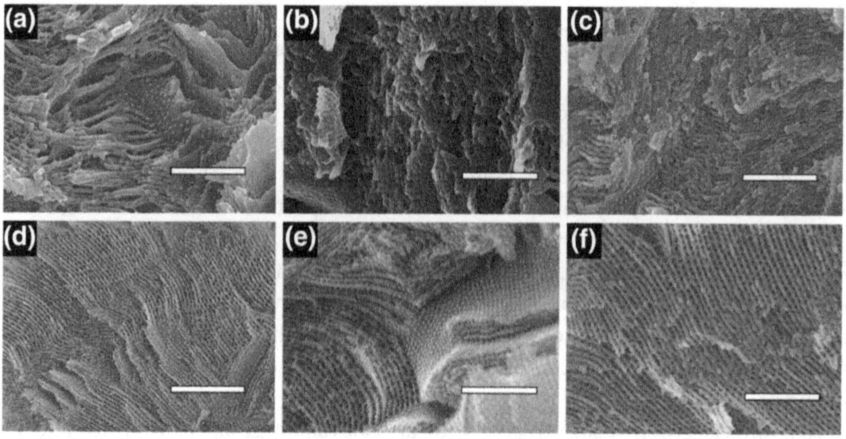

Fig. 4.7 Evolution of the HPL structure from a mixture of PLA-b-poly(styrene-co-2-hydroxyethyl methacrylate) and MDI upon thermal annealing at 150°C [26]. Annealing times were **a** 0 h, **b** 1 h, **c** 2 h, **d** 4 h, **e** 6 h, and **f** 24 h. The images were obtained by SEM after PLA etching. Scale bars represent 200 nm. Adapted with permission of *Macromolecules*. Copyright (2011) American Chemical Society

was specifically selected to exploit interfacial polymerization, which is industrially used to produce thin film composite (TFC) membranes used in reverse osmosis and nanofiltration applications [50]. The TFC membrane is composed of a densely cross-linked polyamide layer on top of a macroporous support, and the preparation involves casting of an organic solution, which consists of multivalent acid chloride molecules, on top of a macroporous support filled with an aqueous solution of *m*-phenylenediamine (MPD). The reaction of the acid chloride with MPD diffusing to the organic layer, produces a cross-linked polyamide thin film at the interface. As the diffusion of the MPD is self-limited by the formation of the film, the film thickness is naturally limited to less than 200 nm. HCl generated as a side product of the reaction is captured with the excess amine in the aqueous layer.

To this end, an acid chloride-containing PLA-b-PS block polymer was synthesized by copolymerizing 4-vinylbenzoyl chloride with styrene by the RAFT process [49]. Then the polymer was used in the interfacial polymerization process, replacing the multivalent acid chloride. By optimizing the organic solvent for the reaction, a TFC membrane consisting of a cross-linked mesoporous PS with cylindrical mesopores was derived, with potential utility for ultrafiltration applications [51].

3.1.3.3 In Situ Cross-Linking by Copolymerization of a Cross-Linker with the Matrix Monomer

Seo and Hillmyer [52] reported a very simple method to produce a cross-linked block polymer precursor with a disordered bicontinuous morphology, by

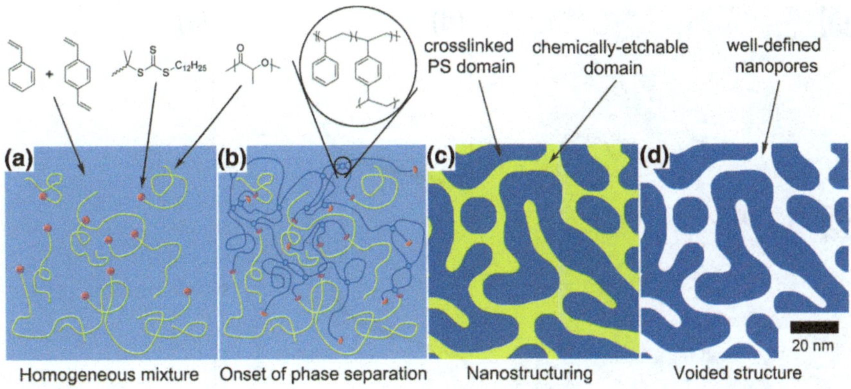

Fig. 4.8 Schematic depiction of cross-linked mesoporous polymer formation by PIMS process [52]. **a** A homogeneous mixture of PLA-CTA (*yellow*) in styrene and DVB (*light blue*). **b** RAFT copolymerization allows the growth of chains (*dark blue*) generating the PLA-b-P(S-co-DVB) block polymer. The growing chains are also being cross-linked in situ. **c** At a critical conversion point, microphase separation occurs and cross-linking arrests the emergent bicontinuous structure at a nanoscopic length scale. **d** Subsequent PLA etching produces percolating mesopores in a cross-linked and mechanically robust matrix. Reproduced with permission of *Science*. Copyright (2012) American Association for the Advancement of Science

copolymerization of styrene and DVB (25 mol% relative to styrene) in the presence of a PLA macro-chain transfer agent (PLA-CTA) in bulk. PLA-CTA allows styrene and DVB to copolymerize via the RAFT mechanism, and the P(S-co-DVB) copolymer grows from the PLA chain end. As polymerization proceeds, the molar mass and volume fraction of P(S-co-DVB) increase in the emergent PLA-b-P(S-co-DVB) block polymer driving the system toward microphase separation. Microphase separation occurs at a critical conversion point, and the emerging bicontinuous morphology is arrested by in situ cross-linking due to the incorporation of DVB in the polymerization mixture. The mechanism is illustrated in Fig. 4.8.

Compared with the previously described PIPS approach where the growth of a third polymer (thermosetting polymer) induces phase separation of the block polymer/thermosetting polymer mixture, this method is purely based on the microphase separation of the block polymer being synthesized. It is designated polymerization-induced microphase separation (PIMS) to clarify the difference. The PIMS method also greatly simplifies synthesis efforts. It does not require copolymerization of the functional monomer with the matrix monomer, or a separate cross-linking reaction because a densely cross-linked block polymer precursor is produced by the copolymerization itself. Purification of the precursor is not necessary as the yield of the copolymerization is almost quantitative and no solvent is used. The precursor can be molded into desired shapes, and further processed by milling and sanding thanks to its mechanical integrity.

Figure 4.9a–c shows SEM images of the cross-linked mesoporous polymers obtained after PLA etching, illustrating the percolating pore structure. As the molar

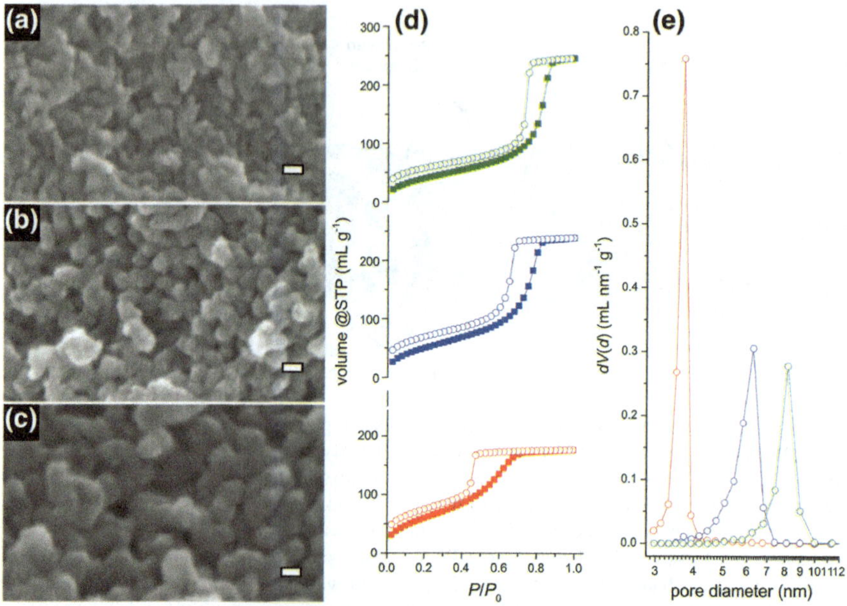

Fig. 4.9 Control over pore size of the cross-linked mesoporous polymer by varying the molar mass of PLA-CTA from 11 (image **a** and *red* data in **d** and **e**) to 22 (image **b** and *blue* data in **d** and **e**) and 41 kg mol^{-1} (image **c** and *green* data in **d** and **e**) [52]. **a–c** SEM images. Scale bars represent 20 nm. **d** Nitrogen adsorption isotherms. **e** Pore size distributions based on Barrett-Joyner-Halenda analysis of the desorption branches. Reproduced with permission of *Science*. Copyright (2012) American Association for the Advancement of Science

mass of the PLA-CTA was increased from 11 to 22 and 41 kg mol^{-1}, the pore size also increased. Nitrogen sorption data fully corroborated that pore size could be controlled by varying the molar mass of the PLA-CTA (Fig. 4.9d, e). When 11 kg mol^{-1} of PLA-CTA was used, the pore size was estimated to be 4 nm according to a Barrett-Joyner-Halenda analysis of the nitrogen sorption isotherm (the desorption branch). This value is remarkably small compared with other mesoporous polymers derived from block polymer self-assembly, and it appears that the densely cross-linked network generated by copolymerization of styrene and DVB via the RAFT process imparted sufficient stability to the pore structure to avoid collapse.

Because the PS matrix was densely cross-linked with DVB, the porous polymer showed no sign of an exothermal transition corresponding to pore collapse up to 150 °C on DSC. Nitrogen sorption analysis confirmed that 83% of the pore volume survived after heating the porous polymer at 100 °C for 1 h in air. The porous polymer also possessed robust mechanical properties, including an ultimate tensile strength of 6.4 MPa and an ultimate elongation of ca. 9%.

The key to the PIMS approach is ensuring that the matrix monomer (i.e., styrene and DVB) is compatible with the sacrificial block (i.e., PLA) in order to produce a homogeneous solution, but also ensuring that the matrix polymer (i.e., P(S-co-DVB)) is not, so that microphase separation can be induced during polymerization. Seo et al. [53] showed that even a sacrificial block can be simultaneously formed in the polymerization mixture to produce PLA-b-P(S-co-DVB) block polymer from a mixture of lactide, styrene, and DVB, literally in one step, if a heterofunctional initiator that can grow PLA on one side and P(S-co-DVB) on the other side is employed. PIMS can be also combined with other phase separation mechanisms such as PIPS to produce a hierarchical structure. Introducing poly(ethylene oxide) (PEO) homopolymer into a polymerization mixture of PLA-CTA, styrene, and DVB resulted ina polymerization-induced macrophase separation (i.e., PIPS) between the PEO and P(S-co-DVB), since PEO is miscible with PLA but incompatible with P (S-co-DVB) [54]. A bicontinuous morphology was also developed inside of the PLA-b-P(S-co-DVB) domains via the PIMS mechanism, and treating the precursor with a basic water/methanol mixture removed the PEO and PLA, forming a hierarchically porous polymer containing macro- and mesopores.

Different matrix monomers which satisfy the compatibility criteria can be used for the PIMS process. Seo et al. [55] investigated vinylbenzyl chloride as a new matrix monomer replacing styrene in PLA-b-P(S-co-DVB). Poly(vinylbenzyl chloride) is a well-known starting material for the synthesis of hyper-cross-linked polymers consisting of micropores smaller than 2 nm. Treating poly(vinylbenzyl chloride) with a Lewis acid such as $FeCl_3$ induces Friedel–Crafts alkylation and forms methylene bridges between the aromatic rings. This permanently fixes the free volume between chains and converts it into micropores [56].

A cross-linked block polymer precursor was successfully prepared from a mixture of PLA-CTA, vinylbenzyl chloride, and DVB via the PIMS process. While treating the precursor with base selectively removed the PLA and produced a mesoporous polymer very comparable to the one obtained from PLA-b-P (S-co-DVB), reacting the precursor with $FeCl_3$ induced hyper-cross-linking of the matrix microdomain and simultaneously degraded PLA, which resulted in a hierarchically porous polymer containing micropores within the mesopore wall. It should be noted that the cross-linked matrix produced by copolymerization with DVB was additionally cross-linked again by hyper-cross-linking, and the pore stability of the mesopore structure was improved even more.

Oh and Seo [57] recently showed that acrylate monomers, and isobornyl acrylate and ethylene glycol diacrylate in particular, can be combined with PLA-CTA in the PIMS process to produce a cross-linked block polymer as a precursor to the corresponding mesoporous polymer. The particular set of acrylate monomers was well suited for the rapid fabrication of a free-standing, cross-linked mesoporous polymer film using the photoinitiated RAFT polymerization process at room temperature. This route avoided the issues of lengthy polymerization time and high vapor pressure that are typical of thermally initiated polymerization methods.

3.1.3.4 Thermosetting Polymerization in the Presence of Block Polymer Precursors

Similar to the PIPS approach described above, where addition polymerization is conducted in the presence of a reactive block polymer precursor, step-growth polymerization, such as epoxy polymerization and phenol/formaldehyde condensation, can be performed with a block polymer precursor to generate thermosets. In this case, the block polymer precursor does not need to possess a functional group that participates in the polymerization. A nanostructured thermoset with the target microphase-separated morphology can be derived if the matrix block is miscible with the emergent thermosetting resin, while the sacrificial block is not.

Using poly(ethylene oxide)-b-poly(propylene oxide)-b-poly(ethylene oxide) (PEO-b-PPO-b-PEO) as the block polymer precursor, phenol/formaldehyde condensation and related reactions have been extensively studied, primarily for the fabrication of mesoporous carbons rather than mesoporous polymers, since the phenol/formaldehyde resin can be readily converted into carbon by thermal treatment. It should be noted that the methodology largely resembles the synthesis of mesoporous silica templating micelles of PEO-b-PPO-b-PEO by the sol-gel reaction of a silica precursor (Zhao et al. 1997). Meng et al. [58] and Tanaka et al. [59] independently reported the synthesis of mesoporous carbon based on this approach. Other polymers, such as PS-b-poly(4-vinylpyridine) [60] and PS-b-PEO [61], have been also used as the templates for mesoporous carbon synthesis.

It has been shown that mild thermal treatment of the cross-linked precursor obtained by phenol/formaldehyde condensation in the presence of PEO-b-PPO-b-PEO can selectively calcine the PPO block and produce a cross-linked mesoporous polymer [58, 62]. Recently, cross-linked mesoporous polyimide with phenol/formaldehyde resin was developed by adding oligomeric poly(amic acid) into a polymerization mixture consisting of PEO-b-PPO-b-PEO, phenol, and formaldehyde [63]. Nitrogen-doped mesoporous carbon could be obtained by further thermal treatment at higher temperature.

3.2 Conclusions

Deriving well-defined mesoporous polymers from block polymer precursors holds great promise, because the approach allows mesoporous polymers to be produced with precisely controlled pore size and structure from highly processable and chemically tunable precursors. For utilization in practical applications, however, the stability of the mesoporous polymers must be guaranteed including thermal stability, mechanical performance, and chemical resistance. While a number of cross-linking methodologies have been discussed in this chapter which can produce robust mesoporous polymers via cross-linking of the matrix block, it is certain that advances in organic and polymer chemistry will provide more refined tools for this purpose contributing to the development of new cross-linked mesoporous polymers in the future.

References

1. Ingham CJ, ter Maat J, de Vos WM (2012) Where bio meets nano: the many uses for nanoporous aluminum oxide in biotechnology. Biotechnol Adv 30:1089–1099
2. Kresge CT, Leonowicz ME, Roth WJ et al (1992) Ordered mesoporous molecular sieves synthesized by a liquid-crystal template mechanism. Nature 359:710–712
3. Beck JS, Vartuli JC, Roth WJ et al (1992) A new family of mesoporous molecular sieves prepared with liquid crystal templates. J Am Chem Soc 114:10834–10843
4. Barton TJ, Bull LM, Klemperer WG et al (1999) Tailored porous materials. Chem Mater 11:2633–2656
5. Zhao D, Feng J, Huo Q et al (1998) Triblock copolymer syntheses of mesoporous silica with periodic 50 to 300 angstrom pores. Science 279:548–552
6. Che S, Liu Z, Ohsuna T et al (2004) Synthesis and characterization of chiral mesoporous silica. Nature 429:281–284
7. Na K, Jo C, Kim J et al (2011) Directing zeolite structures into hierarchically nanoporous architectures. Science 333:328–332
8. Ulbricht M (2006) Advanced functional polymer membranes. Polymer 47:2217–2262
9. Lee JS, Hirao A, Nakahama S (1988) Polymerization of monomers containing functional silyl groups. 5. Synthesis of new porous membranes with functional groups. Macromolecules 21:274–276
10. Lee JS, Hirao A, Nakahama S (1989) Polymerization of monomers containing functional silyl groups. 7. Porous membranes with controlled microstructures. Macromolecules 22:2602–2606
11. Bang J, Jeong U, Ryu DY et al (2009) Block copolymer nanolithography: translation of molecular level control to nanoscale patterns. Adv Mater 21:4769–4792
12. Hillmyer MA (2005) Nanoporous materials from block polymer precursors. Adv Polym Sci 190:137–181
13. Gamys CG, Schumers JM, Mugemana C et al (2013) Pore-functionalized nanoporous materials derived from block copolymers. Macromol Rapid Commun 34:962–982
14. Olson DA, Chen L, Hillmyer MA (2008) Templating nanoporous polymers with ordered block copolymers. Chem Mater 20:869–890
15. Zhang Y, Sargent JL, Boudouris BW et al (2015) Nanoporous membranes generated from self-assembled block polymer precursors: Quo Vadis? J Appl Polym Sci 132:41683/1-17
16. Zalusky AS, Olayo-Valles R, Taylor CJ et al (2001) Mesoporous polystyrene monoliths. J Am Chem Soc 123:1519–1520
17. Todd EM, Hillmyer MA (2011) Porous polymers from self-assembled structures. In: Silverstein MS, Cameron NR, Hillmyer MA (eds) Porous polymers. Wiley, Hoboken, NJ, pp 31–78
18. Bates FS, Fredrickson GH (1999) Block copolymers—designer soft materials. Phys Today 52:32–38
19. Hamley IW (1998) The physics of block copolymers. Oxford University Press Inc, New York
20. Semenov AN (1985) Contribution to the theory of microphase layering in block-copolymer melts. Sov Phys JETP 61:733–742
21. Zalusky AS, Olayo-Valles R, Wolf JH et al (2002) Ordered nanoporous polymers from polystyrene-polylactide block copolymers. J Am Chem Soc 124:12761–12773
22. Meuler AJ, Hillmyer MA, Bates FS (2009) Ordered network mesostructures in block polymer materials. Macromolecules 42:7221–7250
23. Li L, Schulte L, Clausen LD et al (2011) Gyroid nanoporous membranes with tunable permeability. ACS Nano 5:7754–7766
24. Baruth A, Seo M, Lin CH et al (2014) Optimization of long-range order in solvent vapor annealed poly(styrene)-block-poly(lactide) thin films for nanolithography. ACS Appl Mater Interfaces 6:13770–13781

25. Uehara H, Yoshida T, Kakiage M et al (2006) Nanoporous polyethylene film prepared from bicontinuous crystalline/amorphous structure of block copolymer precursor. Macromolecules 39:3971–3974
26. Seo M, Amendt MA, Hillmyer MA (2011) Cross-linked nanoporous materials from reactive and multifunctional block polymers. Macromolecules 44:9310–9318
27. Muralidharan V, Hui CY (2004) Stability of nanoporous materials. Macromol Rapid Commun 25:1487–1490
28. Cavicchi KA, Zalusky AS, Hillmyer MA et al (2004) An ordered nanoporous monolith from an elastomeric crosslinked block copolymer precursor. Macromol Rapid Commun 25:704–709
29. Chan VZH, Hoffman J, Lee VY et al (1999) Ordered bicontinuous nanoporous and nanorelief ceramic films from self assembling polymer precursors. Science 286:1716–1719
30. Liu G, Ding J, Guo A et al (1997) Potential skin layers for membranes with tunable nanochannels. Macromolecules 30:1851–1853
31. Liu G, Ding J (1998) Diblock thin films with densely hexagonally packed nanochannels. Adv Mater 10:69–71
32. Liu G, Ding J, Hashimoto T et al (1999) Thin films with densely, regularly packed nanochannels: preparation, characterization, and applications. Chem Mater 11:2233–2240
33. Thurn-Albrecht T, Steiner R, DeRouchey J et al (2000) Nanoscopic templates from oriented block copolymer films. Adv Mater 12:787–791
34. Sekine R, Sato N, Matsuyama T et al (2007) Radiation-induced fabrication of polymer nanoporous materials from microphase-separated structure of diblock copolymers as a template. J Polym Sci, Part A: Polym Chem 45:5916–5922
35. Hansen MS, Vigild ME, Berg RH et al (2004) Nanoporous crosslinked polyisoprene from polyisoprene-polydimethylsiloxane block copolymer. Polym Bull 51:403–409
36. Szewczykowski PP, Anderson K, Schulte L et al (2009) Elastomers with reversible nanoporosity. Macromolecules 42:5636–5641
37. Schulte L, Grydgaard A, Jakobsen MR et al (2011) Nanoporous materials from stable and metastable structures of 1,2-PB-b-PDMS block copolymers. Polymer 52:422–429
38. Guo F, Andreasen JW, Vigild ME et al (2007) Influence of 1,2-PB matrix cross-linking on structure and properties of selectively etched 1,2-PB-b-PDMS block copolymers. Macromolecules 40:3669–3675
39. Li L, Szewczykowski P, Clausen LD et al (2011) Ultrafiltration by gyroid nanoporous polymer membranes. J Membr Sci 384:126–135
40. Jeong U, Ryu DY, Kim JK et al (2003) Volume contractions induced by crosslinking: a novel route to nanoporous polymer films. Adv Mater 15:1247–1250
41. Okumura A, Nishikawa Y, Hashimoto T (2006) Nano-fabrication of double gyroid network structure via ozonolysis of matrix phase of polyisoprene in poly(2-vinylpyridine)-*block*-polyisoprene films. Polymer 47:7805–7812
42. Drockenmuller E, Li LYT, Ryu DY et al (2005) Covalent stabilization of nanostructures: robust block copolymer templates from novel thermoreactive systems. J Polym Sci, Part A: Polym Chem 43:1028–1037
43. Leiston-Belanger JM, Russell TP, Drockenmuller E et al (2005) A thermal and manufacturable approach to stabilized diblock copolymer templates. Macromolecules 38:7676–7683
44. Harth E, Van Harn B, Lee VY et al (2002) A facile approach to architecturally defined nanoparticles via intramolecular chain collapse. J Am Chem Soc 124:8653–8660
45. Amendt MA, Pitet LM, Moench S et al (2012) Reactive triblock polymers from tandem ring-opening polymerization for nanostructured vinyl thermosets. Polym Chem 3:1827–1837
46. Chen L, Phillip WA, Cussler EL et al (2007) Robust nanoporous membranes templated by a doubly reactive block copolymer. J Am Chem Soc 129:13786–13787
47. Amendt MA, Chen L, Hillmyer MA (2010) Formation of nanostructured poly(dicyclopentadiene) thermosets using reactive block polymers. Macromolecules 43:3924–3934
48. Chen L, Hillmyer MA (2009) Mechanically and thermally robust ordered nanoporous monoliths using norbornene-functional block polymers. Macromolecules 42:4237–4243

49. Seo M, Hillmyer MA (2014) RAFT copolymerization of acid chloride-containing monomers. Polym Chem 5:213–219
50. Petersen RJ (1993) Composite reverse osmosis and nanofiltration membranes. J Membr Sci 83:81–150
51. Seo M, Moll D, Silvis C et al (2014) Interfacial polymerization of reactive block polymers for the preparation of composite ultrafiltration membranes. Ind Eng Chem Res 53:18575–18579
52. Seo M, Hillmyer MA (2012) Reticulated nanoporous polymers by controlled polymerization-induced microphase separation. Science 336:1422–1425
53. Seo M, Murphy CJ, Hillmyer MA (2013) One-step synthesis of cross-linked block polymer precursor to a nanoporous thermoset. ACS Macro Lett 2:617–620
54. Saba SA, Mousavi MPS, Bühlmann P et al (2015) Hierarchically porous polymer monoliths by combining controlled macro- and microphase separation. J Am Chem Soc 137:8896–8899
55. Seo M, Kim S, Oh J et al (2015) Hierarchically porous polymers from hyper-cross-linked block polymer precursors. J Am Chem Soc 137:600–603
56. Zhang X, Shen S, Fan L (2007) Studies progress of preparation, properties and applications of hyper-cross-linked polystyrene networks. J Mater Sci 42:7621–7629
57. Oh J, Seo M (2015) Photoinitiated polymerization-induced microphase separation for the preparation of nanoporous polymer films. ACS Macro Lett 4:1244–1248
58. Meng Y, Gu D, Zhang F et al (2005) Ordered mesoporous polymers and homologous carbon frameworks: amphiphilic surfactant templating and direct transformation. Angew Chem Int Ed 44:7053–7059
59. Tanaka S, Nishiyama N, Egashira Y et al (2005) Synthesis of ordered mesoporous carbons with channel structure from an organic–organic nanocomposite. Chem Commun 2125–2127
60. Kosonen H, Valkama S, Nykänen A et al (2006) Functional porous structures based on the pyrolysis of cured templates of block copolymer and phenolic resin. Adv Mater 18:201–205
61. Hu D, Xu Z, Zeng K et al (2010) From self-organized novolac resins to ordered nanoporous carbons. Macromolecules 43:2960–2969
62. Zhang F, Meng Y, Gu D et al (2005) A facile aqueous route to synthesize highly ordered mesoporous polymers and carbon frameworks with *Ia*3*d* *Ia*3*d* bicontinuous cubic structure. J Am Chem Soc 127:13508–13509
63. Liu Y, Ohnishi K, Sugimoto S et al (2014) Well-ordered mesoporous polymers and carbons based on imide-incorporated soft materials. Polym Chem 5:6452–6460
64. Phillip WA, O'Neill B, Rodwogin M et al (2010) Self-assembled block copolymer thin films as water filtration membranes. ACS Appl Mater Interfaces 2:847–853
65. Wang Z, Guo L, Wang Y (2015) Isoporous membranes with gradient porosity by selective swelling of UV-crosslinked block copolymers. J Membr Sci 476:449–456
66. Deng Y, Yu T, Wan Y et al (2007) Ordered mesoporous silicas and carbons with large accessible pores templated from amphiphilic diblock copolymer poly(ethylene oxide)-b-polystyrene. J Am Chem Soc 129:1690–1697

Chapter 4
Melt-Stretching Polyolefin Microporous Membrane

Lei Caihong and Xu Ruijie

4.1 Introduction

As a prominent part in the lithium battery (LIB), the microporous membrane (also called separator) plays a significant role on the battery's safety and electrochemical property, including ion conductivity, charge and discharge capacity, cycle performance, etc. The pore channel in the microporous membrane affords the passage for the Li ionic from negative to positive electrode, as shown in Fig. 4.1 and is significant for the battery electrochemical performance. Especially, with the development of recent global electrical and hybrid electrical vehicles, the demand for the safety and control of pore structure is more impendent. The main characteristics of microporous separator include porosity, pore size and distribution, and air permeability. The latter is often characterized by Gurley value, which is defined as the time required for a specific amount of air to pass through a specific area of the separator under a specific pressure. The lower the Gurley value is, the better the air permeability is. In addition, high mechanical property and better dimensional stability are also necessary for the application in the field of LIB [1].

Up to now, the industrialized methods for fabricating microporous separator include mono stretching dry method, bi-oriented dry method, and wet method. The corresponding pore structure is shown in Fig. 4.2. Here, polyolefin, as raw materials, is widely used, since it exhibits excellent chemical and electrochemical stability toward the electrolyte and electrode materials.

The wet method is based on thermal-induced phase separation mechanism and is used to prepare high-density polyethylene (HDPE) microporous membrane [2], which has been commercialized by Asahi and Tonen from Japan, Entek from USA,

L. Caihong (✉) · X. Ruijie
Guangdong Provincial Key Laboratory of Functional Soft Condensed Matter,
School of Materials and Energy, Guangdong University of Technology,
Guangzhou 510006, People's Republic of China
e-mail: lch528@gdut.edu.cn

P. Bettotti (ed.), *Submicron Porous Materials*,
DOI 10.1007/978-3-319-53035-2_4

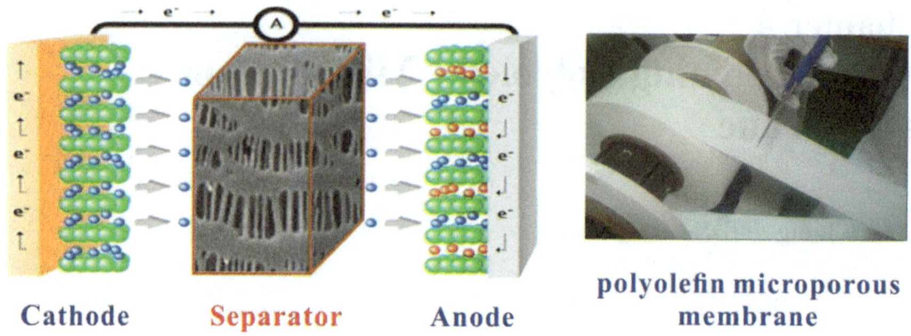

Fig. 4.1 The role of microporous separator in the battery and its appearance

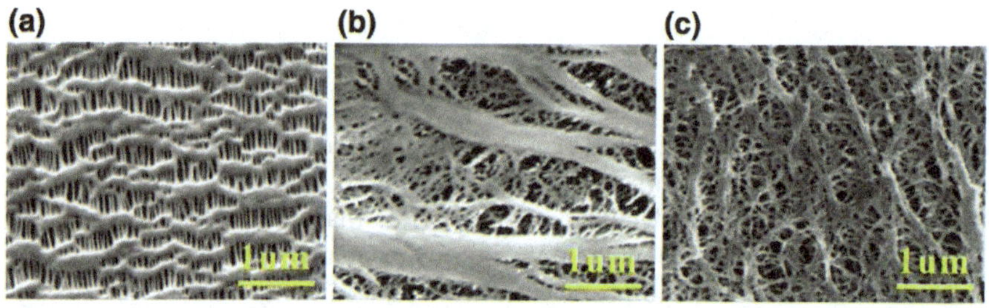

Fig. 4.2 SEM of microporous separator, **a** mono stretching dry method, **b** bi-oriented dry method, **c** bi-oriented wet method

SKC from Korea, Jinhui and Senior from China. During the process, the liquid paraffin is added to HDPE and the obtained compound is extruded into a flat sheet. Then the sheet is bi-oriented. The solvent such as dichloromethane is used to remove the liquid paraffin to form interconnected spherical or elliptical micropores, as shown in Fig. 4.2c. Finally, the obtained pore structure is heat set to improve the dimensional stability. Here, the solvent has to be used for the formation of pore structure, at the same time, it also leads to some environmental problems.

The bi-oriented polypropylene (PP) microporous membrane has also been used in the field of LIB. During the process, first, the initial film with high content of β crystals is prepared and then during the bi-oriented stretching, β crystals are transformed into stable α crystals and pore structure, similar to that in bi-oriented wet method, as shown in Fig. 4.2b, is initiated [3]. This method has been commercialized by Mitsubishi from Japan and Zhongke from China.

The melt-stretching method (mono stretching dry method) using pure polyolefin as raw materials has been used to fabricate PP and HDPE microporous membrane by Celgard (USA), Ube (Japan) and Senior (China) [4]. The process includes three main consecutive stages: (1) production of the precursor film with a row-nucleated lamellar morphology, (2) annealing of the film to thicken the lamellae, and (3) stretching of the film at low temperature to create voids and then stretching at high temperature to enlarge the pores [5, 6]. Slit-like pores are formed by stretching-induced lamellae separation, as shown in Fig. 4.2a. After these stages, the stretched films are heat set to improve their dimensional stability.

Compared with the other two methods, the melt-stretching method has the advantages such as no environmental pollution and uniform pore structure. The first related patent was proposed in 1969 [7]. Although this kind of method has been used to fabricate microporous membrane for more than 20 years by Celgard, the detailed open technical information is very limited. The number of basic research reports about the detailed material–processing–structure–property relationship in the web of science is near zero before 2000. Even now, the related articles in the web of science are less than 60. Our research group has been devoted to clarify the basic scientific and engineering problems during the fabrication since 2007.

Figure 4.3 shows the schematic of pore formation process based on melt-stretching mechanism. First, the pure polyolefin polymer is extruded and the initial row-nucleated crystalline structure is obtained under the applied stretching and temperature field from die to cast roll. Here, the main problems are the control of initial lamellar size and morphology during the stress-induced crystallization. Then annealing is carried out to improve the lamellae periodicity and increase the lamellar thickness. Cold stretching is used to initiate pores and then the pores are enlarged by hot stretching. Apparent connecting bridges among the separated lamellae are formed. Finally, the stretched pore structure is heat set to escape the membrane shrinkage. The main recent works related to this field will be reviewed in the following part.

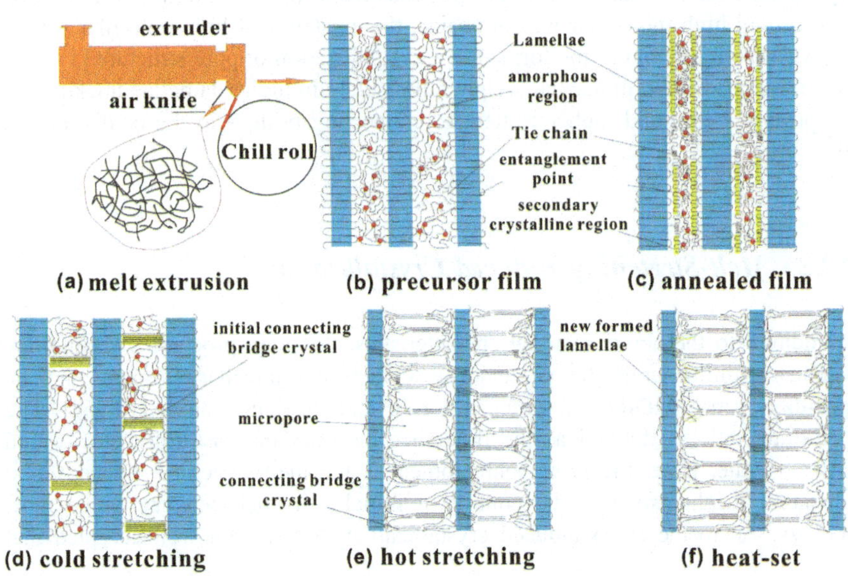

Fig. 4.3 Schematic of pore formation process based on melt-stretching mechanism

4.2 Melt-Stretching Polypropylene Microporous Membrane

4.2.1 *Raw Material Characteristics*

It has been ascertained that the structure of precursor film shows significant influence on the pore structure and properties of finally stretched microporous membrane. One of the main factors influencing the stress-induced crystallization behavior and the precursor row-nucleated lamellar morphology are material characteristics, including molecular weight, molecular weight distribution, branches degree, and isotactic degree. Usually, high molecular weight resins possessing long chains are good candidates for the generation of a proper row-nucleated lamellar since they form long fibrils (threads) that act as sites for lateral lamellae crystallization. Sadeghi et al. [8, 31] found that the resin with a higher molecular weight had a tendency to form a planar crystalline morphology. In another work [9], they reported that the addition of up to 10 wt% of a high molecular weight component to a low molecular weight one enhanced the formation of the row-nucleated structure. The corresponding membrane surface showed more pore density and uniform pore size. [10] have found that for the linear PP a peak in the pore size distribution around 0.10 μm was observed, whereas for the blend membrane with branched PP, the whole curve was shifted to higher values with a peak around 0.15 μm. The porosity was evaluated as 41 and 53% for the linear PP and the blend membranes, respectively.

Usually, the low molecular part is unfavorable for the formation of initial crystalline structure. However, in our previous work, it was found that for PP resin with certain high molecular weight chains, the existence of low molecular weight tail was also important to the formation and stabilization of pore structure [11, 30]. The existence of low molecular weight tail resulted in higher lamellae thickness in the precursor film and higher content of connecting bridges in the cold-stretched film.

4.2.2 *Melt-Stretching-Induced Crystallization*

In addition to the above material characteristics, the main processing technology parameters influencing the precursor row-nucleated lamellar structure include melt-draw ratio (MDR), air knife, cast roll, and die temperature [12, 52]. The melt-draw ratio is defined as the difference between the cast roll velocity and extrusion rate. Here, the prominent problem is the melt-stretching-induced crystallization mechanism and the control of initial row-nucleated crystalline morphology. As to the stress-induced crystallization, it has been well accepted that stress increases the crystallization rate and induces the crystal transformation [13, 14]. Different from usual research in the field of stress-induced crystallization [15, 16], where isothermal crystallization under the stretching field is followed, here, during the preparation of initial precursor film, non-isothermal crystallization during cooling from die to cast roll under stretching field complicates the crystallization behavior. The factors to characterize the initial lamellar structure are lamellae thickness, lamellar lateral size, orientation degree, and crystallinity, which can be characterized by SAXS (small-angle X-ray scattering), WAXS (wide-angle X-ray scattering), FTIR (Fourier transform infrared spectroscopy), and DSC (differential scanning calorimetry). The corresponding typical stress–strain behavior is no necking during stretching and high elastic recovery, the value of which can be more than 80%, whereas it is lower than 20% for usual PP film with spherulite structure.

Some studies have focused on the effect of MDR on the properties of initial film and stretched microporous membrane [8, 12, 17, 31, 52]. Sadeghi [5] indicated that the lamellae stack crystals appeared with increasing MDR. Recently, to build the relationship between lamellae size and the stretched pore structure, the initial film was prepared under different MDRs and the lamellae size was characterized by SAXS [18]. Figure 4.4 shows the SAXS pattern of initial film prepared by different MDRs and the corresponding stretched pore structure. The appearance of typical scattering pattern along meridional direction indicates the formation of oriented crystalline structure. With the increase of MDR, the lamellae thickness is kept at 4.7 nm, whereas the lamellar lateral size is increased from 19.1 to 54.6 nm. The orientation degree is increased from 0.23 to 0.41. The porosity of corresponding stretched membrane is increased from 37.8 to 60.3%. With the increase of MDR, the increase of orientation degree and lamellar lateral size, not lamellae thickness, is beneficial for the improvement of stretched porosity.

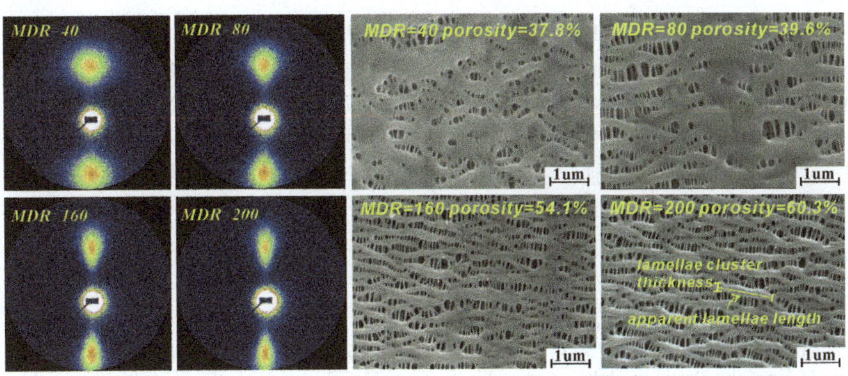

Fig. 4.4 SAXS patterns of initial film and structure of stretched microporous membrane

4.2.3 Annealing

Annealing is carried out to improve the lamellae structure periodicity and increase the lamellae thickness. Usually, annealing time around 3–5 min is needed. The annealing temperature is set between the onset of mobility in the crystalline structure and melting point. For PP microporous membrane, the Gurley value and porosity annealed at temperature of 145 °C are 150 s and 59.5%, respectively, whereas those annealed at temperature of 105 °C are 750 s and 50.8%, respectively [19, 54]. Annealing at temperature of 145 °C is appropriate for the preparation of microporous membrane.

Compared with the precursor film, the significant change after annealing is the appearance of a low temperature endotherm plateau (i.e., annealing peak) in the DSC curves and the decrease of yield stress in the stress–strain curves [19–24, 54, 56]. As to its origin, the accepted view is the crystallization of some chains among the amorphous region—secondary crystallization during annealing. Table 4.1 lists the yield stress, elastic modulus, calculated tie chains, and crystal plane orientation for films annealed at different temperatures. With the increase of annealing temperature, the elastic modulus is decreased and the corresponding calculated tie chains content is decreased from 2.12 to 0.91%. Only the crystal orientation along plane (110) is increased. Combined with the decrease of tie chains content, it could be deduced that during annealing, these disappeared tie chains crystalize along plane (110), leading to the improvement of orientation along plane (110) and the appearance of endotherm plateau in the DSC curves.

In addition to the secondary crystallization, during annealing at temperature such as 145 °C, the melting and recrystallization is inevitable, which has been proved by SAXS and TMDSC (Temperature modulated differential scanning calorimetry) [19,

Table 4.1 The yield stress, tie chains content, and crystal plane orientation

Sample	Precursor film	105 °C	115 °C	125 °C	135 °C	145 °C
Yield stress (MPa)*	35.8	28.1	27.9	26.2	24.7	24.2
Elastic modulus (MPa)*	1658	1129	1103	1006	970	876
Tie chain content (%)**	2.12	1.39	1.32	1.23	1.02	0.91
$A_{040/110+040}$***	0.59	0.55	0.57	0.56	0.57	0.57
$A_{130/110+130}$***	0.38	0.37	0.36	0.37	0.37	0.37
$A_{110/111+110}$***	0.74	0.78	0.81	0.84	0.86	0.88

Note
*obtained from the stress–strain curves of annealed film
**calculated based on $\beta = ((1 - X_c)E)/(E_c - X_c E)$ [25, 26], where E is the elastic modulus of the sample, Ec is the elastic modulus of crystalline region, 41 GPa and Xc is the crystallinity tested by DSC
***obtained from the WAXD testing of annealed films

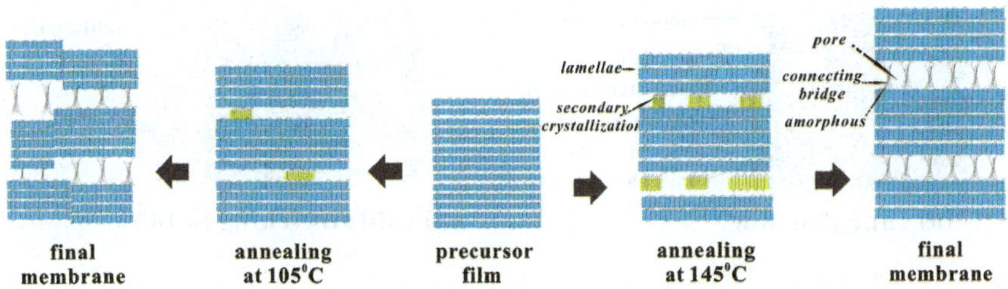

Fig. 4.5 Schematic representation showing the microstructural changes of films before and after stretching, annealed at 105 and 145 °C

23, 54]. Figure 4.5 shows the schematic of annealing-induced microporous structure at different temperatures. At annealing temperature such as 145 °C, the lamellar structure is more perfect due to more melting and recrystallization occurrence, compared with that at 105 °C. In addition, the content of secondary crystals is higher based on the decrease of tie chains content in Table 4.1. During stretching, the initial pore number from the secondary crystals is higher. The perfect lamellar structure can support the scaffold of pore structure. All these result in the uniform connecting bridge structure.

4.2.4 Cold Stretching

It is well accepted that cold stretching is carried out to initiate pores, but how to initiate pores is in fact unclear. In the series work reported by Carreau group, the cold stretching ratio was set at 35–40% [8, 9, 31]. Higher stretching ratio will induce the deformation of lamellae structure. Sadeghi et al. [28] attributed the pore initialization to the stretching and scission of short tie chains. In fact, during cold stretching, the stretching and movement of amorphous region is difficult due to the entanglement. In addition, the deterioration to crystalline structure by stretching is unavoidable. Hence, it is necessary to clarify the origin of initial pores during cold stretching.

The α crystal of PP shows specific mother–daughter crystalline structure [27]. In the annealed film with row-nucleated crystalline structure, main lamellae, melting and recrystallized regions, secondary crystallization from tie chains, daughter crystal and tie chains in the amorphous region coexist. Figure 4.6 gives the schematic of pore formation process [29, 44]. 10% stretching lead to the pronounced increase of amorphous region thickness and the appearance of a few initial connecting bridges. At this time, the daughter crystal content is decreased from 45% to 29% and the recrystallized part is decreased by 77.6%. The stretching of daughter crystal and recrystallized part contributes to the formation of initial bridges. At stretching ratio of 30%, uniform-distributed connecting bridges are observed. Here, except for the stretching of daughter crystal, the stretching of tie chains and

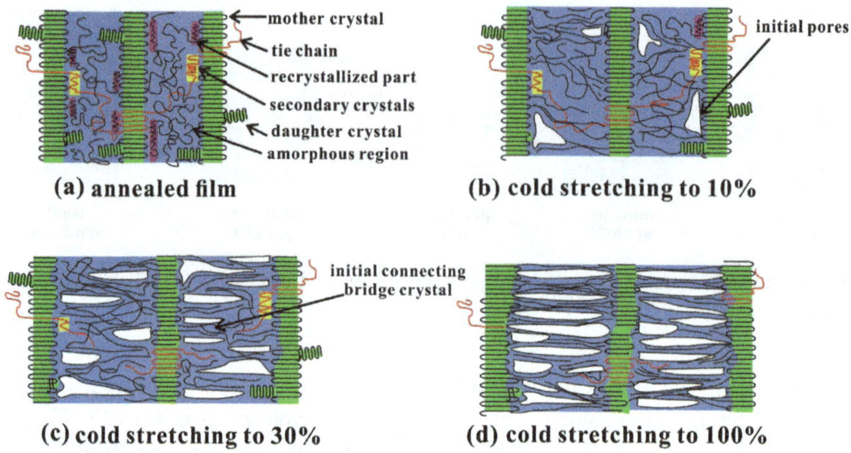

Fig. 4.6 Schematic of initial pore formation during cold stretching

secondary crystals within the amorphous region leads to the formation of more connecting bridges. During cold stretching, the pores are formed progressively and the initial few pores are from the deterioration of crystalline structure, not amorphous regions. At high stretching ratio to 100%, the stretching of main lamellae occurs and lamellae deformation is inevitable. The stretching ratio of 30%, situated at the transition point from plastic plateau to strain-hardening region in the stress–strain curves of annealed film, is appropriate for the initiation of pores.

4.2.5 Hot Stretching

Compared with that only through cold stretching, more stable connecting bridges can be formed after cold and hot stretching [11, 30]. Sadeghi et al. found that for a constant hot stretching level, the air permeability improved when the specific hot stretching temperature was increased [8, 31]. Recently, it was found that compared with cold and hot stretching under the same stretching rate and whole stretching ratio, the sample only by hot stretching showed lower Gurley value and better air permeability [32]. As shown in Fig. 4.7, the membrane only through hot stretching exhibits uniform lamellae arrangement and pore distribution. The corresponding air permeability is better than that of normal stretching process.

Compared with that by only cold stretching, the unrestrained tie chains due to the chain disentanglement at high temperature could be stretched and crystallized into bridges. At the same time, some unstable chains could also be pulled out from the initial lamellae and converted to fiber connecting bridges. All these can contribute to the increase in bridge number after hot stretching. But which one is more prominent to the formation of stable connecting bridges is still unclear. During the

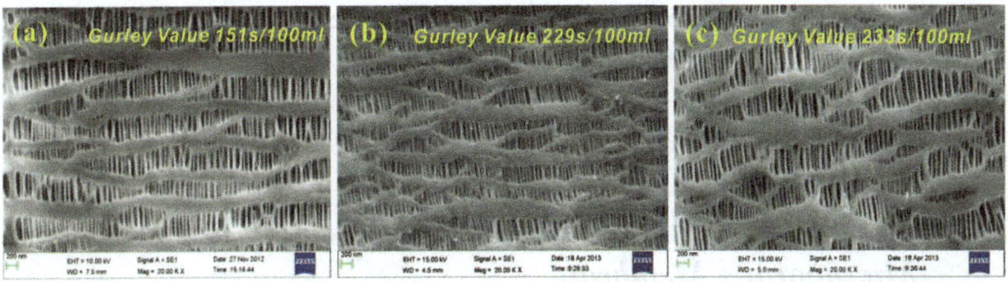

Fig. 4.7 The structure of different membranes **a** the membrane stretched 120% at 105 °C directly. **b** the membrane stretched 15% at room temperature and 105% at 105 °C. **c** the membrane stretched 15% at room temperature and 105% at 145 °C

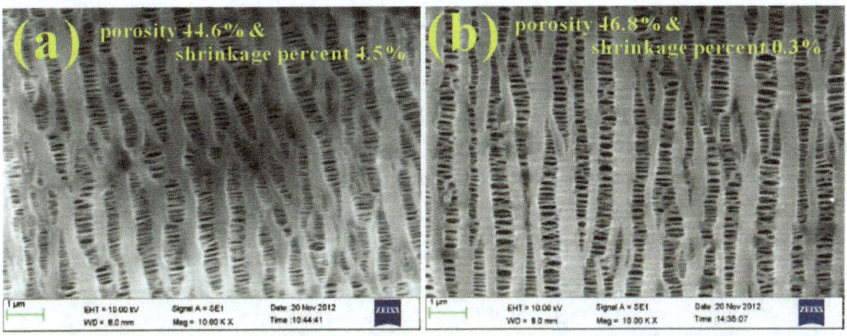

Fig. 4.8 SEM and property of stretched membrane heat set at different temperatures 135 °C (**a**) and 145 °C (**b**)

direct hot stretching, the crystalline part is stronger than that of amorphous entanglement network. Hence, during the stretching under higher temperature, first the tie chains between the lamellae are stretched and converted to connecting bridges. Less deterioration to lamellae structure occurs, resulting in better lamellae structure arrangement after stretching-induced lamellae separation.

4.2.6 Heat Setting

Normally, the heat setting increases the crystallinity, leading to increased tensile strength and modulus. Also the film after heat setting shows better dimensional stability because crystallization occurs [33]. For the preparation of microporous membrane based on melt-stretching mechanism, the heat setting is necessary since the above annealed film shows higher elastic recovery, the value of which is up to 96%. Since this, the annealed film is often called hard elastomer. Without heat setting, the stretched pore structure could not be kept after the stretching stress is relieved. Figure 4.8 shows the SEM micrograph and property of stretched membrane heat set at different temperatures. With the increase of heat setting temperature from 135 to 145 °C, the Gurley value is decreased from 245 to 190 s and the corresponding shrinkage percent under 105 oC for 1 h is decreased from 4.5 to 0.3% [34].

4.2.7 The Whole Change During the Fabrication of PP Microporous Membrane

During the preparation of microporous membrane, the change from initial no-pore structure to uniform-distributed pores is an interesting phenomenon. The change from initial amorphous melt state, to row-nucleated crystalline structure, to final separated lamellae and connecting crystalline bridges is a complex process. Figure 4.9 shows the whole microstructure change during the fabrication. Compared with that of precursor film, the lamellae thickness is increased obviously after annealing under 145 °C. After cold stretching, the lamellae are separated initially and a lot of voids appear. Compared with the only cold-stretched membrane, after hot stretching, the lamellae are separated further and many connecting bridge

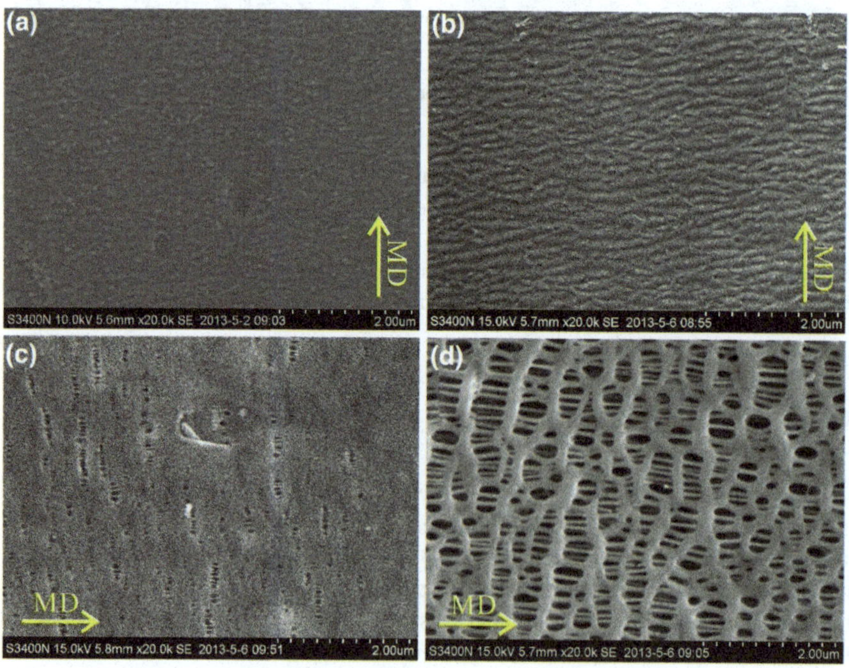

Fig. 4.9 SEM images during different stages of microporous preparation, MD is the machine direction (**a** precursor film, **b** annealed film, **c** cold-stretched membrane, **d** hot-stretched membrane)

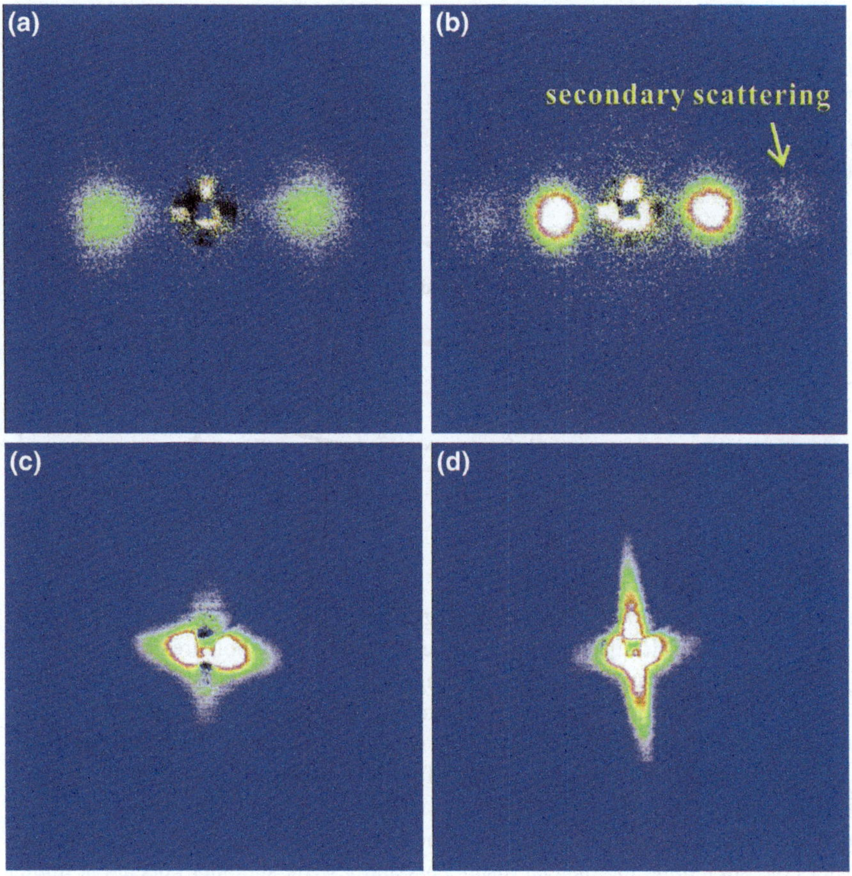

Fig. 4.10 SAXS patterns during different stages of microporous membrane preparation (**a** precursor film, **b** annealed film, **c** cold-stretched membrane, **d** hot-stretched membrane)

crystals emerge between the adjacent lamellae. The pores are enlarged obviously by hot stretching.

Figure 4.10 gives the SAXS patterns during different stages of microporous membrane preparation. For the precursor film, typical equational scattering pattern appears, indicating the formation of high-oriented crystalline structure. After annealing, the scattering pattern moves near to the beam stop, which means the increase of long period is based on the Bragg's law, $L = 2\pi/q_{\max}$, where q is the scattering vector. In addition, the appearance of secondary scattering indicates the improvement of structure periodicity by annealing [35]. After cold stretching, the scattering pattern in the equational direction moves further to the beam stop. The stretching induces the increase of long period, due to the formation of initial pores. At the same time, along the meridional direction, new scattering

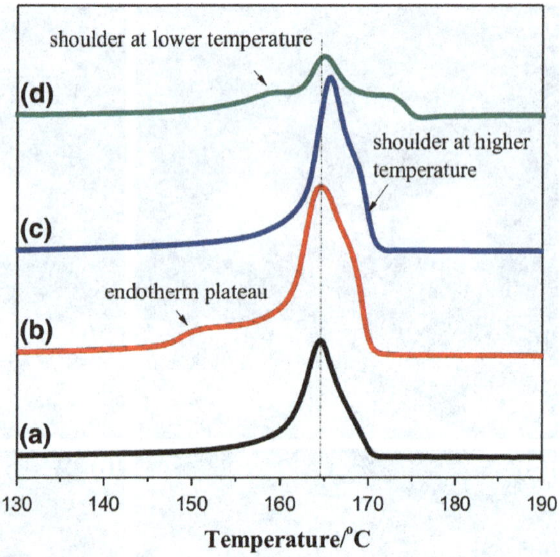

Fig. 4.11 DSC curves during different stages of microporous membrane preparation (**a** precursor film, **b** annealed film, **c** cold-stretched membrane, **d** hot-stretched membrane)

information appears. After hot stretching, these new scattering patterns become strong and sharp. More connecting bridges are formed by hot stretching.

Figure 4.11 presents DSC curves during the micropore formation process. After annealing, it can be seen that there is an endotherm plateau before the main melting point (T_m) and the main melting peak is appreciable enlarged. The plateau is situated around annealing temperature (T_a) and is attributed to tie chain crystallization around initial lamellae [36]. Meanwhile, after annealing, the increase of main melting peak area is ascribed by the lamellae melting and recrystallization. After cold stretching, the plateau disappears and a weak shoulder appears at higher temperature, contributed by the formed initial connecting bridges. Compared with the only cold stretching membrane, the melting peak after hot stretching becomes wide and two shoulders appear at both sides of main melting peak. After hot stretching, more connecting bridge crystals appear and they show more complete structure and higher melting point, resulting in the movement of melting curve to higher temperature [5]. As to the left plateau, on one hand, some chains are pulled out from main lamellae and converted to some connecting bridges during hot stretching, resulting in the decrease of lamellae structure perfectness and the movement of melting curve to lower temperature. On the other hand, the occurrence of crystallization during heat setting leads to the stability of stretched pore structure, at the same time, also leads to the appearance of left endotherm [34].

Figure 4.12 gives the stress–strain curves of different stages during the preparation of microporous membrane. For the initial precursor film, compared with normal PP film with spherulite structure, no necking occurs. After the yield point, a plastic plateau appears before the strain-hardening region. After annealing, the yield stress is decreased and pronounced strain-hardening behavior appears. Here, the decrease of yield stress indicates the decrease of tie chains content, based on the

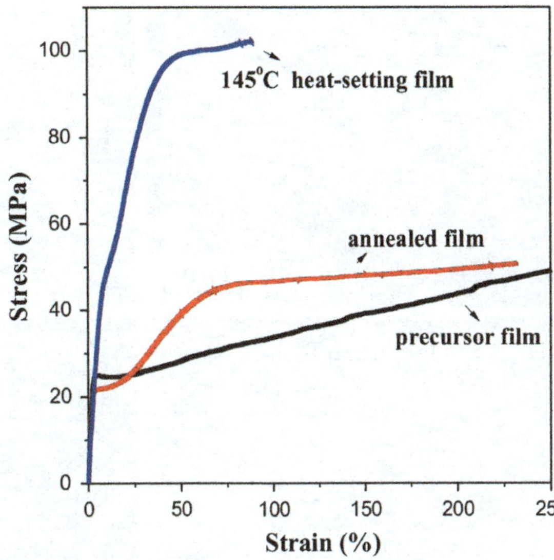

Fig. 4.12 Stress–strain curves during different stage of microporous membrane formation

relationship proposed by Nitta and Takayanagi [37]. Combined with the above-mentioned secondary crystallization during annealing, it can be deduced that these disappeared tie chains go into crystals, which are situated among the amorphous region. After stretching and heat setting, the plastic plateau becomes weak and strain-hardening becomes apparent. Since connecting crystal bridges are formed during stretching, compared with the lamellae structure, the connecting bridges crystal are more perfect, leading to more stress needed to stretching and pronounced strain-hardening phenomenon.

4.3 Melt-Stretching Polyethylene Microporous Membrane

Now, the industrialized PE microporous membrane is fabricated based on thermal-induced phase separation mechanism. However, since the solvent has to be used to produce pores, the environmental pollution and safety problems are inevitable. Compared with this, the melt-stretching PE microporous membrane will not face such problems.

4.3.1 Materials Characteristics

For the preparation of PE microporous membrane, Yu reported that the pore structure was more uniform for the resin of broader molecular weight distribution [38]. Usually, the melt flow index (MFI) is used to characterize the materials. From

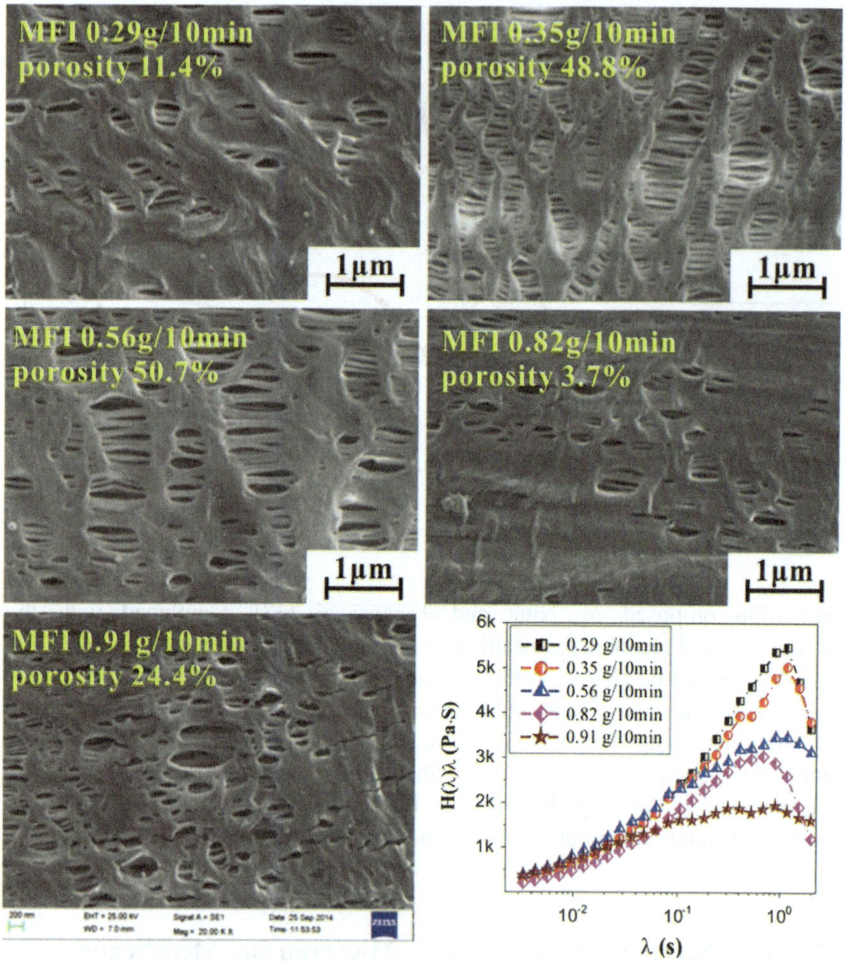

Fig. 4.13 SEM of stretched microporous membrane prepared by PE with different melt flow index and relaxation spectra

the related works, it seems that PE resins with melt flow index from 0.35 to 5.4 g/10 min can be used to fabricate microporous membrane [39–41]. Recently, five resins with different MFI values were chosen to prepare microporous membrane in our group [42]. Their relaxation spectra and corresponding microstructure of stretched microporous membrane are shown in Fig. 4.13. It is apparent that for the membrane prepared by PE with MFI value of 0.35 and 0.56 g/10 min, pronounced pores and connecting bridges appear and the porosity is higher than 45%. The corresponding Gurley value is lower than 220 s/100 mL, whereas it is larger than 3000 s/100 mL for other three materials. For the two materials with MFI value of 0.35 and 0.56 g/10 min, both larger molecular weight tail and smaller molecular weight parts exist and there is a weak relaxation shoulder on the left of main relaxation peak. The larger molecular weight tail affords the formation of long fibrils during extrusion casting. The smaller molecular weight parts support the crystalline ability, showing higher crystalline rate, resulting in larger lamellae thickness. After stretching, the separation of uniform lamellar structure leads to higher porosity. It can be seen that appropriate material characteristics are prominent for the fabrication of microporous membrane.

However, compared with that from thermal-induced phase separation mechanism, the disadvantage of melt-stretching microporous membrane is its lower puncture strength, due to only one-direction orientation. The puncture strength for the membrane with thickness of 18 μm prepared by melt-stretching is only 130 g, whereas the thickness of 16 μm from Asahi company based on thermal-induced phase separation mechanism is 328 g. Therefore, it is necessary to improve the puncture strength of HDPE microporous membrane prepared by melt-stretching. It has been found that the crystallinity of HDPE/LLDPE blend with LLDPE content of 5 wt% is increased from 65.8 to 73.0% due to the existence of cocrystallization behavior [43]. The porosity and air permeability property of stretched blend microporous membrane are decreased by 7.1% and 44.4%, respectively, whereas the puncture resistance is increased by 111.4% [44]. The introduction of LLDPE into HDPE affords a way to improve the puncture resistance, at the same time, not to deteriorate the pore structure seriously.

4.3.2 Annealing

Compared with that without annealing, the porosity of PE microporous membrane annealed at 125 °C for 2 h is increased from 43 to 63% and the corresponding Gurley value is decreased from 430 to 250 s [48]. It can be seen that annealing is an indispensable stage for the preparation of PE microporous membrane.

For PE single crystal or spherulites, Peterlin [45], Statton and Geil [46] and Sanchez et al. [47] have made detailed works about the influence of annealing on the crystalline structure, such as long period, recrystallization, crystal thickening. The above results during annealing process of PP precursor film with row-nucleated crystalline structure have shown that both secondary crystallization and melting and recrystallization phenomena exist. Recently, to clarify their contribution to lamellae thickness, the in situ annealing process under different temperatures using SAXS setup was followed and the calculated long period and lamellae thickness are shown in Fig. 4.14. Compared with that annealed at 105 and 115 °C, annealing at 125 °C induces pronounced long period and crystalline thickness improvement. During annealing, the crystalline thickness is apparently increased before 1800 s due to the secondary crystallization. However, after this stage, the thickening rate becomes

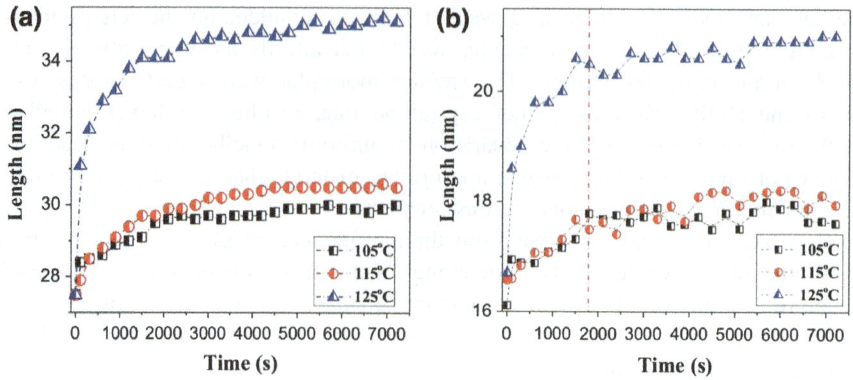

Fig. 4.14 Comparison of long period (**a**) and crystalline thickness (**b**) at annealing temperatures of 105, 115 and 125 °C

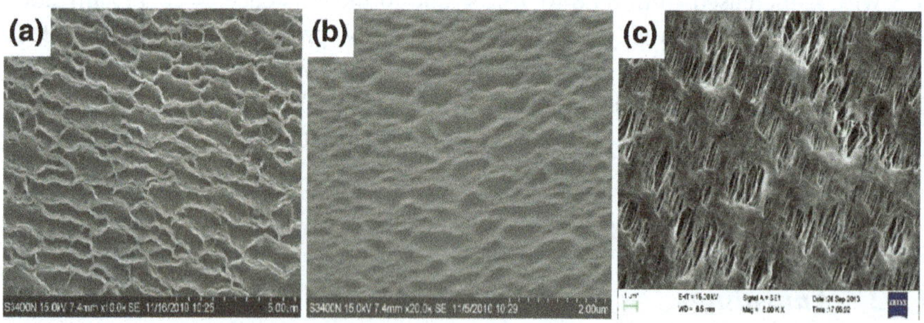

Fig. 4.15 SEM micrographs of precursor film (**a**), annealed film (**b**) and stretched PE microporous membrane (**c**)

very slow and the wave-like-increase appears, indicating the occurrence of melt recrystallization behavior. The increase of lamellar thickness is mainly from the secondary crystallization.

4.3.3 *The Whole Structure Change During the Preparation of PE Microporous Membrane*

Figure 4.15 shows the microstructure change from PE precursor film to stretched microporous membrane. Compared with precursor film, the annealed film shows increased lamellae thickness. Long connecting bridges appear after cold and hot stretching, the length of which is larger than that in PP membrane. But it is apparent that the pore distribution is worse than that in PP membrane.

Figure 4.16 gives the DSC curves of precursor film, annealed film, cold-stretched and hot-stretched membrane. Compared with precursor film, the

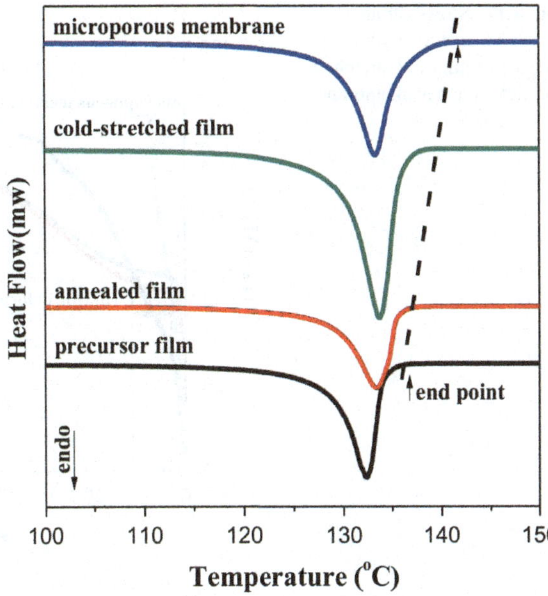

Fig. 4.16 DSC curves of precursor film, annealed film, cold-stretched and hot-stretched microporous membrane

melting peak of annealed film moves to higher temperature. Based on the Thomson equation [49], this indicates the increase of lamellae thickness. Different from that in annealed PP film shown in Fig. 4.11, here no endotherm plateau appears. After cold stretching, the end point moves right. This behavior is more pronounced after hot stretching. The existence of stretched pronounced connecting bridges in Fig. 4.15 results in the movement of end point.

Figure 4.17 gives the stress–strain curves of precursor film, annealed film, cold-stretched and hot-stretched membrane. Similar to that in annealed PP film, here, after annealing, the yield stress is decreased. After cold and hot stretching, the plastic plateau disappears.

4.4 PP/PE/PP Trilayer Microporous Membrane

Compared with pure PP or PE microporous membrane, the trilayer microporous membrane exhibits safety advantages when it is used in the field of LIB. When the temperature in the battery is increased to near the melting point of PE, the middle PE layer will melt and the corresponding pore structure will be closed, leading to the breakage of ion passage. At the same time, the outside PP layer still keeps the membrane size, escaping from the PE-shrinkage-induced short circuit and affording high safety. The trilayer microporous membrane has been commercialized by Celgard (USA). It can be prepared by laminating PP and PE precursor film and then stretching to initiate pores [50]. It can also be prepared by coextrusion and

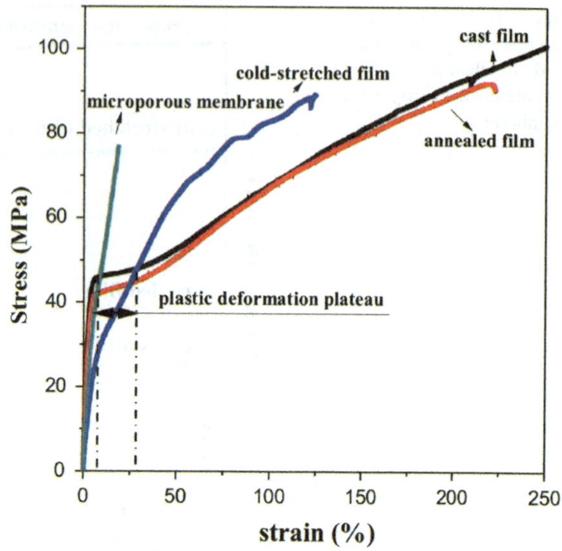

Fig. 4.17 Stress–strain curves of precursor film, annealed film, cold-stretched and hot-stretched membrane

stretching [51]. But since PE and PP exhibit different rheological and crystalline behavior, it is difficult to control the initial row-nucleated crystalline structure in coextruded PP and PE layers. As mentioned above, for the preparation of microporous membrane, different appropriate annealing and stretching temperature window exist for PP and PE, resulting in the difficulty of stretched pore structure formation in the coextruded trilayer membrane. Up to now, few works give the detailed investigation as to the influence of materials characteristics, processing technology including melt-stretching, annealing and cold–hot stretching stage on the structure and property of trilayer microporous membrane [12, 52].

4.5 Compound Microporous Membrane

Even though the stretched microporous membrane shows some orientation degree due to the contribution of stretching field during the whole process, its tensile strength and modulus are still unsatisfactory due to the existence of about 40% porosity. Recently, a low thickness with a tendency toward 10 μm is required for high energy and power densities of LIB. However, this adversely affects the mechanical strength and the battery safety. The nano-reinforcement has been widely used in the normal polymer modification. However, for the preparation of microporous membrane based on melt-stretching mechanism, the introduction of nano-filler may destroy the initial row-nucleated crystalline structure, due to its nucleating effect. In addition, the formation of some pores induced by the interphase separation of nano-filler with polymer matrix during stretching may also destroy the pore size distribution. Few works have been carried out to reinforce the microporous membrane by using nano-fillers.

Recently, nano-meter silicon dioxide was introduced into PP and the compound membrane was prepared. It was found that when the silicon dioxide content was 2 wt%, the mean pore diameter of the stretched microporous membrane was increased from 28.3 to 32.5 nm. The Gurley value only decreased from 200 to 194 s. The tensile strength and elastic modulus were increased by 33.3 MPa and 72.0 MPa, respectively [53]. Similar phenomena were observed for the compound membrane of PP/Al_2O_3 [19, 54]. The introduction of some inorganic nano-fillers leads to the improvement of mechanical property, at the same time, the electrolyte uptake could also be increased [22, 55, 56].

4.6 Ceramic-Coated Membrane

With the development of electric vehicles (EVs) and energy storage systems (EESs) in the world, the demand for LIB with higher safety is increasing. One of the important factors influencing the safety of LIB is the dimensional stability of separator under thermal shock. Normally, PP and PE microporous membranes exhibit large thermal shrinkage at high temperatures, which causes a short circuit between electrodes in cases of unusual heat generation.

To address this issue, many efforts have been devoted to minimize the separator shrinkage upon thermal shock, amongst which coating ceramic layer onto polyolefin membranes has been demonstrated [57, 58]. The introducing coating layer will effectively decrease the thermal shrinkage at high temperature and increase the electrolyte uptake and retention by the separator and show better electrochemical performance. Nowadays, the separator with alumina (Al_2O_3) coating layer is the most successfully commercialized product [59]. Some separators with new doped ceramic powders are in-depth studied, such as titania, silica oxide, magnesium hydrate, silica tube [60–63].

Boehmite, a flake-like powder, is an aluminum oxide hydroxide (γ-AlO(OH)) mineral and is a component of the aluminum ore bauxite. Figure 4.18 shows the microstructure and property of coated membrane using polyvinyl alcohol (PVA) as the binder. The shrinkage onset temperature for the separator coated by boehmite with a diameter of 0.78 μm is up to 170 °C and the shrinkage percent at 200 °C is lower than 5%. The coated sample displays higher ionic conductivity (0.72×10^{-4} S cm^{-1}) than PP (0.42×10^{-4} S cm^{-1}). The capacity retention is close to 99% after 50 cycles at 1C charge/discharge when using coated membrane as separator. When charged at 0.2 C and discharged at 0.1 C, 0.2 C, 0.5 C, and 1 C and 0.1 C, the capacity retentions of $LiFePO_4$ tested with PP membrane are 127.5, 125.7, 121.7, and 111.3 $mAhg^{-1}$, respectively, which are smaller than those for the corresponding coated membrane (140.3, 138.2, 133.9, and 123.9 $mAhg^{-1}$ for 0.1 C, 0.2 C, 0.5 C and 1.0 C, respectively). Compared with pure PP membrane, the coating layer containing the boehmite particles provides better electrolyte

Fig. 4.18 Surface and cross-section morphology of membranes before and after coating, rate capabilities of cells assembled with coated membrane at various current densities

wettability and electrolyte uptake. The higher electrolyte uptake induces Li^+ transport rate increase and the interface resistance between the electrode and membrane turns small.

4.7 Polymer Powder Coated Membrane

Except for the above ceramic coating, some works directly apply polymers onto the surface of polyolefin membrane to improve the electrochemical performance, such as cellulose, polyimide, polyvinylidene, and fluoride [64–66]. Normally, to apply polymer onto the surface, polymer solution has to be prepared and coated. Here, the organic solvent has to be used, resulting in some environmental problem. In addition, the solution coated polymer layer is structurally dense, which inevitably blocks the penetration of liquid electrolyte into the pores of the microporous membrane. Similar to the above ceramic coating, the polymer powders are proposed to directly coat on the surface of membrane using some binders such as PVA. Figure 4.19 gives the coated microstructure and corresponding electrochemical performance of PVDF-powder coated membrane. The ionic conductivity of coated membrane is 0.59×10^{-4} S cm^{-1}, much higher than that of pure PP membrane (0.42×10^{-4} S cm^{-1}). The liquid electrolyte contact angle is decreased from 99° to 0°. The discharge capacities with PVDF separators are higher than those of cells with PP separator for all C rates tested. This superior rate capability can be ascribed not only to the higher ion conductivity of PVDF separator, but also to lower interfacial resistance due to better interfacial contacts between the PVDF separator and electrodes in the lithium-ion assembled cell.

4.8 Future Prospective

With the development of EV and HEV, the demand for high property separator is increasing. But for a long time, the microporous membranes prepared by mono stretching dry method, bi-oriented dry method, and bi-oriented wet method are still

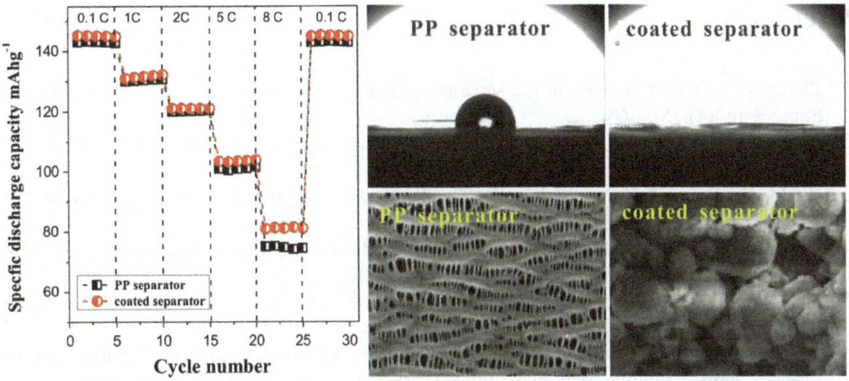

Fig. 4.19 Discharge capacities of the lithium-ion batteries assembled with various separators as a function of C rate, the liquid electrolyte contact angle and microstructure of coated separator

during the rapid development stage, except that new separators such as solid electrolyte can be commercialized.

Although the microporous membrane prepared based on melt-stretching mechanism has been commercialized for more than 20 years, the basic scientific and engineering problems during the preparation are still unclear, including the stress-induced crystallization during the preparation of precursor film, stretching-induced deterioration and lamellae to fiber connecting bridge crystals transformation during cold and hot stretching. The further research of these problems will no doubt be beneficial to improve the pore control in the stretched microporous membrane. In fact, clarifying these basic problems is also prominent for the fabrication of high property fiber and films.

The coated membranes by using inorganic filler show higher dimensional stability and better initial cycle performance. Although this kind of new separator has been commercialized recently and widely used in the LIB, the long-term battery performance is still in doubt. In addition, the liquid electrolyte in the battery is inevitable to induce some safety problems. Hence, it is expected that in the near future the full solid electrolyte can be widely used in the battery.

Acknowledgements The authors would like to thank National Science Foundation of China under Grant No. 51003017, the Project of High Level Talents in Higher School, Guangdong Province Major Key Projects of Applied Research and Development of Science and Technology (2015B090925021), Guangdong Province Science and Technology Plan Project (2016A010103030), Guangdong Province Ordinary University Innovation Project, PhD Start-up Project of Guangdong Province Natural Science Foundation (2016A030310344) and Postdoctoral Science Foundation Project (2016M592461). They also want to thank Shenzhen Senior Materials Company, Ltd., for generously supplying raw materials.

References

1. Zhang SS (2007) A review on the separators of liquid electrolyte Li-ion batteries. J Power Sources 164(1):351–364
2. Hasegawa T, Kondo T (2000) Short circuit-resistant polyethylene microporous film. US. patent 6054498
3. Xu M, Hu S, Guan J, Sun X, Wu W, Zhu W (1992) Polypropylene microporous film. US patent 5134174
4. Hashimoto A, Yagi K, Mantoku H (2000) Porous film of high molecular weight polyolefin and process for producing same. US. patent 6048607
5. Sadeghi F (2007) Developing of microporous polypropylene by stretching. PhD Thesis, Ecole Polytechnique de Montreal, Canada
6. Johnson MB (2000) Investigations of the processing-structure–property relationship of selected semi-crystalline polymers. PhD Thesis, Virginia Polytechnic Institute and State University, United States
7. Zimmerman D (1969) Production of novel open-celled microporous film. U.S. patent 3801692
8. Sadeghi F, Ajji A, Carreau PJ (2007) Analysis of row nucleated lamellar morphology of polypropylene obtained from the cast film process: effect of melt rheology and process conditions. Polym Eng Sci 47(7):1170–1178
9. Sadeghi F, Ajji A, Carreau PJ (2008) Microporous membranes obtained from polypropylene blends with superior permeability properties. J Polym Sci Part B Polym Phys 46(2):148–157
10. Tabatabaei SH, Carreau PJ, Ajji A (2008) Microporous membranes obtained from polypropylene blend films by stretching. J Membr Sci 325(2):772–782
11. Lei CH, Wu SQ, Xu RJ, Peng XL, Shi WQ, Hu B (2013) Influence of low molecular weight tail of polypropylene resin on the pore structure by room-temperature stretching. Polym Eng Sci 53(12):2594–2602
12. Tabatabaei SH, Carreau PJ, Ajji A (2009) Effect of processing on the crystalline orientation, morphology and mechanical properties of polypropylene cast films and microporous membrane formation. Polymer 50:4228–4240
13. Keller A, Kolnaar HWH (1997) Flow induced orientation and structure formation, vol 18. VCH, New York
14. Flory PJ (1947) Thermodynamics of crystallization in high polymers. I. Crystallization induced by stretching. J Chem Phys 15:397–408
15. Liu D, Tian N, Huang ND, Cui KP, Wang Z, Hu TT, Yang HR, Li XY, Li LB (2014) Correlation between flow-induced nucleation morphologies and strain in polyethylene: from un-correlated point nuclei, scaffold-network, micro-shish to shish. Macromolecules 47:6813–6823
16. Nogales A, Hsiao BS, Somani RH, Srinivas S, Tsou AH, Balta-Calleja FJ, Ezquerra TA (2001) Shear-induced crystallization of isotactic polypropylene with different molecular weight distribution: in situ small- and wide-angle X-ray scattering studies. Polymer 42 (12):5247–5256
17. Lin Y, Meng L, Wu L, Li X, Chen X, Zhang Q et al (2015) A semi-quantitative deformation model for pore formation in isotactic polypropylene microporous membrane. Polymer 80:214–227
18. Xu RJ, Chen XD, Xie JY, Cai Q, Lei CH (2015) Influence of melt-draw ratio on the crystalline structure and properties of polypropylene cast film and stretched microporous membrane. Ind Eng Chem Res 54(11):2991–2999
19. Cai Q, Xu RJ, Lei CH, Li LB (2015) Influence of annealing temperature on the lamellae and connecting bridge structure of stretched polypropylene microporous membrane. Polym Int 64 (3):446–452

20. Ding ZT, Bao RY, Zhao B, Yan J, Liu ZY, Yang MB (2013) Effects of annealing on structure and deformation mechanism of isotactic polypropylene film with row-nucleated lamellar structure. J Appl Polym Sci 130(3):1659–1666
21. Liu DM, Kang J, Xiang M, Cao Y (2013) Effect of annealing on phase structure and mechanical behaviours of polypropylene hard elastic films. J Polym Res 20(5):1–7
22. Saffar A, Ajji A, Carreau PJ, Kamal MR (2014) The impact of new crystalline lamellae formation during annealing on the properties of polypropylene based films and membranes. Polymer 55(14):3156–3167
23. Jiang ZY, Tang YJ, Rieger J, Enderle HF, Lilge D, Roth SV, Gehrke R, Wu ZH, Li ZH, Li XH, Men YF (2010) Structural evolution of melt-drawn transparent high-density polyethylene during heating and annealing: Synchrotron small-angle X-ray scattering study. Eur Polymer J 46(9):1866–1877
24. Lei CH, Huang WL, Xu RJ, Xu YQ (2012) The correlation between the lower temperature melting plateau endotherm and the stretching-induced pore formation in annealed polypropylene films. J Plast Film Sheeting 28:151–164
25. Takayanagi M, Imada K, Kajiyama T (1967) Mechanical properties and fine structure of drawn polymers. J Polym Sci Part C Polym Symp 15(1):263–281
26. Ushui K, Kondo Y, Hatada K, Gima S (1996) Effect of tie molecules on the craze strength of polypropylene. Polymer 37(24):5375–5379
27. Lovinger AJ (1983) Microstructure and unit-cell orientation in α-polypropylene. J Polym Sci Polym Phys Ed 21(1):97–110
28. Sadeghi F, Ajji A, Carreau PJ (2005) Study of polypropylene morphology to obtain porous membranes by stretching. In: Proceedings of SPE ANTEC, vol 1 pp 163–167
29. Chen XD, Xu RJ, Xie JY, Lei CH (2016) The study of room-temperature stretching of annealed polypropylene cast film with row-nucleated crystalline structure. Polymer 94:31–42
30. Lei CH, Wu SQ, Xu RJ, Cai Q, Hu B, Peng XL, Shi WQ (2013) Formation of stable crystalline connecting bridges during hot stretching of polypropylene hard elastomer film. Polym Bull 70(4):1353–1366
31. Sadeghi F, Ajji A, Carreau PJ (2007) Analysis of microporous membranes obtained from polypropylene films by stretching. J Membr Sci 292(1–2):62–71
32. Wu SQ, Lei CH, Cai Q, Xu RJ, Hu B, Shi WQ, Peng XL (2014) Study of structure and properties of polypropylene microporous membrane only by hot stretching. Polym Bull 71 (9):2205–2217
33. Greener J, Tsou AH, Blanton TN (1999) Physical and microstructural effects of heat setting in polyester films. Polym Eng Sci 39(12):2403–2418
34. Lei CH, Wu SQ, Cai Q, Xu RJ, Hu B, Shi WQ (2014) Influence of heat-setting temperature on properties of stretched polypropylene microporous membrane. Polym Int 63(3):584–588
35. Guinier A, Fournet G (1955) Small-angle scattering of X-rays. Wiley, New York
36. Dudic D, Kostoski D, Djokovic V (2002) Formation and behavior of low-temperature melting peak of quenched and annealed isotactic polypropylene. Polym Int 51(2):111–116
37. Nitta KH, Takayanagi M (2000) Tensile yield of isotactic polypropylene in terms of a lamellar-cluster model. J Polym Sci Part B: Polym Phys 38(8):1037–1044
38. Yu TH, Wilkes GL (1996) Orientation determination and morphological study of high density polyethylene (HDPE) extruded tubular films: effect of processing variables and molecular weight distribution. Polymer 37(21):4675–4687
39. Lee SY, Park SY, Song HS (2006) Lamellar crystalline structure of hard elastic HDPE films and its influence on microporous membrane formation. Polymer 47(10):3540–3547
40. Kim J, Kim SS, Park M, Jang M (2008) Effects of precursor properties on the preparation of polyethylene hollow fiber membranes by stretching. J Membr Sci 318(1):201–209
41. Shen LQ, Xu ZK, Xu YY (2002) Preparation and characterization of microporous polyethylene hollow fiber membranes. J Appl Polym Sci 84(1):203–210

42. Xie JY, Xu RJ, Chen CB, Chen XD, Zhang F, Lei CH, Lin YF, Li LB Influence of material characteristics on the structure and properties of high-density polyethylene microporous membrane. RSC Advances. doi:10.1039/C6RA05133B
43. Hu SR, Kyu T, Stein RS (1987) Characterization and properties of polyethylene blends I. Linear low-density polyethylene with high-density polyethylene. J Polym Sci Part B: Polym Phys 25(1):71–87
44. Chen CB, Xu RJ, Chen XD, Xie JY, Zhang F, Yang Y, Lei CH (2016) Influence of cocrystallization behaviour on structure and properties of HDPE/LLDPE microporous membrane. J Polym Res 23(3):1–9
45. Peterlin A (1965) Crystalline character in polymers. J Polym Sci Part C Polym Symp 9(1):61–89
46. Statton WO, Geil PH (1960) Recrystallization of polyethylene during annealing. J Appl Polym Sci 3(9):357–361
47. Sanchez IC, Colson JP, Eby RK (1973) Theory and observations of polymer crystal thickening. J Appl Phys 44(10):4332–4339
48. Chen CB, Lei CH, Cai Q, Mo HB, Xu RJ (2015) Influence of annealing time on the structure and properties of HDPE microporous membrane. J Plast Film Sheeting 31(1):78–95
49. Wlochowicz A, Eder M (1984) Distribution of lamella thicknesses in isothermally crystallized polypropylene and polyethylene by differential scanning calorimetry. Polymer 25:1268–1270
50. Yu WC (1997) Shutdown trilayer battery separator. U.S. patent 5691077
51. Yu TH (2000) Trilayer battery separator. U.S. patent 6080507
52. Tabatabaei SH, Carreau PJ, Ajji A (2009) Microporous membranes obtained from PP/HDPE multilayer films by stretching. J Membr Sci 345(1–2):148–159
53. Cai Q, Xu R J, Chen X D, Chen C B, Mo H B, Lei C H Structure and properties of melt-stretching polypropylene/silicon dioxide compound microporous membrane, Polym Compos. doi:10.1002/pc.23462
54. Cai Q, Xu RJ, Xu LF, Chen XD, Lei CH (2015) The preparation and study of melt-stretched polypropylene/Al_2O_3 compound microporous membrane. J Funct Mater 46(22):22145–22148
55. Lei C, Cai Q, Xu R J, Chen X D, Xie J Y (2016) Influence of magnesium sulfate whiskers on the structure and properties of melt-stretching polypropylene microporous membranes. J Appl Polym Sci. doi:10.1002/app.43884
56. Saffar A, Carreau PJ, Kamal MR, Ajji A (2014) Hydrophilic modification of polypropylene microporous membranes by grafting TiO_2 nanoparticles with acrylic acid groups on the surface. Polymer 55(23):6069–6075
57. Jung YS, Cavanagh AS, Gedvilas L, Widjonarko NE, Scott ID, Lee SH, Kim GH, George SM, Dillon AC (2012) Poly (aryl ether ketone) composite membrane as a high-performance lithium-ion batteries separator. Adv Energy Matter 2:1022–1027
58. Jeong HS, Hong SC, Lee SY (2010) Effect of microporous structure on thermal shrinkage and electrochemical performance of Al_2O_3/poly (vinylidene fluoride-hexafluoropropylene) composite separators for lithium-ion batteries. J Membr Sci 364(1):177–182
59. Miao YE, Zhu GN, Hou H, Xia YY, Liu T (2013) Electrospun polyimide nanofiber-based nonwoven separators for lithium-ion batteries. J Power Sources 226:82–86
60. Juang RS, Hsieh CT, Chen PA, Chen YF (2015) Microwave-assisted synthesis of titania coating onto polymeric separators for improved lithium-ion battery performance. J Power Sources 286:526–533
61. Zhang P, Chen L, Shi C, Yang P, Zhao J (2015) Development and characterization of silica tube-coated separator for lithium ion batteries. J Power Sources 284:10–15
62. Kim M, Park JH (2012) Inorganic thin layer coated porous separator with high thermal stability for safety reinforced Li-ion battery. J Power Sources 212:22–27
63. Yeon D, Lee Y, Ryou MH, Lee YM (2015) New flame-retardant composite separators based on metal hydroxides for lithium-ion batteries. Electrochim Acta 157:282–289

64. Liao H, Zhang H, Hong H, Li Z, Qin G, Zhu H, Lin Y (2016) Novel cellulose aerogel coated on polypropylene separators as gel polymer electrolyte with high ionic conductivity for lithium-ion batteries. J Membr Sci 514:332–339
65. Zhang H, Lin CE, Zhou MY, Angelin E, Zhu BK (2016) High thermal resistance polyimide separators prepared via soluble precursor and non-solvent induced phase separation process for lithium ion batteries. Electrochim Acta 187:125–133
66. Heo J, Choi Y, Chung KY, Park J (2016) Controlled pore evolution during phase inversion from the combinatorial non-solvent approach: application to battery separators. J Mater Chem A 4:9496–9501

Chapter 5
AAO Templates with Different Patterns and Channel Shapes

Grzegorz D. Sulka, Agnieszka Brzózka, Leszek Zaraska, Ewa Wierzbicka and Anna Brudzisz

5.1 Introduction

Recently, there has been a great deal of interest in both fundamental research and practical applications of nanostructured materials. In particular, intensive research attention has been focused on the fabrication of periodic nanostructures including highly ordered and densely packed arrays of nanopores, nanodots, nanotubes, and nanowires with interspace distance below 100 nm. The underlying reasons for the widespread interest in studying nanostructured materials are their enhanced performance and unique catalytic, electronic, magnetic, optical, thermoelectric, and mechanical properties that can be tuned by varying their size and/or interspace separation distance [1–3].

A rapidly growing involvement in nanotechnology research has been manifested through an increased number of new synthetic techniques with varying degrees of speed, cost, and structural quality, developed for fabrication of nanostructured materials. A wide variety of bottom-up and top-down methods, including conventional and unconventional fabrications routes, like X-ray [4], electron beam [5], ion beam [6] and interference [7, 8] lithography techniques, dip-pen nanolithography [9], nanosphere lithography [10, 11], nanoimprint lithography [12, 13], microcontact printing [14, 15], wet-chemical processes [16, 17], and methods based on self-organization [18–20] have already been used for generating arrays of nanodots, nanowires, and nanotubes. From the viewpoint of nanotechnology, process simplicity, low equipment costs, and a good reproducibility of the synthetic method are essential for a high-throughput fabrication of ordered arrays. Most of

G.D. Sulka (✉) · A. Brzózka · L. Zaraska · E. Wierzbicka · A. Brudzisz
Faculty of Chemistry, Department of Physical Chemistry & Electrochemistry, Jagiellonian University in Krakow, Ingardena 3, 30060 Krakow, Poland
e-mail: sulka@chemia.uj.edu.pl

P. Bettotti (ed.), *Submicron Porous Materials*,
DOI 10.1007/978-3-319-53035-2_5

the commonly used fabricating techniques (e.g., conventional lithography techniques) are expensive and time-consuming. Therefore, their transfer to industry, to large-scale industrial implementation, remains a great challenge. In contrast to the complex technical demands of conventional methods, next-generation unconventional lithography techniques are remarkably simple and straightforward. Although unconventional patterning techniques are well-suited for a precise replication of nanoscale patterns over large surface areas, their usefulness is generally limited to the fabrication of small classes of materials with specific shapes such as holes, lines, prisms, and rings. On the other hand, significant drawbacks of numerous bottom-up approaches are a poor control over the nanostructure dimensions and lack of a long-range order in the formed nanostructure even though a short-range order is kept and satisfying.

Among the different bottom-up strategies, developed recently for fabrication of nanostructured materials, the templating methods are one of the most attractive, convenient, and versatile for generating diverse nanostructures. In this alternative approach, the templates that differ in material, pattern, range of order, periodicity, feature size, and overall size, simply serve as the scaffolds in or around which materials are deposited. The structural features of the template used for the replication directly affect the structure of the resulting array. Using templating methods, large-scale arrays of ordered nanostructures with desired composition, size, and aspect ratio can be fabricated on various substrates. A variety of alternative templates have been successfully used for synthesis of metallic, semiconductor, and polymer arrays of nanodots, nanowires, and nanotubes. Notable examples of templates include highly oriented pyrolytic graphite (HOPG) [21–23], track-etched mica [24, 25], porous silicon [26–30], block copolymers [31–36], protein [37, 38], DNA [39, 40], and virus templates [41, 42]. In addition, considerable attention has been recently focused on two types of porous hard-templates: track-etched polycarbonate (PC) and porous anodic aluminum oxide (AAO) membranes. Both types of membranes offer a parallel arrangement of cylindrical channels with a pore diameter ranging from a few nm to over μm, accurately controlled pore size distribution, and desired hydrophilic properties for aqueous environments. The versatility of commercially available PC membranes has allowed a broad range of materials (e.g., metals, metal oxides, semiconductors, carbons, polymers, proteins, and others) to be fabricated via chemical or electrochemical methods, and the size of resulting nanowires and nanotubes can be easily tuned by choosing the appropriate template structure [43–51].

The main drawback of PC templates is the lack of a regular nanopore arrangement that limits considerably their ability to create ordered nanostructured arrays. In a similar way to PC membranes, porous AAO membranes are used for fabrication of nanostructured materials, but a key advantage of using AAO is a highly ordered and close-packed distribution of pores within the template. In addition, fabrication of porous AAO membrane via electrochemical anodization is relatively simple and inexpensive. Furthermore, all structural features of oxide templates such as diameter of the pores, pore-to-pore distance, thickness of the template, pore density, porosity, etc., can be easily controlled by adjusting anodizing conditions, especially type of electrolyte, potential difference applied during electrolysis, temperature and duration of the process. Another great advantage of porous AAO membranes is possibility of fabrication of membranes with not only straight but also branched nanochannels or nanochannels with specific shapes. Finally, posttreatment procedures (e.g., chemical etching) can be also applied to obtain anodic alumina templates with desired morphology. Therefore, significant progress has been recently made toward the development of new templating strategies employing porous AAO membranes.

This paper intends to review the current status of different fabrication strategies that have been successfully developed to generate nanopore arrays of AAO with different patterns and pore shapes.

5.2 AAO Templates

Porous AAO templates with nominal pore sizes of 20, 100, and 200 nm, as 60-μm-thick through-hole membranes are commercially available from Whatman Co. (Anopore™, Anodisc™). The standard sizes of membrane circles are 13, 25, and 47 mm. The crucial faults of these membranes are limited total size of the membrane, higher pore diameters than specified by the manufacturer, broad pore size distribution, and lack of order in the pore arrangement due to the fabrication procedure based on a one-step anodization. A wider range of anodic alumina-based nanotemplates, as free-standing membranes or Si, glass and Al-supported templates with various pore sizes (18–200 nm) and template thicknesses (15–150 μm) is offered by Synkera Technologies, INC. These templates are characterized by much higher degree or pore ordering and narrower pore size distribution. Three types of nanoporous AAO membranes are offered by SmartMembranes GmbH. The SmartPor membranes exhibit circular, hexagonally arranged pores with a uniform pore diameter (25, 35, or 180 nm) and interpore spacing (63, 105, and 500 nm). On the other hand, the FlexiPor membranes consist of constant diameter nanochannels from the top side and smaller pores from the bottom side of the membrane. The pore diameter at the bottom side of the membrane can be adjusted to the needs of the customer. SmartMembranes GmbH offers also QualiPor membranes with an extremely uniform hexagonal pore arrangement and pore diameter (deviation lower than 5%) synthesized using pre-patterning methods prior anodization. Ultrathin AAO templates with a uniform pore diameter in the range of 25–400 nm and interpore spacing of 65, 125, and 450 nm are offered by Topmembranes Technology Ltd. The thickness of AAO membranes does not exceed 1000 nm.

However, it should be mentioned, that AAO membranes with only a circular and straight shape of pores and a certain range of pore diameters are commercially available. Consequently, many researchers prepare their homemade AAO membranes by adjusting anodizing conditions (anodizing voltage, current density, type of electrolyte, and temperature) or by changing procedures of the pre- or post-treatment (pre-patterning of Al substrate or pore widening). Figure 5.1 shows diverse types of homemade AAO templates fabricated by anodizing of aluminum. A typical structure of porous AAO formed by anodization of aluminum consists of close-packed array of columnar hexagonal cells with nanopores at their centers (Fig. 5.1a). The AAO template with unidirectional channels provides an excellent architecture that can be characterized by structural features such as pore diameter, interpore distance, pore density, porosity, and aspect ratio, being a ratio between pore depth and pore diameter. These structural features of porous AAO templates are conveniently controlled by changing anodizing conditions.

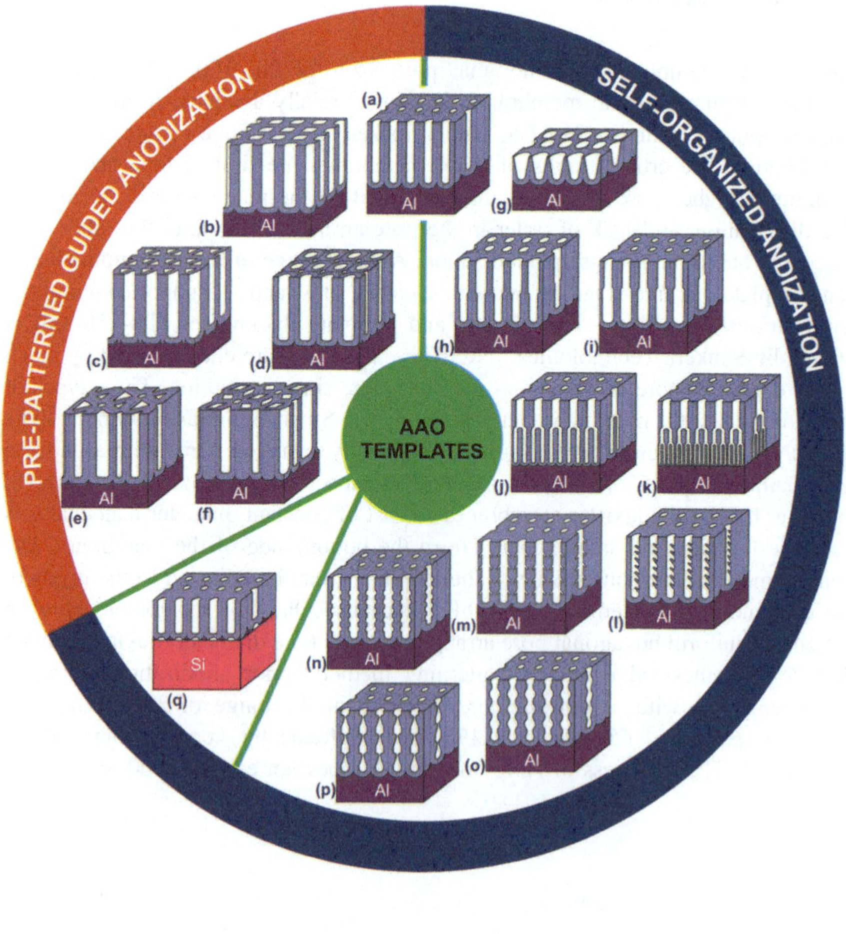

Fig. 5.1 Various types of AAO templates fabricated by pre-patterned guided anodizations and self-organized anodizations. **a** hexagonal cells with circular pores formed by anodizing of pre-textured Al or by self-organized two-step anodizing, **b** square cells by anodizing pre-textured Al, **c** triangle cells by anodizing pre-textured Al, **d** checkerboard cell pattern by anodizing of pre-textured Al with alternating defects in the array of convexes in a square lattice, **e** triangle–circle cell pattern, and **f** complex triangle–diamond cell pattern developed by anodization from square pre-patterns on Al, **g** conical pores fabricated by a few alternating cycles of anodization and pore widening processes, **h** step-shaped nanochannels fabricated by two anodizations performed at the same voltage separated by pore widening process or by two anodizations carried out at the same voltage in different electrolytes (phosphoric acid as a first anodizing electrolyte and oxalic acid as a second one), **i** reversed step-shaped nanochannels fabricated by two anodizations at the same voltage in different electrolytes (inverse sequence of used electrolytes comparing to (h) or by mild anodization (MA) at lower voltages followed by hard anodization (HA) at higher voltages in the same electrolyte, **j** Y-branched nanochannels fabricated by anodization at constant voltages with a reduction of anodizing voltage by a factor of $1/\sqrt{2}$; for n-branched nanochannels the anodizing voltage should be reduced by a factor of $1/\sqrt{n}$, **k** multilevel branched nanochannels by multi-step reduction of anodizing voltage by a factor of $1/\sqrt{2}$, **l** parallel serrated nanochannels formed by anodization with the generation of oxygen gas bubbles, **m** periodically branched pores by potentiostatic anodization with periodically changing anodizing voltage, **n** periodically modulated pore diameter by potentiostatic pulse anodization (subsequent combination of mild anodization (MA) and hard anodization (HA) processes), **o** periodically modulated symmetrical pore diameter by galvanostatic cyclic anodization with a sinusoidal current signal or by potentiostatic HA with spontaneous symmetric sinusoidal current oscillations, **p** periodically modulated ratchet-type pore diameter by galvanostatic cyclic anodization with an asymmetrical current signal (exponential saw-tooth with two different amplitudes) or by potentiostatic HA with spontaneous asymmetric current oscillations, and **q** hexagonally arranged pores with an arched barrier layer fabricated on a Si substrate

For the preparation of porous AAO templates, two different anodizing approaches can be distinguished: a pre-patterned guided anodization (Fig. 5.1a–f) and a self-organized two-step anodization (Fig. 5.1a, g–p). In the first case, a pre-patterned (pre-textured) aluminum substrate is used for anodization, while the latter is based on a self-organized pre-patterning of Al during the first anodizing step and the removal of irregular oxide layer before the final (second) anodization.

5.3 Pre-patterned Guided Anodization

In the prepatterning method, electrochemically polished high purity Al foils or thin films deposited on various types of substrates (mostly SiO_2/Si, less frequently glass or Ti) are used. Since the lattice constant of the pre-patterned concaves on Al substrate is matched with the applied anodizing voltage, a perfectly ordered nanopore arrangement over the previously pre-textured surface area is observed. The pre-patterning of Al substrate can be realized by a direct indentation of the surface (Fig. 5.2a), for example with a tip of the scanning probe microscope or focused-ion beam, and by an indirect method applying an imprint mold to impress a pattern on the Al surface (Fig. 5.2b).

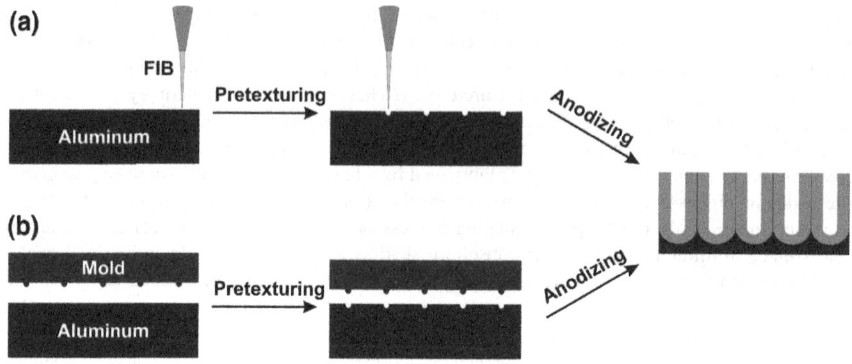

Fig. 5.2 Fabrication of perfectly ordered arrays of pores by direct indentation (**a**) using a focused-ion beam (FIB) and indirect nanoimprinting (**b**) by using a mold followed by anodizing

5.3.1 Direct Indentation Methods

Among methods that have been already employed for a direct pre-texturing of Al surface, the most commonly used are atomic force microscopy (AFM) [52–55] and focused-ion beam (FIB) lithography [56–64], where the Al surface is directly patterned by a tip of the scanning probe microscope or Ga ion beam, respectively.

A key factor in pre-texturing methods is an appropriate selection of amount of energy/force that should be necessarily applied to create nanoconcaves on the metal surface. The depth of created surface concaves (holes) should be enough to develop well-ordered pores in AAO during the subsequent anodization process. In addition, the minimum nanohole depth strongly depends on the surface roughness, which is also related to the type of the applied (electropolished or sputter deposited) Al substrate. For instance, AFM indentation technique requires the use of a force from the range between about 25 and 40 μN [52]. It was proved that in case when the indentation force was lower than about 20 μN for the root mean square of roughness of 2 nm, as formed concaves were too shallow (depth between 1 and 2.5 nm), and the morphology of porous alumina film grown during anodization did not show clear periodicity [55]. On the contrary, when the force of 40 μN was applied for the indentation, concaves depth reached 15–18 nm, and subsequent anodic oxidation resulted in formation of the AAO layer with a very good periodicity [54].

For the pre-patterning method using focused-ion beam (FIB) lithography, relationships between ion dose/dwell time, ion energy, and dimensions of concaves (mainly size and depth) formed on Al surface have been already established [56, 57], however, it was found that the surface roughness is still a crucial parameter affecting the formation of well-ordered AAO arrays. For instance, when the root mean square roughness of Al surface is lower than 1 nm, anodic formation of well-ordered alumina layers is achieved for concaves deeper than 3 nm [56]. On the other hand, when the Al surface roughness is 2.7 nm, the concave depth has to be about 15 nm to observe the guiding process [65]. The FIB prepatterning allows for obtaining of anodic alumina with ultrasmall pores (about 10 nm before a post-anodizing etching treatment) with a lattice constant between 40 and 50 nm [58]. At the same time, statistical analysis comparing the guided anodization with the self-organized process showed a significant increase in the long-range order and the uniformity of the pore size as well as the interpore spacing for the pre-textured anodized samples [66].

Recently, some more advanced techniques such as interference lithography [67], Ar-ion beam etching [68, 69], and reactive ion etching (RIE) [70, 71] through nanoporous PMMA [69] and AAO masks [68, 70] or through a thin film of poly (styrene-b-4-vinylpyridine) (PS-b-P4VP) diblock copolymer [71] have been also employed for a direct pre-texturing of Al substrates before anodization. In addition, pre-texturing of Al can be also accomplished by resist-assisted lithographic processes followed by wet chemical etching of aluminum. Examples include PMMA-assisted electron beam lithography [72], resist-assisted focused-ion beam lithography [73], and resist-assisted direct writing laser (DWL) lithography [74]. Moreover, the step and flash imprinting lithography (SFIL) followed by wet chemical etching was also used for the pre-patterning of aluminum surface before anodization [75].

Although creating an ordered pattern on the Al substrate by direct indentation is effective, the procedure is time-consuming, expensive, and requires high precision in alignment. Therefore, direct indentation methods are rarely used in laboratories.

5.3.2 Indirect Indentation Methods

An interesting alternative approach to direct indentation methods used for the pre-patterning of aluminum offers a nanoimprint lithography. Prior to anodization, an imprint mold with the ordered array of convexes is used to impress patterns on the Al surface. By using the mold, shallow indentations on the Al surface are created, and their depth depends on the applied pressure. The generated array of concaves on the aluminum surface is a negative replica of convexes of the master mold, and each concave is an initiation site for the pore nucleation during anodization. A great advantage of mold-assisted techniques is mold reusability.

The materials used for a mold preparation have to be harder than Al in order to effectively imprint the surface. Hard molds made of SiC [76–94], Si_3N_4 on Si [95–102], Si [103, 104], TiN [105], Ni [106–111], and W [112] with various convex shapes and dimensions have been already used for nanoindentation of aluminum. The increased hardness of the mold requires to use for its preparation of expensive lithography methods (e.g., e-beam lithography). Moreover, such molds are prone to cracking, when very high pressure is applied during stamping. The problem of insufficient mechanical stability of the mold can be significant especially for high pattern densities, where higher pressure is required to create concaves with suitable size and depth. The appropriate pressure, at which the formed concaves were deep enough to guide the anodization, is typically in the range of 2.8–5 ton cm^{-2}. The relationship between the applied pressure and depth of formed concaves have been also investigated in detail, however, the results were strongly dependent on the shape and dimensions of mold convexes [82, 83, 96, 103, 105–107]. It should be mentioned that the roughness of the Al surface is also reduced during stamping process. As expected, it is beneficial for creation of highly ordered AAO during anodization [112].

Recently, a commercially available optical diffraction grating (5 × 5 mm with 1200 lines mm^{-2}) [113, 114] have been employed to transfer of the relief of the grating to the aluminum surface. A two-step press-in procedure, used for the generation of a triangular lattice of concaves on the Al surface, requires rotation of the diffraction grating by 60° around the axis perpendicular to the Al surface before applying the second press forming procedure.

Another intriguing approach toward the pre-patterning of Al surface, based on nanosphere lithography (NSL) with 2D hexagonal close-packed arrays of polystyrene spheres [115, 116], silica nanospheres [117, 118], Fe_2O_3 spheres [119], and glass beads [120] has been demonstrated as well. The spheres are immobilized onto Si, mica or a glass substrate and used as a mold. Before anodization, spheres are removed from the aluminum surface. The advantage of this method is the possibility to form structures with various interpore distances by appropriate selection of the size of spheres. Molding with particles having diameter of 13–500 nm has been successfully applied for creating patterns on a large surface area of Al. Unlike other methods, this approach does not require a clean room or sophisticated equipment for the synthesis of mold. Moreover, nanobeads with continuously tunable diameters are commercially available. Considering the use of monolayer of nanoparticles as a master mold, the most problematic issue is polydispersion of particles, which can lead to poor ordering of pores grown during anodization [116–119].

Although methods described above allow for the formation of highly ordered AAOs with a tunable interpore distance by anodization in sulfuric [83], oxalic [82, 83, 102, 104], and mostly phosphoric [76, 77, 118] acids, some disadvantages and limitations can be also identified. Pre-patterning seems to be the most applicable in the case of anodic oxidation of Al deposited on Si and glass substrates, where two-step anodization can be often hardly applied due to the limited thickness of the metallic layer. However, when the mechanical pressure is applied during pattern transfer, a fragile supporting substrate is vulnerable to damage. The main obstacle to using nanoimprint lithography methods for a large-scale pre-patterning of Al surface, is a small surface area of imprint molds, usually not exceeding 1 mm^2. Consequently, the perfect order in the pore arrangement can be obtained only over a limited surface area of AAO.

5.3.3 AAO Layers Grown on the Pre-textured Al

A key advantage of using the pre-pattern aluminum substrate is a perfect order in the hexagonal arrangement of circular pores observed after anodization [76–90,

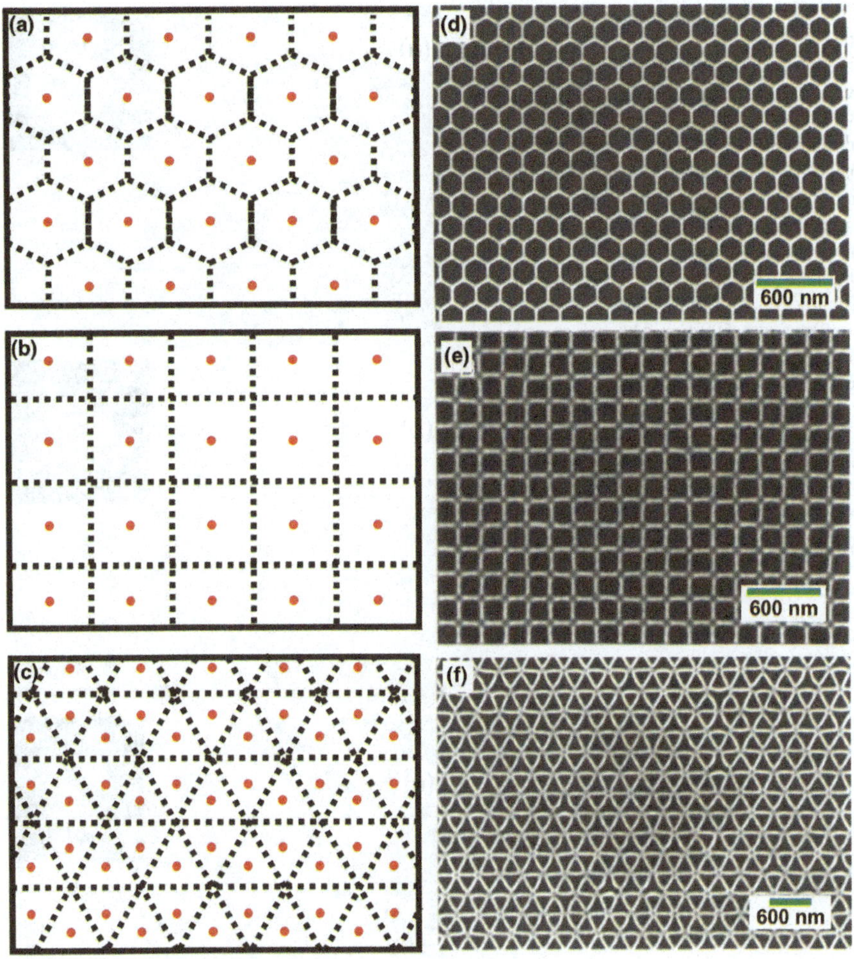

Fig. 5.3 Nanoindentation patterns with the interval of initiation sites of 200 nm (**a–c**). *Dot* in each cell represent a pore generation site. The triangular (**d**), square (**e**) and graphite (**f**) pore arrangements in AAOs obtained after anodization of prepatterned Al in 0.05 M oxalic acid at 17 °C at 80 V for 6 min. Adapted with permission from Ref. [90]. Copyright 2001 WILEY-VCH Verlag GmbH, D-69469 Weinheim

96–100, 105–109]. Moreover, the pre-pattern guided anodization can result in pore shapes and cell configurations different from the pattern with the circular pores within hexagonally arranged cells, typically observed in AAO templates (Fig. 5.3). By transferring different patterns from the resist to Al surface or by using master molds with convexes arranged in a square and graphite structure lattices, square- (Figs. 5.1b and 5.3e) [65, 90–93, 102, 109] and triangular-shaped (Figs. 5.1c and 5.3f) [90, 91, 94, 95] pores and cells in AAO have been obtained. The shape of formed pores obtained during anodization is not determined by the shape of indentation, but by the pattern of indentation [90–92, 100].

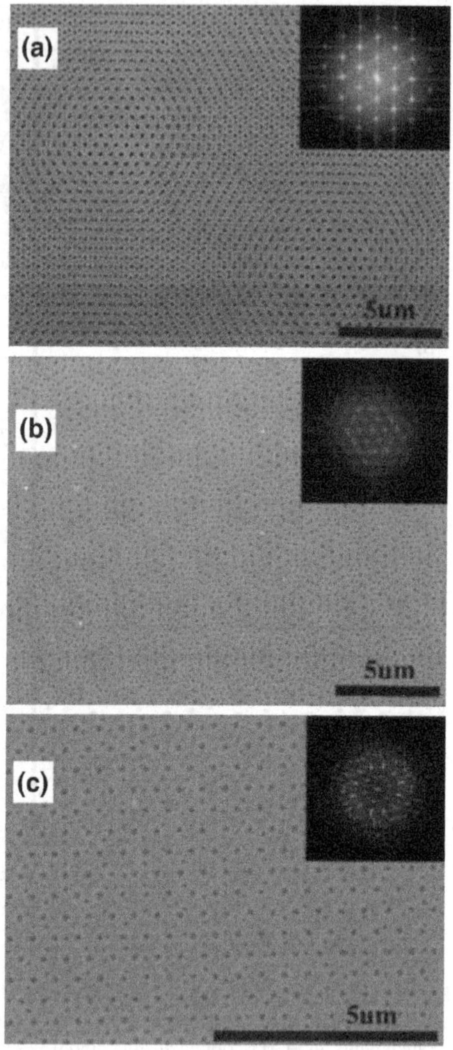

Fig. 5.4 Scanning electron microscope (SEM) images of the porous alumina surface with Moiré patterns together with their fast Fourier transforms (insets). The patterns were formed by anodization of aluminum imprinted twice with a Ni_3N_4 mold rotation angle of 2.2° (**a**), 11.5° (**b**), and 30° (**c**). The interval between the mold convexes was 500 nm. The anodizations were carried out in 10 wt% H_3PO_4 at 2 mA cm^{-2} and 3 °C for 1–60 min. Reprinted with permission from Ref. [101]. Copyright 2003 WILEY-VCH Verlag GmbH & Co. KGaA, Weinheim

A porous AAO array with a Moiré pattern (Fig. 5.4) has been successfully created by FIB indentation of two overlaid square concave patterns at 45° [62] or by double mold stamping with different rotation angles [101].

It is very important, that a non-hexagonal ordering of pores can be maintained only for ca. 500 nm thick of AAO films. Under a prolonged anodization, the pore ordering either changes to hexagonal or no longer exhibits any significant pore arrangement depending on the imprinted pattern and anodization conditions [90, 102]. The formation of nanochannels on the indented metallic substrate is a result of the presence of both the electrical field and the mechanical stress field around the concaves [62]. Consequently, some new pores can nucleate and develop outside the textured pattern, and the growth of other pores can be hindered. For this reason, AAO templates with non-hexagonal patterns and non-circular pores are obtained only with low aspect ratio, however they can be used, e.g., as ultrathin masks for evaporation or deposition [90, 111].

The effect of pore rearrangement occurring during anodization can be minimized by appropriate selection of the anodizing potential [54, 121]. In general, for conventional anodization of Al resulting in the hexagonal pore arrangement, the interpore distance is dependent on the applied anodizing voltage with a linear proportional constant of 2.5 nm V^{-1} which is known as voltage limitation [52]. On the other hand, for anodization of Al films with square patterns and pre-textured via AFM imprinting Fois et al. proposed another empirical relation between the optimal applied voltage (U) and interpore distance (D_{int}) in nm [121].

$$D_{int} = 1.04\,U + 33 \tag{5.1}$$

A different empirical relation for the hexagonal pore arrangement was established by Robinson et al. for anodization of the Al film pre-textured by FIB [65].

$$D_{int} = 2.17\,U + 15.8 \tag{5.2}$$

The authors also suggested that for a given voltage, the interpore distance for the square array is lower than that for the hexagonal arrangement by a factor of $(\sin 60)^{0.5}$ due to a less dense packing factor (pore density) compared to the hexagonal array. It is generally accepted that the voltage limitation is not only connected with the interpore distance, but also defines the pore diameter ($D_p \approx 0.9$ nm V^{-1}) and wall thickness ($d_w \approx 0.8$ nm V^{-1}). For exactly tailored anodizing potential or current conditions, oxide walls can fully fill the space between neighboring pores during the oxide growth process without affecting cell boundaries (Fig. 5.5b). In consequence, clear thin lines at cell boundaries which reflect the net of ordered cells are visible. When the interpore distance is lower from the empirical rule of 2.5 nm V^{-1}, the oxide walls of two neighboring nanopores overlap before pores expand to their final size (Fig. 5.5a). Consequently, the boundary lines are invisible. The overlapping of cells leads to a lower field strength due to the reduced curvature of the barrier layer. Therefore, the oxide dissolution rate becomes slower, and consequently, the oxidation rate and the wall thickness between the pores increase what results in a reduced nanopore diameter. On the other hand, if the interpore distance of concaves is significantly larger (Fig. 5.5c), new self-ordering pores are developed. It is a result of inhomogeneous current distribution, breakdown of the barrier layer, and gas evolution on protrusions of the Al substrate that are present at the corner of each cell. The processes are accompanied by the incorporation of crystalline particles of aluminum into the oxide film [93]. The formation of new pores can be explained in detail as follows. The oxide layer grown at the wedge of the indented pits (so-called protective oxide) suppresses the creation of new pores,

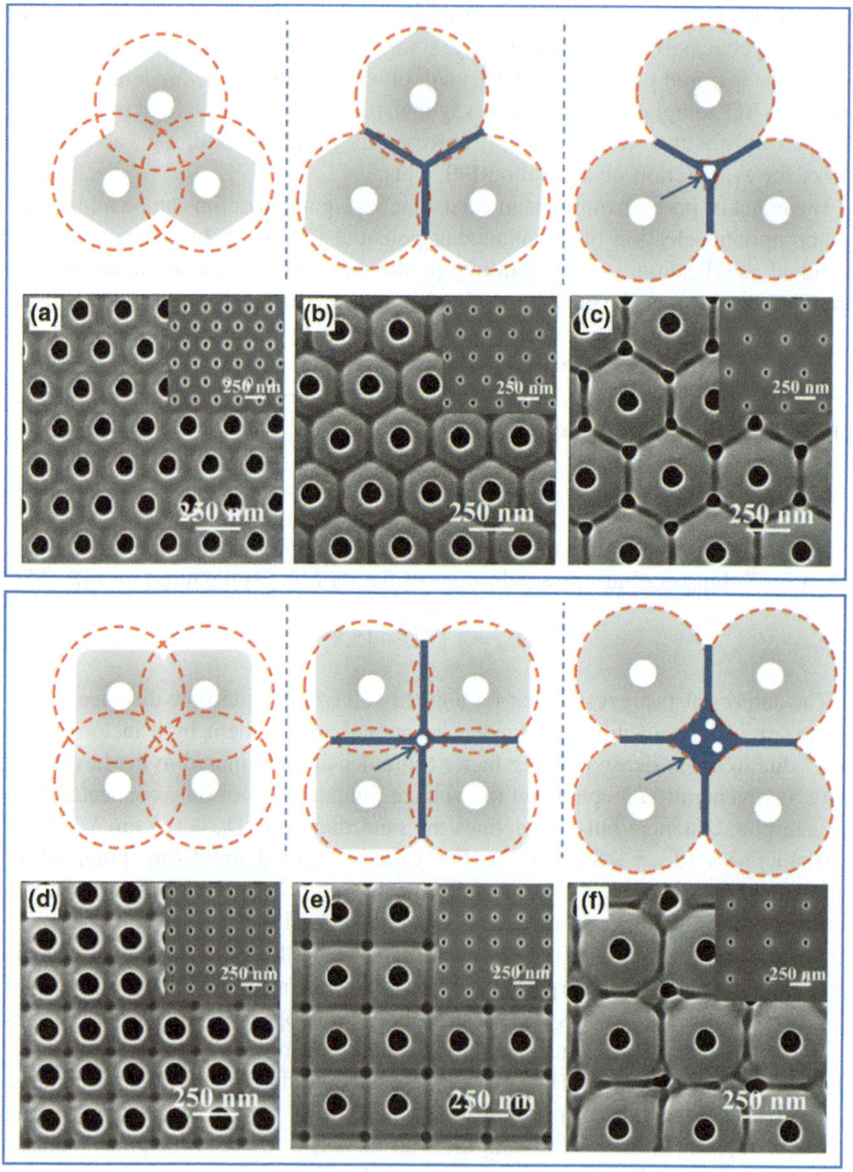
(a)
250 nm
250 nm
(b)
250 nm
250 nm
(c)
250 nm
250 nm
(d)
250 nm
250 nm
(e)
250 nm
250 nm
(f)
250 nm
250 nm

Fig. 5.5 SEM images of AAO layers formed after the FIB guided anodization of hexagonal- (**a**–**c**) and square- (**d**–**f**) patterned Al with small (**a** and **d**), medium (**b** and **e**), and large (**c** and **f**) interpore distances. The corresponding schematic representations show the effect of interpore distance on the interaction of neighboring pore walls and new pore formation. The *arrow* indicates the surface area where new pores are formed. For the square patterns, the surface area at the cell junctions is larger than that for the hexagonal patterns with the same final pore size and interpore distance. As a result, new pores can be formed and developed. The hexagonal and square pore patterns were created with the FIB guided anodization. All samples were anodized in 0.3 M H_3PO_4 at 0 °C and 20 mA cm^{-2} for 6 min. The insets show the SEM images of Al surface after FIB patterning. Adapted from Ref. [62] with the permission of AIP Publishing

due to the hindrance of the field-enhanced alumina dissolution at wedges. Consequently, if the diameter of protective oxide is low and not-protected area is too broad, nucleation of new pores occur between the pits. However, if the protected area is larger, pore nucleation is restricted only to the center of the nanoindented pits. Depending on the size of the not-protective area, new pores can be generated in the ordered or non-ordered pattern [95].

Masuda et al. [77] for the first time reported a generation of new self-ordering pores during anodization of Al in sites non-defined by the indentation pattern, known as a self-healing process. To demonstrate it, a hexagonally patterned master mold with some defects in the convex arrangement was used for the prepatterning of Al surface. After anodization, the pore development was observed even at defect sites (non-patterned sites). If the distance between the non-patterned and the nearest patterned sites is 2.5 times larger than applied anodizing potential, the highly ordered hexagonal pattern was obtained in AAO. The missing pores are generated at sites equaled to the interpore distance of the lattice. The ordered configuration of pores is maintained only up to 2 μm of the oxide thickness, due to a different growth rate of pores at patterned and non-patterned sites [77].

The discovery of the self-healing effect, lead to the development of smart-nanoimprint technology, which allows for creating versatile structures, difficult to be prepared by other conventional methods. Some examples of such structures are presented in Fig. 5.6. New pores are guided by at least three pre-patterned sites, therefore indentation patterns with higher lattice constants are sufficient for the preparation of AAO with much lower interpore distances [122]. This approach can overcome some limitations of lithographic methods. For the hexagonal concave arrangement with the interpore distance D_{int}, the potential for the guided anodization should be reduced by a factor of $\sqrt{3}$ in order to generate new self-ordering pores. The lattice constant of new pores, formed in the free spaces between the patterned concaves, is reduced by the factor of $\sqrt{3}D_{int}/3$ [61, 63, 81, 97, 99, 123]. For the square arrangement of patterned concaves, the spontaneous formation of new pores at the distance of $\sqrt{2}D_{int}/2$ occurs when the applied anodizing potential is reduced by the factor of $\sqrt{2}$. At the initial stage of guided anodization, the size of pores formed at missing sites is smaller than the diameter of pores formed at the patterned sites, and consequently, the cell size observed at the bottom side of the oxide layer is not uniform [122]. This effect was used for the

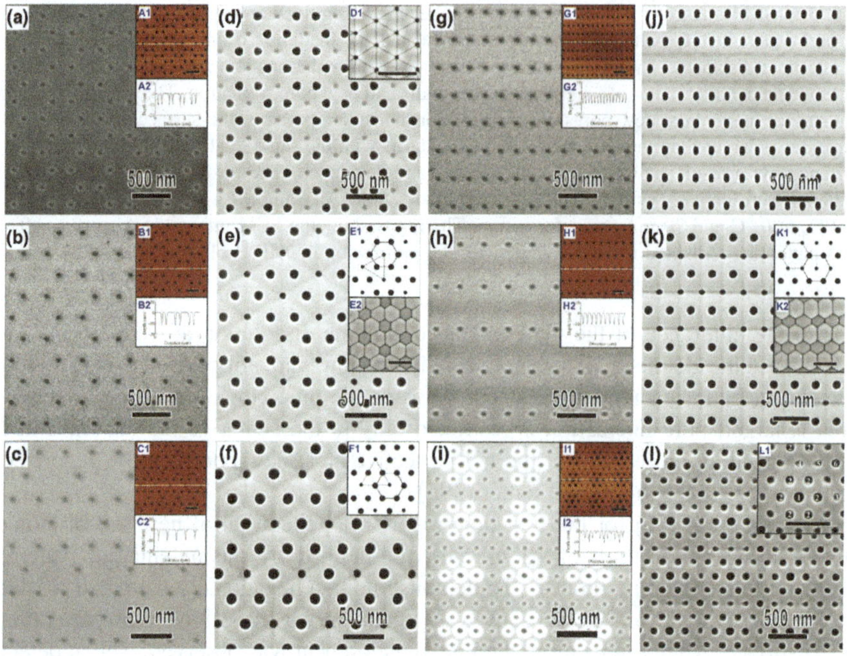

Fig. 5.6 SEM images of Al surfaces patterned with FIB (**a**–**c** and **g**–**i**) and corresponding AAOs after anodizations (**d**–**f** and **j**–**l**) in 0.3 M phosphoric acid at 11.4 mA cm^{-2} and 0 °C for 5 min. During anodizations, the steady-state potential was ~ 140 V. The FIB concaves were formed in: a graphite lattice arrangement with the interpore distance 250 nm (**a**) and 300 nm (**b**), hexagonal arrangement with some defect sites and interpore distance of 300 nm (**c**), rectangular arrangement with the shorter (horizontal) interpore distance of 200 nm (**g**) and 300 nm (**h**), and hexagonal arrangement with the interpore distance of 200 nm (**i**) and different concave diameters (80, 65, and 45 nm) and depths (10, 6, and 3 nm). Insets: the AFM images (A1–C1 and G1–I1) and corresponding surface topologies (A2–C2 and G2–I2) along the line marked in AFM images, schematic representations of the alternating nanopore arrangement (E1, F1, and K1), bottom views of formed cells in AAOs (D1, E2, and K2), and the enlarged SEM image of L with different pores labeled (L1). All the insert scale bars are 500 nm. Adapted with permission from Ref. [63]. Copyright 2010 American Chemical Society

preparation of a nontypical, regular checkerboard pattern (Fig. 5.1d and 5.7) [111], and other mosaic patterns [81, 85] (Fig. 5.8).

For instance, modification of the imprinted pattern by new generated pores can result in more complex nontypical regular patterns on AAO, such as triangular-circle (Figs. 5.1e and 5.8), and hybrid triangular-diamond patterns (Figs. 5.1f and 5.8) [72]. It is worth mentioning that the shape of formed pores is not circular.

Noncircular pores in AAO can be also formed by anodization of the Al surface with concaves having different sizes [62, 63]. The concave size can affect the shape of the nanopores, and consequently, shape of cells, due to asymmetric electric field caused by the differences between small and large pore bottoms (small and large imprinted concaves). For the alternating-sized concave pattern with 65 and 45 nm

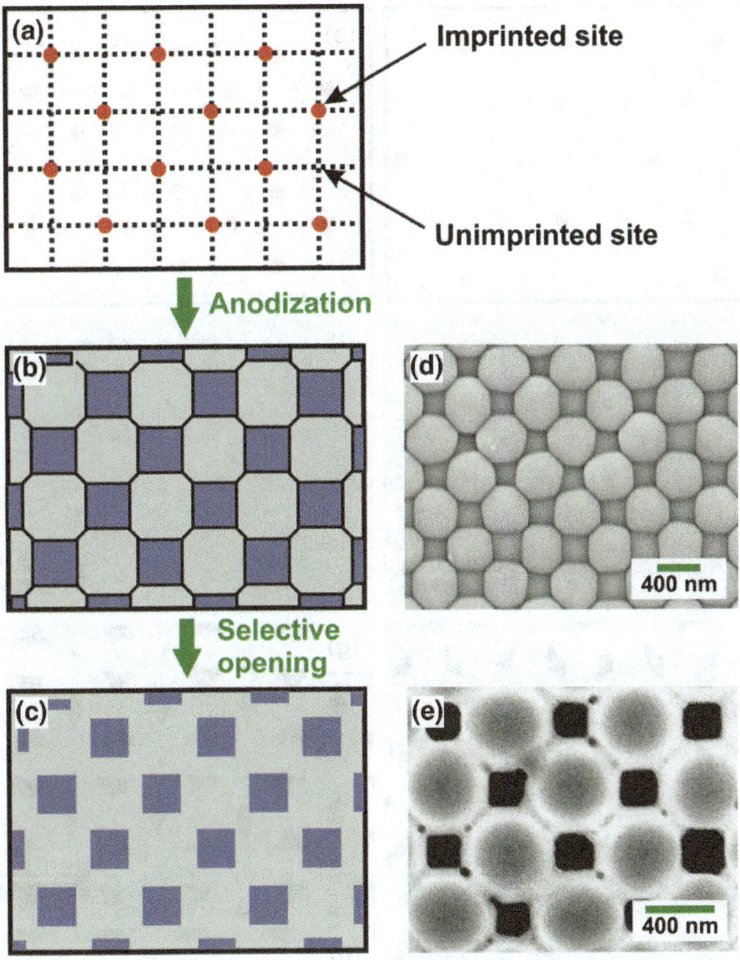

Fig. 5.7 Schematic for the preparation of AAO with the checkerboard pattern: square lattice with alternating defects patterned on the Al surface (**a**), bottom view of cell arrangement (**b**) and pore bottoms after a selective opening by chemical etching (**c**). The interpore distance between imprinted pores in the square lattice was 400 nm. The bottom views SEM images of AAO after anodization in 0.1 M H_3PO_4 at 200 V and 0 °C for 30 min (**d**) followed by a selective opening of self-ordered pores (**e**). Adapted with permission from Ref. [111]. Copyright 2010 The Japan Society of Applied Physics

pore sizes and 200 nm interpore distance formed over the Al surface, the small pores are elongated in the direction perpendicular to the large cell bordering lines after anodization (Fig. 5.9a). On the other hand, elliptical pores in AAO are formed not only by spontaneous deformation of cells, but also from the elongated guiding concaves created by the FIB patterning (Fig. 5.9b) even in the case of the square

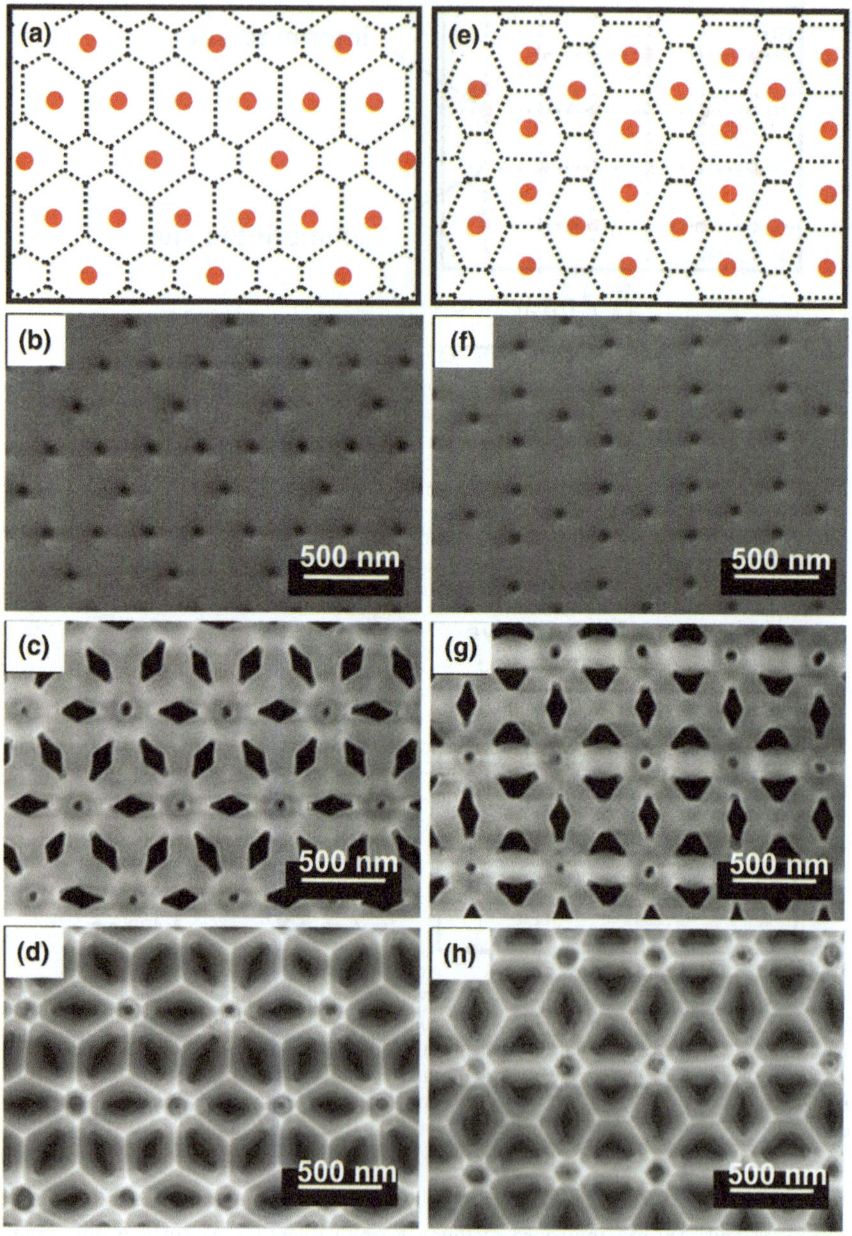
(a)
(e)
(b)
500 nm
(f)
500 nm
(c)
500 nm
(g)
500 nm
(d)
500 nm
(h)
500 nm

Fig. 5.8 Schematic representations of the diamond (**a**) and triangle–diamond (**e**) patterns used for the polymethylmethacrylate (PMMA)-assisted patterning of Al surface. *Red dots* represent the imprinted sites, and dashed lines signify the location of cell walls. The interpore distance between imprinted sites was 300 nm. The SEM images of the PMMA surface after e-beam lithography (**b** and **f**), resulting top-views of AAOs (**c** and **g**), and bottom views of AAOs (**d** and **h**) after chemical etching of the barrier layer. The anodizations were carried out at 120 V in a 0.3 M H_3PO_4 electrolyte at 5 °C for 30 min. Adapted from Ref. [72] with the permission of AIP Publishing.

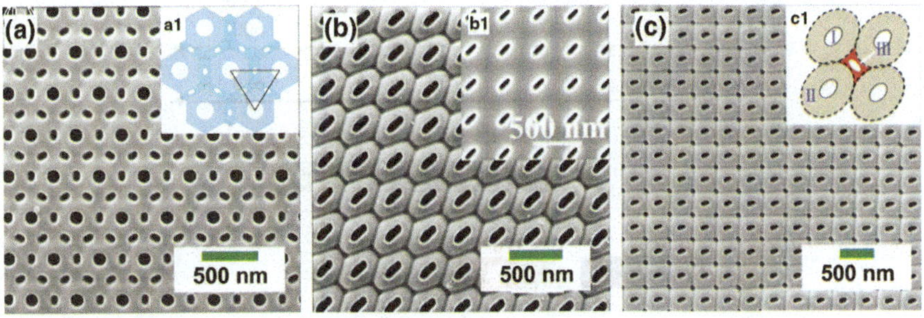

Fig. 5.9 SEM image of AAO with the alternating FIB pattern (**a**) after anodization carried out at 11.4 mA cm^{-2} in 0.3 M phosphoric acid at 0 °C for 5 min. The inset (a1) shows the schematic of the pore shape and oxide wall shape development from the small and large FIB concaves. The SEM image of the square arrangement of elliptical pores in AAO (**b**) after the anodization at 20 mA cm^{-2} for 6 min in 0.3 M phosphoric acid at 0 °C. The inset (b1) shows the FIB patterned pores on the Al surface. The interpore distance between FIB imprinted sites was 350 nm, and the long and short axes for the elliptical pores were150 nm and 50 nm, respectively. The SEM image of the AAO (**c**) grown from the square arrangement of elliptical pores with the reduced the long axis (100 nm). The interpore distance between FIB imprinted sites was 350 nm, and the short axis for the elliptical pores was 50 nm. The inset (c1) shows the formation process of the new, small pores at cell junctions. The areas I, II, and III represent the anodized pore, cell wall, and the self-generated pore, respectively. Figure 5.9a reprinted from Ref. [62] with the permission of AIP Publishing. Figure 5.9b reprinted with permission from Ref. [63]. Copyright 2010 American Chemical Society

cells arrangement [62]. When the size of elongated concaves created over the Al surface is reduced (e.g., along the longer axis), new self-ordering pores can be formed during anodization (Fig. 5.9c).

5.4 Two-Step Self-organized Anodization

The problem with anodizing of aluminum over a small surface area can be overcome by using a self-organized two-step anodization. The two-step anodizing procedure, developed by Masuda and Satoh [124], starts form an initial anodization of the substrate at the constant cell voltage followed by chemical etching of the grown oxide layer. The objective of the first-step anodizing is an electrochemical pre-patterning of Al prior the final anodization (a second step). After dissolution of the oxide layer, a periodic array of concaves, observed on the aluminum surface,

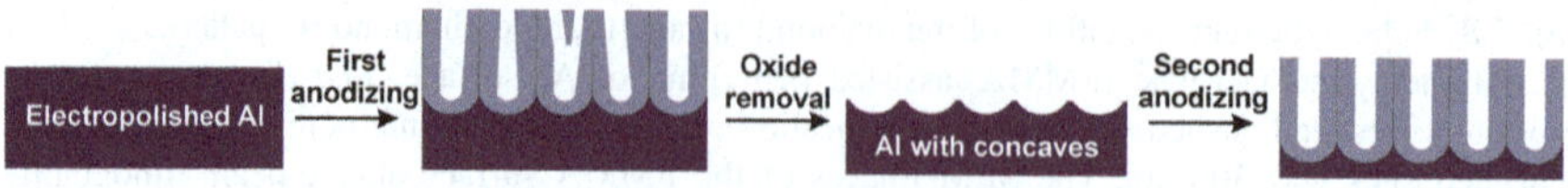

Fig. 5.10 Schematic representation of the two-step anodizing of aluminum

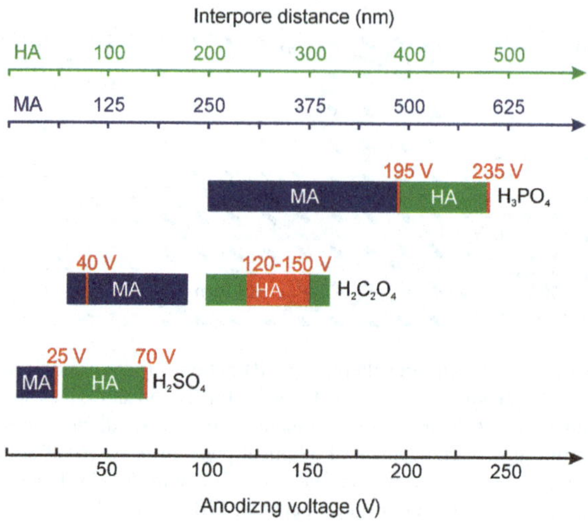

Fig. 5.11 Applicable anodizing voltages for mild anodization (MA) and hard anodization (HA) processes with indicated self-ordering regimes (in *red*) for typical anodizing electrolytes. The effect of anodizing voltage (U) on the interpore distance (D_{int}) for a mild anodization (MA, $D_{int}^{MA} = 2.5$ U) and hard anodization (HA, $D_{int}^{HA} = 2.0$ U) processes

serves as nucleation sites for the ordered pore formation during the second anodization (Fig. 5.10) [125].

The two-step anodization of aluminum, leading to a quasi-monodomain nanopore structure, is usually carried out in acidic electrolytes under constant voltage or constant current conditions. Since a linear relationship between anodizing potential and interpore distance in porous AAO was found, the potentiostatic anodization has been more frequently used. The nanoporous AAO templates with a close-packed hexagonal arrangement of cell (Fig. 5.1a), fabricated by self-organized process, are observed after anodizations carried out in a sulfuric [e.g., 125–131], oxalic [e.g., 125, 132–136] and phosphoric acid solutions [e.g., 125, 137, 138]. The fabrication procedure of AAO can be based on a traditional mild anodization (MA) [126–140] or hard anodization (HA) also known as high-field anodization [108, 141–146]. For mild anodizing processes, a window of applicable voltages is rather narrow and correlated with a type of used electrolyte (Fig. 5.11).

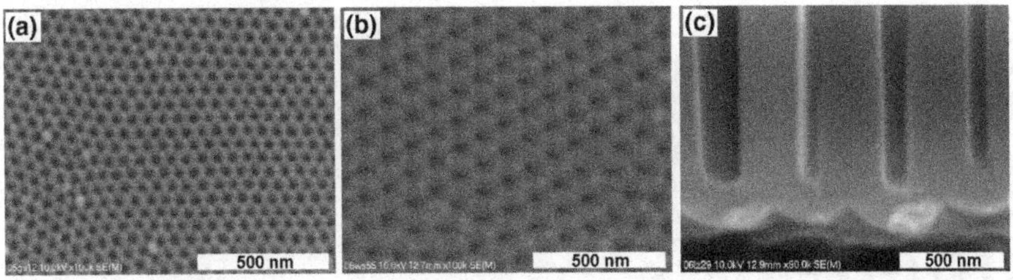

Fig. 5.12 SEM top-views (**a** and **b**) and cross-sectional view (**c**) of AAO template formed by the self-organized two-step anodization carried out in **a** 2.3 M H_2SO_4 at 25 V and –8 °C, **b** 0.3 M $H_2C_2O_4$ at 40 V and 3 °C, **c** 0.3 M H_3PO_4 at 160 V and 0 °C. The duration of the first and second anodizations was **a** 10 min and 45 min, **b** and **c** 14 h and 14 h, respectively

The best hexagonal arrangement of nanopores is observed only at a certain value of anodizing voltage (a self-ordering regime) that was found to be 25 V, 40 V, and 195 V for sulfuric acid, oxalic acid, and phosphoric acid, respectively. Figure 5.12 shows typical SEM top-views of AAO films anodized in 2.3 M sulfuric acid at 25 V and −8 °C (Fig. 5.12a), and in 0.3 M oxalic acid at 40 V and 3 °C (Fig. 5.12b). The cross-sectional SEM image (Fig. 5.12c) shows a hemispherical scallop of the barrier layer that is formed during anodizing of aluminum foil in 0.3 M H_3PO_4 at 160 V and 0 °C.

The ranges of applicable voltages are a bit wider for hard anodization processes, and the self-ordering regimes are different from those observed in MA processes. The interpore distance of the porous lattice depends linearly on the anodizing voltage with a proportionality constant of about 2.5 nm V^{-1} and 2.0 nm V^{-1} for MA and HA processes, respectively [108, 125]. Although the pore diameter of the AAO template depends linearly on anodizing voltage with a proportionality constant between 0.8 and 1.2 nm V^{-1} for MA and about 0.4 nm V^{-1} for HA [108, 125], the pore sizes can be easily adjust by wet chemical etching (pore widening) after anodizing. By the pore widening process, the pore size of the AAO template can be extended to 80–85% of the cell size (interpore distance) [133].

A typical two-step anodization procedure, carried out at a specified conditions, results in porous AAO layers with straight cylindrical nanochannels parallel to each other. However, a great variety of the procedure modifications, which allow for the fabrication of anodic alumina films with a more complex internal architecture and various pore shapes have been already proposed in the literature.

5.4.1 AAO with Conical Channels

The alternating anodization and pore widening treatments were used for fabricating porous alumina with conical channels (Fig. 5.1g) [147–163]. An example of such kind of AAO layer is shown in Fig. 5.13.

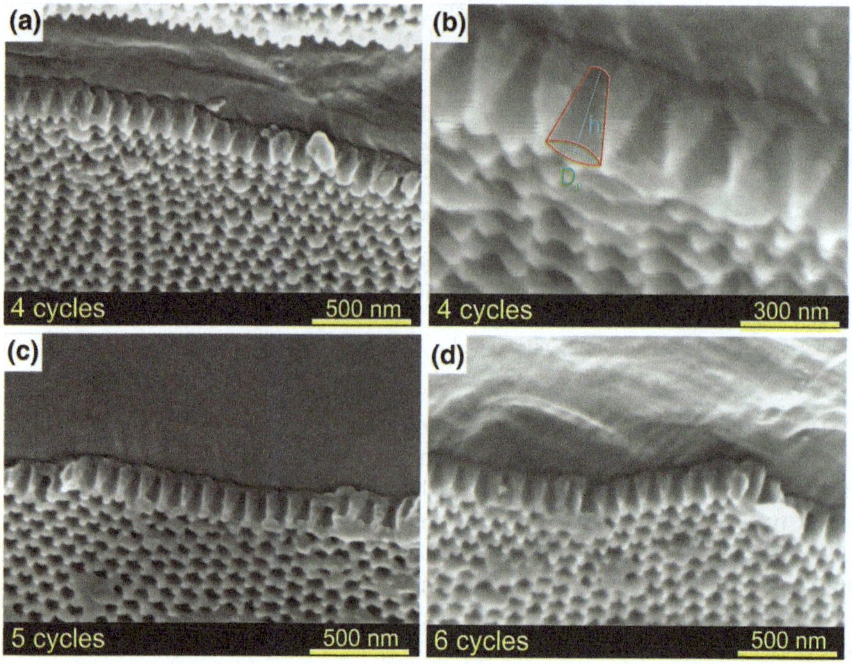

Fig. 5.13 SEM cross-sectional views of AAOs with conical nanopores after different number of anodization and pore widening cycles: **a** 4, **c** 5, **d** 6. **b** SEM micrograph of AAO with marked pore height (h) and pore diameter (D_p). Reprinted with permission from Ref. [163]. Copyright 2016 Elsevier.

The alternating cycles of anodization and chemical etching were typically repeated 3–9 times. A main disadvantage of this procedure is a low aspect ratio of formed AAO films, varying usually between 1 and 10. The shape and dimensions of the pores of AAO depends on anodization and etching conditions. Many scientists studied the influence of experimental conditions on the pore diameter [156, 159, 163], interpore distance [157], pore depth [149, 154, 158, 159, 163], pore shape [153, 156–158, 161], and aspect ratio [149, 154, 163] of AAO with conical nanopores. The AAO layer with conical nanopores can be also obtained by a post-anodization treatment of anodic alumina membranes. For instance, Cui et al. [164] received AAO membranes with a conical pore shape by anodization of aluminum in oxalic acid followed by a subsequent dissolution of the barrier layer in a phosphoric acid solution. During the dissolution of the barrier layer, etching of pore walls inside the pores occurs. The formation of pores with conical shapes is a result of the presence of diffusion gradient of H^+ ions along the pore length. Consequently, the etching rate inside the pores, close to the pore bottoms, is lower than that observed at the pore mouths close to the AAO surface.

Porous AAOs with conical nanopores have been used as templates for the synthesis of the high-performance antireflection polymethylmethacrylate (PMMA) films [155], polystyrene (PS) nanocone array [158], Ni hollow-nanocones, and taper-nanotubes [160], Ni nanocone array [148, 150, 154], Cu nanocone array electrode with electrocatalytic properties [163], and cone-shaped gold nanowire array [164]. Furthermore, Li et al. [158] have studied a controlled diffusion-limited chemical etching of the polystyrene-filled AAO templates for releasing polystyrene nanocone arrays from recyclable alumina templates. It was found that the AAO template with conical pores can be used for a structure replication more than once.

5.4.2 AAO with Step-Shaped Nanochannels

A procedure, consisting of two MA processes carried out at the same voltage and separated by chemical etching, was used for fabrication of AAO films with step-shaped nanochannels (Figs. 5.1h and 5.14a) [147, 162, 165–175]. On the other hand, by combination of MA and HA processes the step-shaped pores can be also created when the pre-patterned Al substrate is anodized [108]. The step-shaped nanopores in AAO might be successfully formed by a potentiostatic anodization with an anticipated exchange of the anodizing electrolyte [176, 177]. At the beginning of the process, phosphoric acid with a lower pH is used, then oxalic acid is poured into the cell. Zhao et al. [171] have found that step-shaped nanopores can be also created after the anodizing voltage increase for a very short time by a factor of $\sqrt{3}$. The pore narrowing after voltage increase was attributed to the enhanced activity of H^+ ions at the original pore bottoms.

The AAO layer with step-shaped nanopores can be used as a template for the synthesis of step-shaped carbon nanotubes [170, 171], carbon nanotubes decorated with magnetic (Ni) particles [171], bismuth nanowires with metal–semiconductor junction characteristic [165], copper nanotubes [166], and nickel–iron ($Ni_{80}Fe_{20}$) nanowires with adjustable magnetic properties [173]. The AAO with step-shaped

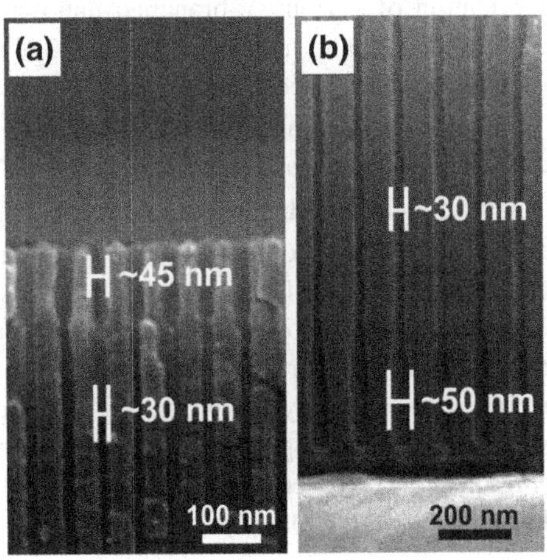

Fig. 5.14 SEM images of AAO layers with step-shaped nanopores **a** formed by 10 min of anodization in 0.3 M H_3PO_4 followed by 5 min anodizing in a 0.3 M $H_2C_2O_4$, and reversed step-shaped nanopores **b** formed after changing the order of the used electrolytes. All anodizations were carried out at 45 V and 20 °C

nanopores can be also used as a superhydrophobic surface [169], and as a label-free interferometric biosensor [172].

By changing the order of used electrolytes: (i) oxalic acid before phosphoric acid [162, 176, 177], (ii) phosphoric acid before tartaric acid [162, 178] or by applying two different anodizing steps [162, 179, 180], reversed step-shaped nanopores in AAOs were obtained (Figs. 5.1i and 5.14b). The similar AAO structure with reversed step-shaped pores can be engineered by combining the MA and HA processes performed at different voltages in the same electrolyte (e.g., MA at 40 V and HA at 110 V in 0.3 M oxalic acid) [171, 180–182].

Such kind of AAOs with reversed step-shaped nanopores were successfully employed as templates for the synthesis of magnetic nanowires (e.g., Ni, Co) [181, 182], carbon nanotubes and carbon nanotubes decorated with Ni particles [171].

5.4.3 AAO Layers with Branched Nanochannels

Most of the studies on AAO templates have been focused on fabricating linear nanopores in the alumina films. Recently, AAO templates with more complex-shaped nanopores were successfully synthesized by anodization. Since the cell diameter of porous anodic alumina is proportional to the anodizing voltage, reducing the applied voltage by a factor of $1/\sqrt{2}$ results in twice as many pores appearing on the surface. Consequently, nearly all pores branch into two smaller diameter pores (a symmetrically divided Y) in order to maintain the original total area of the template. Therefore, the AAO layer with Y-branched nanopores (Fig. 5.1j) can be obtained by reducing or increasing the applied anodizing voltage by a factor of $1/\sqrt{2}$ [166, 167, 183–197] and by anodization of aluminum in two different electrolytes, i.e., in oxalic acid at 40 V and then in sulfuric acid at 25 V [198, 199] or in oxalic acid at 40 V and then phosphoric acid at 80 V [200].

The porous alumina templates with Y-branched nanochannels were used for fabrication of various Y-branched nanowires [166, 183, 184, 188–190, 192, 194, 196–198, 201] and nanotubes [166, 183, 185, 187, 189, 190, 200, 202]. Such kind of membranes has been also tested for ultrafiltration applications [186, 191].

A stepwise increasing or reduction of the anodizing voltage by a factor of $1/\sqrt{n}$ creates n-branched nanochannels in the AAO template [189, 195–197, 203, 204]. A few sequential reduction of anodizing voltage, each time by a factor of $1/\sqrt{2}$ and $1/\sqrt{n}$, generates a complex multilevel Y-branched and multi-branched pore structure, respectively [189]. Using a stepwise reduction [183, 189, 195, 205] or exponential reduction [206–211] techniques of potentiostatic anodization, AAO templates with several levels of multiply branched channels (Fig. 5.1k) were fabricated. The AAO layers with multiply branched nanopores can be also produced by sequentially stepping down the anodization potential for each anodization step (Fig. 5.15). However, after each anodizing step etching or thinning of the barrier layer is required [205].

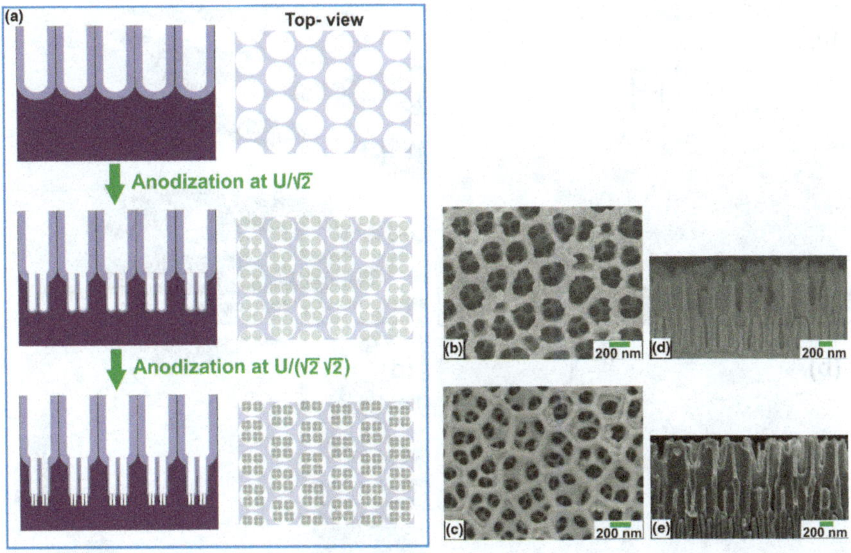

Fig. 5.15 **a** Schematic for fabrication of multilevel Y-branched nanopores in AAO by a sequential reduction of anodizing potential. Top (**b** and **c**) and cross-sectional (**d** and **e**) SEM images of AAOs formed by anodization with a single (**b** and **d**) and double (**c** and **e**) reduction of anodizing potential by a factor of $1/\sqrt{2}$. The initial anodization was carried out in 0.3 M H_3PO_4 at 130 V for 35 min, the barrier layer was thinned for 150 min in 5 wt% H_3PO_4. After the first reduction of potential (**b** and **d**), anodization was performed in 0.15 M $C_2H_2O_4$ at 80 V for 3.5 min, and the barrier layer was thinned for 90 min. After the second reduction of potential (**c** and **e**), anodization was performed in 0.3 M $C_2H_2O_4$ at 50 V for 15 min, and the barrier layer was thinned for 40 min. Adapted with permission from Ref. [205]. Copyright 2008 WILEY-VCH Verlag GmbH & Co. KGaA, Weinheim

The AAO with multilevel-branched nanostructures can be also obtained by four sequential anodization steps performed in different electrolytes at sequentially increasing anodizing potential. The first, second, third, and fourth anodizing steps were conducted in 0.3 M sulfuric acid at 25 V, 0.3 M oxalic acid at 40 V, 1.67 M malonic acid at 108 V, and 1 M tartaric acid at 195 V, respectively [178].

Most often, AAO with multiply branched nanochannels are used as templates for the synthesis of branched nanowires [189, 196, 197, 206–211] and nanotubes [189]. Figure 5.16 shows Y-branched and two level-branched Au and polystyrene nanowires synthesized by electrodeposition of Au in AAO and infiltration of AAO templates, respectively.

5.4.4 AAO with Serrated Nanochannels

Anodic aluminum oxide templates with periodic serrated nanochannels (Fig. 5.11) have been also successfully fabricated by self-organized anodizations [212–218]

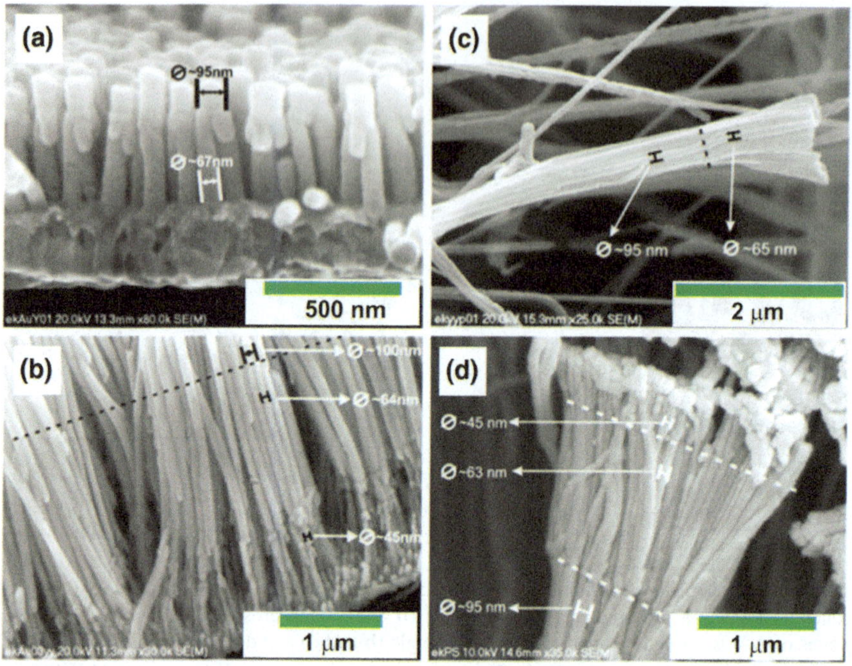

Fig. 5.16 SEM images of Y-branched Au (**a**) and polystyrene (**c**) nanowires and two level-branched Au (**b**) and polystyrene (**d**) nanowires after the removal of alumina templates. The AAO templates were prepared by anodization in 0.3 M oxalic acid at 60 V for 120 min followed by anodization at the reduced potential of 42 V (Y-branched template) for 30 min and at 30 V (two level-branched) for 30 min. Reprinted with permission from Ref. [197]

and used for a template-assisted fabrication of various nanomaterials, e.g., carbon nanotubes [212, 213], Pt nanowires [215] and Co nanowires [216]. The serrated pores in AAO were observed for the first time after one-step anodization performed in oxalic acid at 60 V. It was found that in as grown anodic films about 90% of channels were straight, however 10% exhibited randomly distributed periodical branches [212, 213]. The most popular method for the preparation of AAO with serrated pores is anodization performed in phosphoric acid at much lower anodizing voltage (about 10–80 V) than that required for the self-ordering regime (195 V). Usually, the Al substrate is pre-textured with hexagonally arranged concaves during the first anodization carried out in oxalic acid or sulfuric acid at a suitable self-ordering regime [214–218]. During the second anodization, the hexagonal geometry is sustained at the top surface of AAO (Fig. 5.17a), however after prolonged oxidation the ordered pore arrangement in the inner oxide layer is lost and serrated channels appear in the AAO structure (Fig. 5.17b, c) [218].

It is worth mentioning that by applying sequentially changed conditions (anodization at 45 V in oxalic acid at 5 °C followed by anodization in phosphoric acid

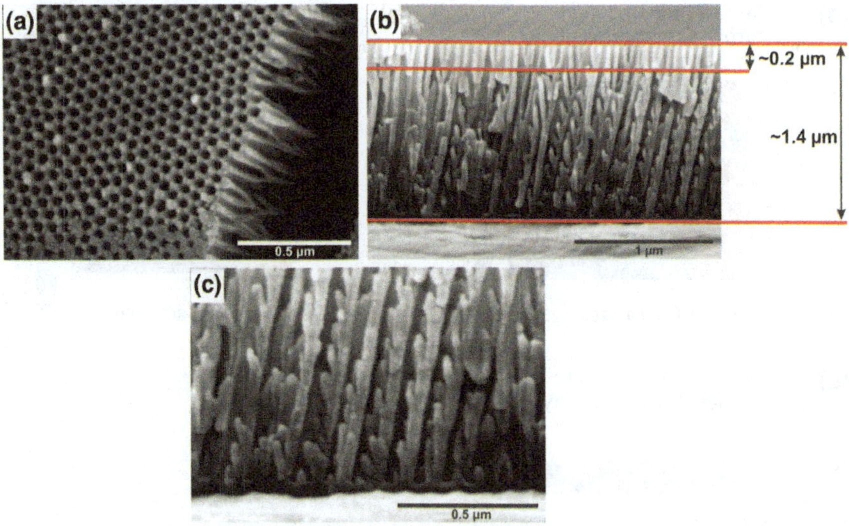

Fig. 5.17 Side (**a**) and cross-sectional (**b** and **c**) SEM images of AAO with serrated channels after the second anodization performed for 8 h in 0.3 M H_3PO_4 at 25 V and 1 °C. The sample was pre-textured for 16 h in 0.3 M H_2SO_4 at 25 V and 1 °C. Reprinted with permission from Ref. [218]. Copyright 2016 Elsevier

at 22 °C), a multilayered (straight and serrated layers) hybrid anodic oxide film can be also created [215].

The formation of serrated channels is attributed to the generation of oxygen gas bubbles from a lower side of the stem pore followed by plastic deformation of the oxide film caused by compressive stress accumulated from the evolution of gas bubbles [215, 216]. Because the electric field strength is highest at the pore bottom, O^{2-} are easily oxidized and oxygen bubbles are formed (Fig. 5.18). Initially formed small bubbles become trapped inside the barrier layer, however further oxidation of O^{2-} causes expansion of such bubbles, instead of generation of new ones. The gas bubble inside the barrier layer deforms the electric field, influencing current distribution (Fig. 5.18a). The bubble region suppresses the ionic transport and current density becomes concentrated around the oxygen bubble. This leads to enhanced aluminum oxide formation at the local points (Fig. 5.18b). Once the bubble size reaches the thickness of the barrier layer, oxygen is released, new equilibrium of electric field is established and the barrier layer is deformed (Fig. 5.18c). As-formed protuberance is pushed upward and a serrated channel is generated.

The periodicity in serrated channel appearance was associated with periodical release of oxygen bubbles [215, 216]. It should be mentioned that serrated channels are aligned with an inclination angle of ~25°.

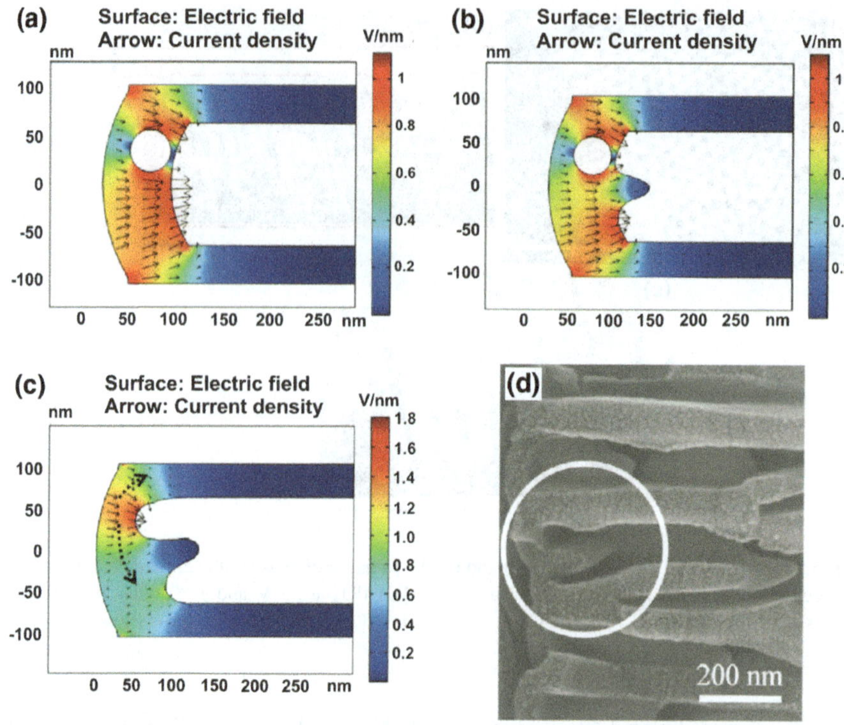

Fig. 5.18 Schematic diagram for the formation of AAO with serrated channels. The field and current distributions: in the presence of gas bubble in the barrier layer (**a**), after the formation of the protuberance at the pore bottom (**b**), and after the release of oxygen bubble (**c**). The SEM image of AAO with serrated channels formed at 60 V in 6 wt% phosphoric acid (**d**). Reprinted with permission from Ref. [216]. Copyright 2010 American Chemical Society

5.4.5 AAO with Periodically Branched Channels

A potentiostatic anodization with periodic changes in the anodizing voltage [162, 219–231] or current density [232, 233] was used for the fabrication of periodically branched nanoporous structures (Fig. 5.1m). Figure 5.19 shows the schematic illustration of the typical voltage signal applied during the periodic anodization of aluminum together with the grown AAO structure and SEM cross-sectional view images of the AAO with periodically branched nanochannels.

A typical synthesis procedure involves a linear decrease of the anodization voltage to a certain value, and then, anodization voltage increases sinusoidally to the initial value. By repeating of this process, a periodic structure with a main (stem) channel and branched channels is formed. The stem channels are formed during high-voltage duration (t_1 in Fig. 5.19a), and each stem channel branches into several small channels during a low-voltage duration (t_2 in Fig. 5.19a). The AAO film with periodically branched nanopores possesses very interesting properties and

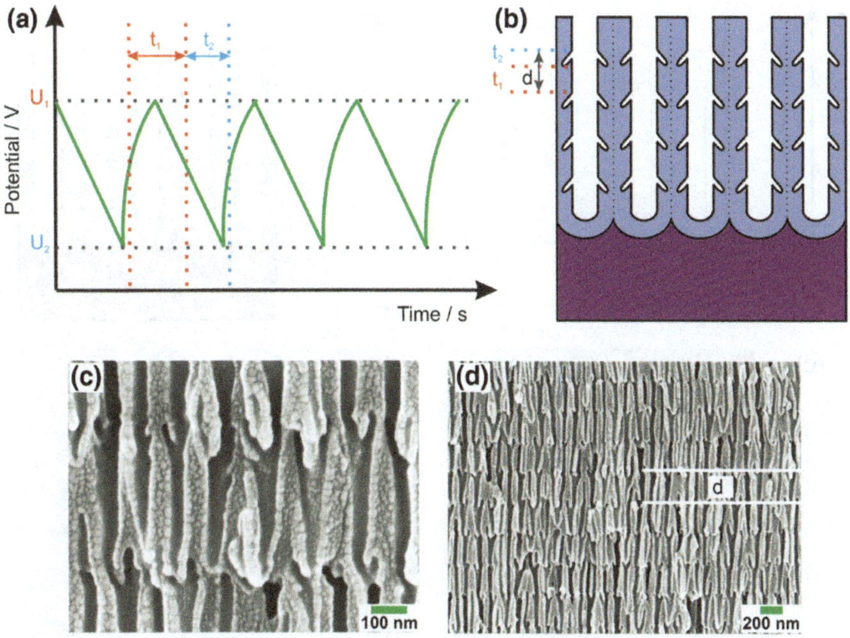

Fig. 5.19 Schematic illustration of the typical voltage signal applied during the periodic anodization of aluminum (**a**) together with the formed AAO layer (**b**). The cross-sectional SEM images of prepared AAO layers with periodically branched nanopores (**c** and **d**). The anodization was carried out at 18 °C in 0.3 M oxalic acid. The applied anodizing potential decreases linearly from 53 V to 23 V in 2.5 min, and then increases sinusoidally from 23 V to 53 V in 15 s. The thickness of the single layer (d) was about 290 nm. **c** and **d** reprinted with permission from [225]. Copyright 2009 Chinese Physical Society and IOP Publishing Ltd

applications. For example, its optical properties can be tuned by anodization conditions [222, 224, 227, 231, 232] and post-anodization treatment [219, 221, 224–226], and therefore, the periodic AAO structure can be used as photonic crystals [219, 221–227, 229–232]. The AAO films with periodically branched nanopores were also used as templates for fabrication of three-dimensionally interconnected porous carbons [228] and as chemical sensors [232].

5.4.6 AAO with Modulated Pore Diameters

Over the past decade, AAO templates with periodically modulated pore diameter (Fig. 5.1n) have been synthesized by: (a) pulse anodization that combines subsequent mild anodization (MA) and hard anodization (HA) processes performed at different voltages [108, 234–239] or different current density [240], (b) hard-pulse anodization technique that combines two different HA potential pulses [241], (c) pulse changes of electrolyte flow [242–244], (d) anodization in sequentially

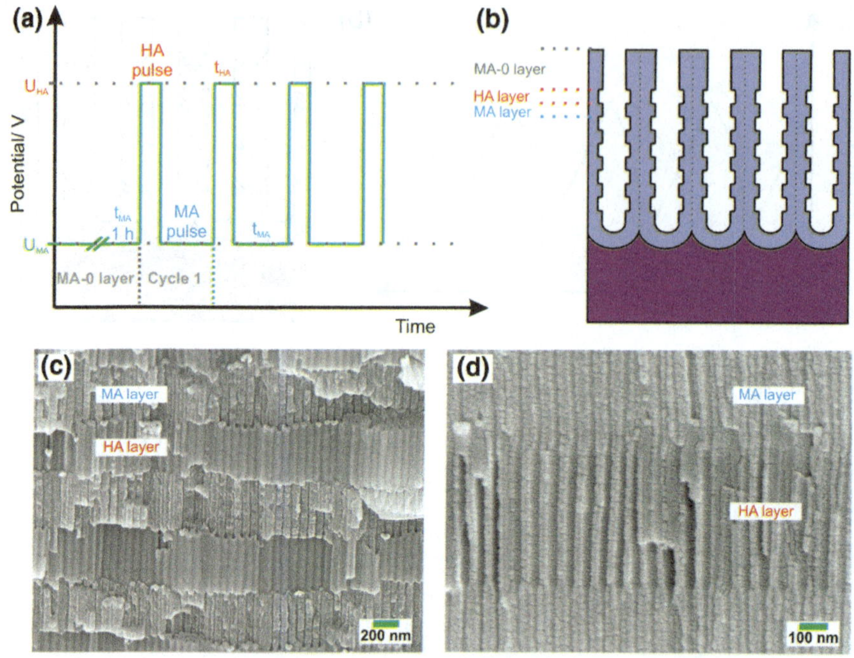

Fig. 5.20 Pulse anodization of aluminum for pore diameter modulations. **a** A typical voltage signal applied during the pulse anodization of aluminum, where U_{MA} and U_{HA} are the anodizing potentials used to achieve mild anodizing (MA) and hard anodizing (HA) conditions, respectively. **b** A schematic representation of the AAO structure with a modulated pore diameter grown after a few HA and MA pulses. (**c** and **d**)The SEM cross-sectional views of AAOs with modulated pore diameters by pulse anodization in 0.3 M H_2SO_4 at 1 °C. Each cycle consisted of the MA pulse (U_{MA} = 25 V, t_{MA} = 180 s) and HA pulse (U_{HA} = 35 V, t_{HA} = 0.5 s)

changed electrolytes but at the same anodizing current density [178] or anodizing potential [177], (e) controlling spontaneous current oscillations occurring during potentiostatic hard anodization in an unstirred solution [245] or switching the anodization voltage between the spontaneous oscillatory and non-oscillatory behaviors [246], and finally (f) discontinuous anodization in phosphoric acid [247]. Figure 5.20 shows an example of fabrication of AAO with modulated pore diameters by the pulse anodization carried out in sulfuric acid at the MA and HA potential of 25 V and 35 V, respectively.

For the fabrication of AAO with advanced structural pore modulations, Losic et al. [248, 249] have developed a cyclic anodization, in which the anodizing current in the form of periodic waves with different shapes (profiles), amplitudes, and periods was applied. By applying a symmetrical sinusoidal and asymmetrical current signals, AAO templates with a symmetrical geometry (Fig. 5.1o) and a ratchet-type asymmetric geometry along the pore axis (Fig. 5.1p) were prepared, respectively. Other advanced pore modulations were obtained by applying more complex current signals during a galvanostatic cyclic anodization [248]. The AAO layer with a modulated symmetrical pore diameter (Fig. 5.1o) has been also prepared by HA carried out in sulfuric acid at anodization voltages ranging between 27 and 32 V [146] and in oxalic acid under unstirred electrolyte conditions [245]. It was found that the modulated pore structure is a result of spontaneous symmetric sinusoidal current oscillations observed during the potentiostatic HA process. On the other hand, a ratchet-type asymmetric pores in AAO templates (Fig. 5.1p) was also observed for asymmetric current oscillations [245].

The AAO membrane with modulated pore diameter have been successfully used as a template for the synthesis of Ag nanowires [238], Au nanowires [237], nanoporous Au nanowires [238], CoFe nanowires [241], Ni nanowires [238, 250], Fe_3O_4 nanotubes [235] with magnetic properties, and PMMA nanotubes [244]. The AAO with modulated pore diameter was also used as photonic crystals [239, 251].

5.5 AAO Layers Grown on Supporting Substrates

Another interesting example of the AAO template currently being explored for the preparation of various nanostructured materials is a template grown by anodization of aluminum sputtered on a supporting substrate (Fig. 5.1q) including Si [e.g. 205, 252–259], GaAs [260], InP [261], Ti [262], ITO glass [e.g., 263–271] and others [272–277]. The main difference between anodized layers, formed by anodizing of Al foils and sputtered Al on supporting substrates, is their thickness. When aluminum foils are anodized, the oxide layer thickness can range from a few hundredths of nanometers to a few dozens of micrometers. On the other hand, the thickness of the oxide layer formed on a supporting substrate is limited by the thickness of the sputtered Al layer and usually does not exceed a few microns. The fabrication of ultrathin AAO membranes with a thickness up to 50 nm on Si substrate was also reported [278–280].

Due to a relatively low thickness of the Al layer in most cases the one-step anodization is applied to obtain AAO films on supporting substrates resulting in formation of poorly ordered nanoporous structures. In addition, the evaporated Al films consist of small and randomly distributed grains with densely spaced grain boundaries. In consequence, the pore arrangement and uniformity of formed anodic alumina films is much worse than those grown on Al foils. Although some examples of the two-step anodization procedure have been already reported [e.g. 255, 260, 281–287], no significant improvement in the nanopore order was achieved by applying this strategy due to the extremely limited duration of the first anodizing step which is a key factor for formation of the regular array of nanoconcaves on the Al surface (see paragraph 5.4).

As it was already mentioned in paragraph 5.3.3, a significant improvement in pore order can be achieved by an appropriate pre-texturing of the substrate before anodization. This approach has been also employed for the fabrication of well-ordered AAO layers on various supporting substrates (namely Si) [57, 68, 75,

78, 104, 118, 288, 289]. A successful fabrication of anodic alumina with square cells on Si has been also reported [57, 75, 289]. However, it should be mentioned that the nanoindentation of Al is performed by using a hard mold under high pressure, the crack of substrates having low mechanical strength can be expected [68]. Therefore, in most cases "soft imprinting" methods are used for the pre-patterning of Al on supporting substrates (e.g., [68, 118, 288]). A very interesting approach allowing for the fabrication of almost ideally ordered AAO on Si substrate was proposed by Nasir et al. [290]. The technique is based on formation of an array of conductive holes on a SiO_2-covered Si substrate by its imprinting followed by Al evaporation instead of direct indentation of the metal. Further anodization creates a porous alumina layer with pore positions correlated with the previously imprinted holes, since the electric field is guided by the array of holes created on the conductive substrate [290].

It should be mentioned that AAO layers with a complex internal structure can be also successfully obtained on supporting substrates. For instance, Ho et al. reported synthesis of multilevel-branched anodic alumina by a sequential decrease of voltage during anodization and etching of the barrier layer (see paragraph 5.4.3) [205].

A very serious problem that should be taken into consideration during anodization of Al on various supporting substrates is a relatively poor adhesion of the Al film often resulting in its peeling-off from the substrate. To overcome this problem, thin layer of other metals such as Ta [291, 292], Ti [268, 293], Nb [294], W [270], and others is usually sputtered on the substrate before deposition of aluminum. If it is necessary, a highly conductive layer of Au or Pt can be also sputtered beneath the Al layer to ensure after anodization the electrical contact at the pore bottoms [292, 295]. However, it should be remembered that the presence of such kind of the interstitial layer can influence the structure of barrier layer formed during anodization [295].

Another significant difference between anodic porous alumina layers, formed by anodizing of Al foils and Al sputtered on supporting substrates, is the shape of the barrier layer formed at the bottom of pores. In a typical AAO film, the barrier layer has a hemispherical and scalloped geometry (see Fig. 5.12c) while a unique arched structure, often with a void beneath the barrier layer, is formed when the aluminum layer on the supporting substrate is fully converted into aluminum oxide (Fig. 5.21). The barrier layer rearrangement and formation of characteristic voids beneath the pores was observed for different supporting substrates such as Si [253, 254, 257, 286, 295–297], glass [263–265, 298] and others [299].

In general, at the beginning of anodization a typical hemispherical structure of the barrier layer is observed, even when the oxide/Al interface reached the surface of the supporting substrate (Fig. 5.21a, b). If the anodizing process is continued further, a gradual change of the barrier layer structure from hemispherical to inverted arched is observed (Fig. 5.21c, d). Simultaneously, hemispherical pore bottoms altered drastically their shape into rectangular and finally into arched. It is generally accepted that the rearrangement of morphology of the barrier layer and formation of such kind of voids is a result of the mechanical stress generated at the AAO/substrate interface [286, 297].

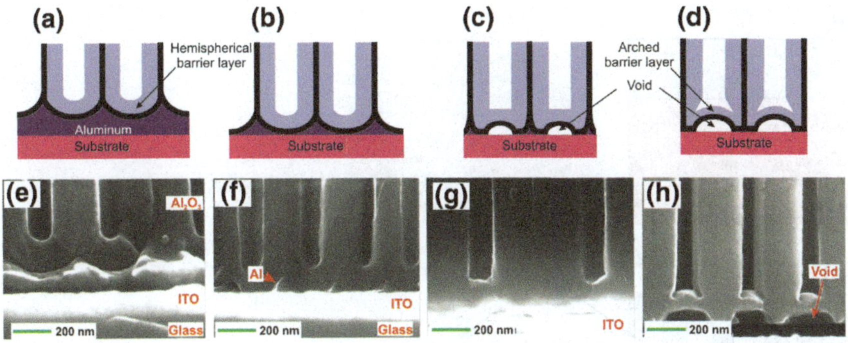

Fig. 5.21 Formation of the arched barrier layer structure in AAO on a supporting ITO/glass substrate at different stages (**a**–**d**), and corresponding SEM images of fracture sections near the barrier layer of AAO after 38 min (**e**), 45 min (**f**), 50 min (**g**), and 60 min (**h**) of anodizing in 1.1 M H_3PO_4 at 130 V and 2 °C. Reproduced in part with permission from Ref. [263]. Copyright 2002 The Electrochemical Society

A further practical application of as-formed AAO, e.g., as a template for deposition of various nanomaterials, often requires a complete removal of the dielectric barrier layer to form a through-hole alumina layer and ensure a direct contact of electrolyte with the supporting substrate. Fortunately, the barrier layer in AAO films obtained by anodization of Al sputtered on supporting substrates is significantly thinner than pore walls that allows for its selective removal through a chemical dissolution usually performed in a phosphoric acid solution [78, 252, 255, 265, 293, 300–303]. However, some alternative techniques for the barrier layer removal including hydrogen plasma treatment [304], cathodic polarization [253, 295, 305], and stepwise reduction of anodizing potential followed by chemical etching in H_3PO_4 [306] have been already proposed. An interesting method based on anodization of the Al/W bilayer followed by a selective removal of tungsten oxide has been also proposed for perforation of the barrier layer by Oh and Thompson [277].

Many promising applications of AAO formed by anodization of aluminum deposited on various supporting substrates have been already reported. For instance, an ordered pattern can be successfully transferred from anodic alumina into the underlying Si layer to form an array of concaves with desired size, shape, and depth [78, 252, 300, 307–310]. In addition, such kind of porous silicon can be also oxidized to form a nanostructured layer of SiO_2 [258]. A similar approach has been also employed for the pattern transfer into the Fe-doped InP substrate [311]. On the other hand, when anodization of Al on Si is continued after a complete conversion of metallic Al into porous alumina, an electrolyte can reach the substrate what results in oxidation of silicon and formation of the two-dimensional array of SiO_2 nanodots (see Fig. 5.22) [257, 258, 296, 312]. The most important advantage of this method is its low-cost and possibility to be used in large areas. Moreover, the size and density of pores in the template can be easily tuned by appropriate adjusting of anodizing conditions. It is obvious that the density of as-formed silicon

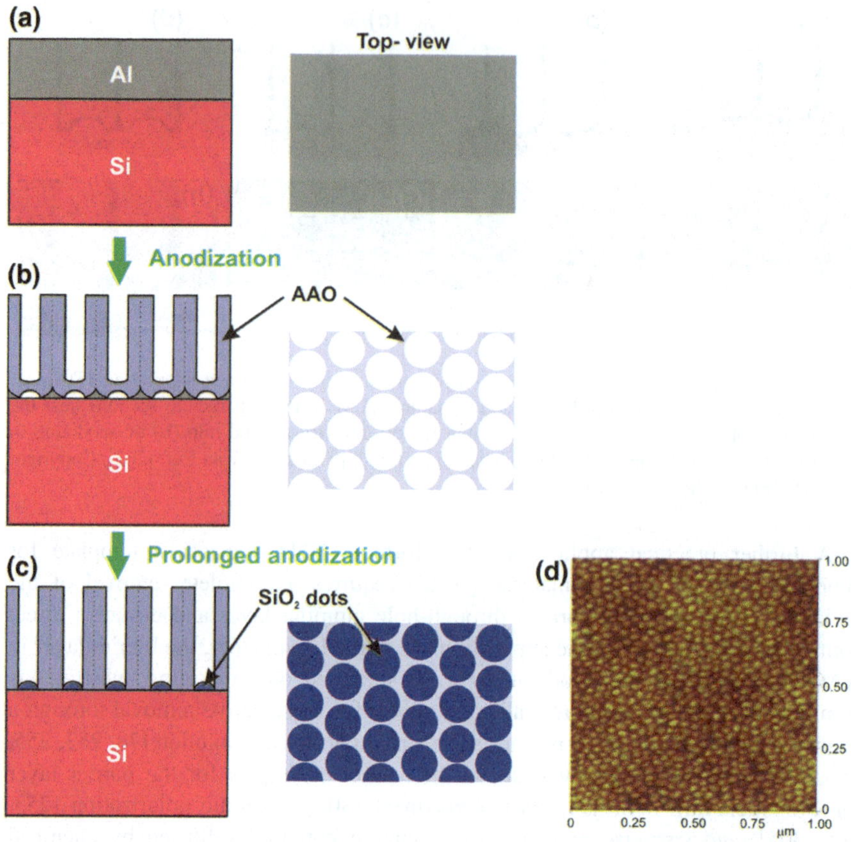

Fig. 5.22 Schematic for fabrication of the SiO_2 dot array by anodization of the Al thin film on Si substrate (**a–c**). The anodization was carried out in 6 vol.% H_2SO_4 at 20 V. The pore diameter and thickness of formed AAO was 10–13 nm and 56 nm, respectively (**b**). The prolonged anodization results in oxidation of Si substrate at each pore bottom and formation of SiO_2 nanodots (**c**). The size of SiO_2 dot arrays was about 30–35 nm in diameter and 8–10 nm in height. The AFM image of SiO_2 dots after the AAO template dissolution (**d**). Figure 5.22d reprinted with permission from Ref. [312]. Copyright 2008 WILEY-VCH Verlag GmbH & Co. KGaA, Weinheim

oxide nanoislands is determined by the pore density in AAO template since a single dot is formed at each pore tip.

Recently, it was also proved that anodic oxidation of Al deposited on a layer of other metal, e.g., Ta, Nb, W, or Ti can result in the formation of nanostructures of the corresponding metal oxide beneath and within the pores of the formed AAO layer. The main idea of this method, often called a through-mask anodization technique, is schematically shown in Fig. 5.23.

In the first step, metallic Al layer (Fig. 5.23a) is anodized in commonly used electrolytes (sulfuric, oxalic, or phosphoric acid) under appropriate conditions to form the oxide film with desired nanoporous morphology (Fig. 5.23b). When the

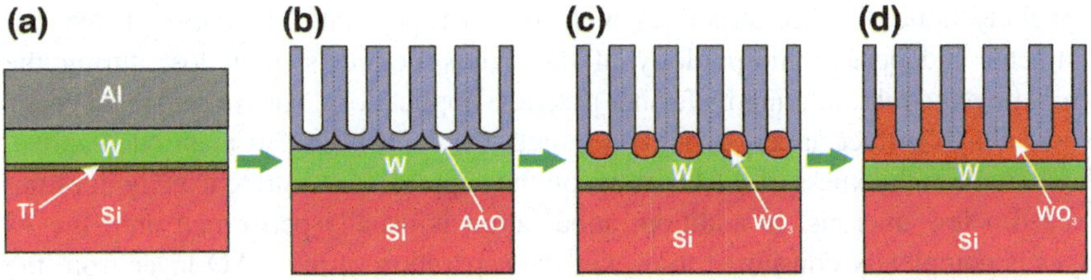

Fig. 5.23 Schematic for fabrication of the nanostructured tungsten oxide film: sputter-deposition of the Al/W/Ti layer on Si (**a**), formation of nanoporous AAO by anodization of Al (**b**), anodization of the W layer through nanoporous AAO (**c**), and prolonged anodization or re-anodization of the W layer at high voltages (**d**)

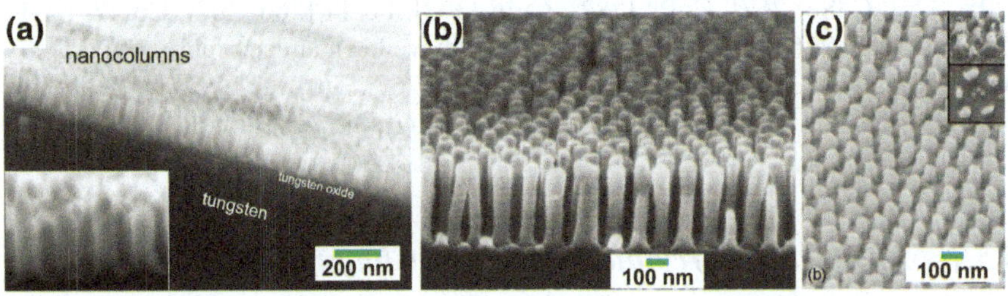

Fig. 5.24 SEM images of column-like structures of WO_3 (**a**), Ta_2O_5 (**b**), and TiO_2 (**c**) formed by the through-mask anodization technique. Reprinted with permission from [274] (**a**), [273] (**b**) and [316] (**c**). (a) Copyright 2008 American Chemical Society. (b) Copyright 2004 The Electrochemical Society. (c) Copyright 2009 IOP Publishing

barrier layer of alumina reaches the underlying metal, nanosized regions of the corresponding metal oxide are raised and penetrate the AAO layer beneath the pores (Fig. 5.23c). The further re-anodizing of the specimen at higher anodic voltages results in formation of anodic metal oxide which gradually fills the pores of the alumina template (Fig. 5.23d). Finally, the ordered array of transition metal oxide nanostructures is obtained after a selective dissolution of the alumina template. Up to now, this promising approach was successfully employed for the fabrication of Ta_2O_5 [104, 272, 273, 275, 313, 314], Nb_2O_5 [272, 276], WO_3 [274, 315], and TiO_2 [299, 316] nanostructures. Some examples of these materials are shown in Fig. 5.24.

A great advantage of such kind of porous AAOs grown on supporting substrates is the possibility to use them as templates for fabrication of various nanomaterials, e.g., nanowire and nanodot arrays. It is obvious that practical applications of these nanostructures, e.g., for a device fabrication, require a complete removal of the alumina template. However, typical high aspect ratio AAO membranes formed on the surface of Al foils exhibit a relatively poor mechanical stability after removal of the remaining substrate and pore opening procedure [140, 317]. Moreover, collapse of the nanostructure often occurs during chemical dissolution of the template causing the aggregation of nanowires into bundles [140]. This phenomenon, being the result of the existence of capillary forces between individual nanowires is especially noticeable for structures with a high aspect ratio. In consequence, the character and desired morphology of the nanostructure is often lost during the template removal hindering its further practical applications, e.g., in sensing devices where a high surface area is strongly required. On the contrary, as mentioned above, the maximum thickness of AAO grown on the supporting substrate usually does not exceed a few microns. In addition, anodization is usually performed until the Al layer is completely consumed, however the separation of the AAO layer from the supporting material does not occur. It allows for the fabrication of mechanically stable nanostructures even after a final dissolution of the template.

A great variety of metallic or semiconducting nanostructures (nanowires, nanorods, nanodots) such as: Al [259], Au [103, 253, 278, 292, 293, 301, 318], Cu [270, 319–322], Ni [281, 291, 292, 302, 306, 319, 323, 324], Pd [304, 325], Pt [295, 326], CdSe [269], CoPd [282], CuIn [303], Fe–Pt [266, 267], Co–Pt, Ni–Pt [267], Si [255, 256, 327–329], and ZnO [330] have been already fabricated directly on Si (e.g., [293, 295, 320]), ITO (e.g., [264–267, 270]) or even on a flexible polymer substrate [325, 326] by deposition through the AAO mask.

5.5.1 AAO with Horizontally Aligned Channels

As mentioned above, typically, the pores in alumina templates are aligned perpendicular to the surface of the substrate. However, from the point of view of potential applications of 1D nanomaterials (e.g., nanowires of nanotubes) grown inside the template, AAO membranes with horizontally aligned nanochannels are also strongly desirable [331–338]. The possibility of fabrication of such anodic alumina layers was reported for the first time in 1993 by Masuda et al. [331]. The authors performed the anodization of the Al film deposited on a glass substrate and covered with a sputtered layer of Al_2O_3 in oxalic acid at the potential of 100 V at 17 °C for 3 min. After 40 min of the pore widening carried out in a 5 wt% phosphoric acid solution at 30 °C, an array of horizontal nanochannels with an average diameter of about 80 nm and depth of about 1 μm was achieved [331]. In 2005, Cojocaru and co-workers reported that a simple anodic oxidation of the Al film sandwiched between two insulating layers of SiO_2 at very low anodizing potentials (3–5 V) can lead to formation of horizontally aligned arrays of pores with an extremely small size (3–4 nm in diameter) and a length of about 200–300 nm [332]. A similar approach was successfully employed by several research groups to obtain anodic alumina with horizontal pores with controlled pore diameters ranging from 10 to 130 nm and pore-to-pore distances between 20 and 275 nm, depending on the type of used electrolyte and anodizing conditions [333–336]. The SEM images of such kind of AAO layers with various pore sizes are shown in Fig. 5.25. Furthermore, Xiang et al., have observed that if the interpore distance is about one fourth of the film thickness, a change in a pore arrangement from honeycomb structure to quadratic occurs [334]. Finally, Zhang et al. proved that it is possible to obtain an array in the form of 1D row of nanopores [333].

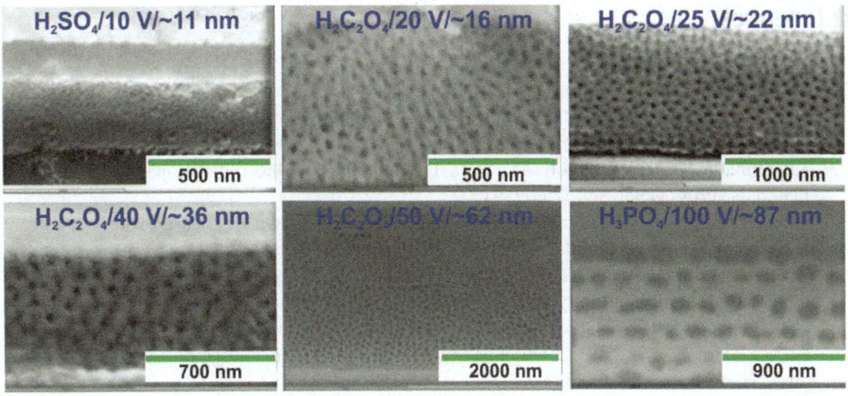

Fig. 5.25 SEM images of the lateral AAO layers with different pore diameters obtained at various anodizing potentials and electrolytes. The types of used acid/anodizing potential/average pore diameter are shown at top part of each image. Adapted with permission from [335]. Copyright 2008 IOP Publishing

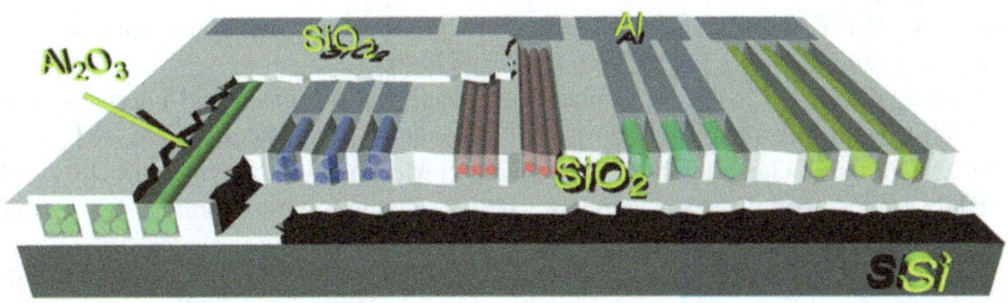

Fig. 5.26 Schematic representation of a multicontact horizontal AAO device with deposited different nanowires. A horizontal porous alumina structure containing different pore morphologies (pore diameter, interpore distances, and pore length) was obtained by individual anodization of each contact at different conditions. Reprinted with permission from [337]. Copyright 2010 American Chemical Society

A very interesting concept of multicontact anodization on a single chip was reported by Xiang et al. [337]. The main idea of this procedure is schematically shown in Fig. 5.26. As can be seen, Al fingers with various sizes are individually contacted and separated from each other that allow for their individual anodization in various electrolytes at different anodizing conditions. As a result, a complex system composed of alumina segments with different channel lengths, diameters and pore densities on a single chip can be obtained [337].

Recently, Bolger et al. have found that the pore morphology is also strongly influenced by mechanical constraint caused by the insulating layers surrounding the individual Al fingers. As a result, the flow of oxide is restricted at the interfaces and dendritic pore morphology can be created [338]. The AAO layers with horizontally aligned channels have been also successfully employed as templates for fabrication of metallic [334, 337] and semiconducting [332, 337] nanowires as well as carbon nanotubes [336].

References

1. Moriarty P (2001) Nanostructured materials. Rep Prog Phys 64:297–381
2. Dresselhaus MS, Lin YM, Rabin O et al (2004) Nanowires. In: Bushan B (ed) Springer Handbook of Nanotechnology. Springer, Heidelberg, pp 99–145
3. Kuchibhatla SVNT, Karakoti AS, Bera D et al (2007) One dimensional nanostructured materials. Prog Mater Sci 52:699–913
4. Klauser R, Huang ML, Wang SC et al (2004) Lithography with a focused soft X-ray beam and a monomolecular resist. Langmuir 20:2050–2053
5. Chen A, Chua SJ, Chen P et al (2006) Fabrication of sub-100 nm patterns in SiO_2 templates by electron-beamlithography for the growth of periodic III–V semiconductor nanostructures. Nanotechnology 17:3903–3908
6. Choi Y, Hong S, Lee LP (2009) Shadow overlap ion-beam lithography for nanoarchitectures. Nano Lett 9:3726–3731
7. Campbell M, Sharp DN, Harrison MT et al (2000) Fabrication of photonic crystals for the visible spectrum by holographic lithography. Nature 404:53–56
8. Choi WK, Liew TH, Dawood MK et al (2008) Synthesis of silicon nanowires and nanofin arrays using interference lithography and catalytic etching. Nano Lett 8:3799–3802
9. Senesi AJ, Rozkiewicz DI, Reinhoudt DN et al (2009) Agarose-assisted dip-pen nanolithography of oligonucleotides and proteins. ACS Nano 3:2394–2402
10. Cong CX, Yu T, Ni ZH et al (2009) Fabrication of graphene nanodisk arrays using nanosphere lithography. J Phys Chem C 113:6529–6532
11. Li C, Hong G, Qi L (2010) Nanosphere lithography at the gas/liquid interface: A general approach toward free-standing high-quality nanonets. Chem Mater 22:476–481
12. Chen X, Thomas J, Gangopadhyay P et al (2009) Modification of symmetrically substituted phthalocyanines using click chemistry: phthalocyanine nanostructures by nanoimprint lithography. J Am Chem Soc 131:13840–13843
13. Ell JR, Crosby TA, Peterson JJ et al (2010) Formation of SiO_2 air-gap patterns through $scCO_2$ infusion of NIL patterned PHEMA. Chem Mater 22:1445–1451
14. Weibel DB, DiLuzio WR, Whitesides GM (2007) Microfabrication meets microbiology. Nature Rev Microbiol 5:209–218
15. Campos LM, Truong TT, Shim DE et al (2009) Applications of photocurable PMMS thiol-ene stamps in soft lithography. Chem Mater 21:5319–5326
16. Zhang H, Shen Ch, Chen S et al (2005) Morphologies and microstructures of nano-sized Cu_2O particles using acetyltrimethylammonium template. Nanotechnology 16:267–272
17. Xiao W, Wang D, Lou XW (2010) Shape-controlled synthesis of MnO_2 nanostructures with enhanced electrocatalytic activity for oxygen reduction. J Phys Chem C 114:1694–1700
18. Lu MH, Zhang Y (2006) Microbead patterning on porous films with ordered arrays of pores. Adv Mater 18:3094–3098
19. Lee SH, Bantz KC, Lindquist NC et al (2009) Self-assembled plasmonic nanohole arrays. Langmuir 25:13685–13693
20. Fang ZL, Li Y, Chen C et al (2010) pH-induced simultaneous synthesis and self-assembly of 3D layered β-FeOOH nanorods. Langmuir 26:2745–2750
21. Zach MP, Ng KH, Penner RM (2000) Molybdenum nanowires by electrodeposition. Science 290:2120–2123
22. Penner RM (2002) Mesoscopic metal particles and wires by electrodeposition. J Phys Chem B 106:3339–3353
23. Menke EJ, Li Q, Penner RM (2004) Bismuth telluride (Bi_2Te_3) nanowires synthesized by cyclic electrodeposition/stripping coupled with step edge decoration. Nano Lett 4:2009–2014
24. Sun L, Searson PC, Chien CL (1999) Electrochemical deposition of nickel nanowire arrays in single-crystal mica films. Appl Phys Lett 74:2803–2805

25. Sun L, Searson PC, Chien CL (2000) Finite-size effects in nickel nanowire arrays. Phys Rev B 61:R6463–R6466
26. Matthias S, Schilling J, Nielsch K et al (2002) Monodisperse diameter-modulated gold microwires. Adv Mater 14:1618–1621
27. Steinhart M, Wendorff JH, Greiner A et al (2002) Polymer nanotubes by wetting of ordered porous templates. Science 296:1997
28. Steinhart M, Wehrspohn RB, Gösele U et al (2004) Nanotubes by template wetting: A modular assembly system. Angew Chem Int Ed 43:1334–1344
29. Chen X, Steinhart M, Hess C et al (2006) Ordered arrays of mesoporous microrods from recyclable macroporous silicon templates. Adv Mater 18:2153–2156
30. Chen C, Loo J, Deng M et al (2009) Hollow platinum nanoshell tube arrays: Fabrication and characterization. J Phys Chem C 113:5472–5477
31. Hamley IW (2003) Nanostructure fabrication using block copolymers. Nanotechnology 14: R39–R54
32. Lazzari M, López-Quintela MA (2003) Block copolymers as a tool for nanomaterial fabrication. Adv Mater 15:1583–1594
33. Cheng JY, Mayes AM, Ross CA (2004) Nanostructure engineering by templated self-assembly of block copolymers. Nat Mater 3:823–828
34. Kim HC, Park SM, Hinsberg WD (2010) Block copolymer based nanostructures: materials, processes, and applications to electronics. Chem Rev 110:146–177
35. Li M, Ober CK (2006) Block copolymer patterns and templates. Mater Today 9:30–39
36. Aizawa M, Buriak JM (2007) Block copolymer templated chemistry for the formation of metallic nanoparticle arrays on semiconductor surfaces. Chem Mater 19:5090–5101
37. Allred DB, Cheng A, Sarikaya M et al (2008) Three-dimensional architecture of inorganic nanoarrays electrodeposited through a surface-layer protein mask. Nano Lett 8:1434–1438
38. Liu J, Mao Y, Lan E et al (2008) Generation of oxide nanopatterns by combining self-assembly of S-layer proteins and area-selective atomic layer deposition. J Am Chem Soc 130:16908–16913
39. Deng Z, Mao C (2003) DNA-templated fabrication of 1D parallel and 2D crossed metallic nanowire arrays. Nano Lett 3:1545–1548
40. Gu Q, Cheng C, Gonela R et al (2006) DNA Nanowire Fabr Nanotechnol 17:R14–R25
41. Knez M, Kadri A, Wege C et al (2006) Atomic layer deposition on biological macromolecules: metal oxide coating of tobacco mosaic virus and ferritin. Nano Lett 6:1172–1177
42. Sotiropoulou S, Sierra-Sastre Y, Mark SS et al (2008) Biotemplated nanostructured materials. Chem Mater 20:821–834
43. Owrutsky JC, Pomfret MB, Brown DJ (2009) Coherent acoustic oscillations of nanorods composed of various metals. J Phys Chem C 113:10947–10955
44. Sieb NR, Wu NC, Majidi E et al (2009) Hollow metal nanorods with tunable dimensions, porosity, and photonic properties. ACS Nano 6:1365–1372
45. Mourachkine A, Yazyev OV, Ducati C et al (2008) Template nanowires for spintronics applications: nanomagnet microwave resonators functioning in zero applied magnetic field. Nano Lett 8:3683–3687
46. Callegari V, Gence L, Melinte S et al (2009) Electrochemically template-grown multi-segmented gold-conducting polymer nanowires with tunable electronic behavior. Chem Mater 21:4241–4247
47. Menaker A, Syritski V, Reut J et al (2009) Electrosynthesized surface-imprinted conducting polymer microrods for selective protein recognition. Adv Mater 21:2271–2275
48. Zhou H, Wong SS (2008) A facile and mild synthesis of 1-D ZnO, CuO, and α-Fe2O3 nanostructures and nanostructured arrays. ACS Nano 2:944–958
49. Mo D, Liu J, Yao HJ et al (2008) Preparation and characterization of CdS nanotubes and nanowires by electrochemical synthesis in ion-track templates. J Cryst Growth 310:612–616
50. Qu X, Komatsu T (2010) Molecular capture in protein nanotubes. ACS Nano 4:563–573

51. Savariar EN, Sochat MM, Klaikherd A et al (2009) Functional group density and recognition in polymer nanotubes. Angew Chem Int Ed 48:110–114
52. Masuda H, Kanezawa K, Nishio K (2002) Fabrication of ideally ordered nanohole arrays in anodic porous alumina based on nanoindentation using scanning probe microscope. Chem Lett 31:1218–1219
53. Shingubara S, Murakami Y, Morimoto K et al (2003) Formation of aluminum nanodot array by combination of nanoindentation and anodic oxidation of aluminum. Surf Sci 532–535:317–323
54. Jaafar M, Navas D, Hernández-Vélez M et al (2009) Nanoporous alumina membrane prepared by nanoindentation and anodic oxidation. Surf Sci 603:3155–3159
55. Shingubara S, Murakami Y, Sakaue H et al (2003) Aluminum nanodot array formed by anodic oxidation and its conduction properties. Proc of SPIE 4999:387–395
56. Liu CY, Datta A, Wang YL (2001) Ordered anodic alumina nanochannels on focused-ion-beam-prepatterned aluminum surfaces. Appl Phys Lett 78:120–122
57. Hong J, Kim K, Heo J et al (2010) Formation of square matrix pores on Al film utilizing focused ion beam milled indent. Thin Solid Films 518:4572–4577
58. Peng CY, Liu CY, Liu NW et al (2005) Ideally ordered 10 nm channel arrays grown by anodization of focused-ion-beam patterned aluminum. J Vac Sci Technol B 23:559–562
59. Lin MN, Liu CY, Liu NW et al (2006) Fabrication of an ordered nanoparticle array with a nanoaperture membrane used as a contact-mask. Nanotechnology 17:315–319
60. Chen B, Lu K, Tian Z (2010) Gradient and alternating diameter nanopore templates by focused ion beam guided anodization. Electrochim Acta 56:435–440
61. Tian ZP, Lu K, Chen B (2010) Unique nanopore pattern formation by focused ion beam guided anodization. Nanotechnology 21:405301
62. Tian Z, Lu K, Chen B (2010) Fundamental mechanisms of focused ion beam guided anodization. J Appl Phys 108:094306
63. Chen B, Lu K, Tian Z (2011) Novel patterns by focused ion beam guided anodization. Langmuir 27:800–808
64. Lu K, Zhao J (2010) Focused ion beam lithography and anodization combined nanopore patterning. J Nanosci Nanotechnol 10:6760–6768
65. Robinson AP, Burnell G, Hu M et al (2007) Controlled, perfect ordering in ultrathin anodic aluminum oxide templates on silicon. Appl Phys Lett 91:143123
66. Lin MN, Lin MT, Liu CY et al (2005) Long-range ordered nanoaperture array with uniform diameter and interpore spacing. Appl Phys Lett 87:173116
67. Sun Z, Kim HK (2002) Growth of ordered, single-domain, alumina nanopore arrays with holographically patterned aluminum films. Appl Phys Lett 81:3458–3460
68. Maria Chong AS, Tan LK, Deng J et al (2007) Soft imprinting: Creating highly ordered porous anodic alumina templates on substrates for nanofabrication. Adv Funct Mater 17:1629–1635
69. Oshima K, Kikuchi H, Nakao H et al (2007) Detecting dynamic signals of ideally ordered nanohole patterned disk media fabricated using nanoimprint lithography. Appl Phys Lett 91:022508
70. Yin A, Tzolov M, Cardimona D et al (2007) Fabrication of highly ordered anodic aluminium oxide templates on silicon substrates. IET Circ Dev Syst 1:205–209
71. Kim B, Park S, McCarthy TJ et al (2007) Fabrication of ordered anodic aluminum oxide using a solvent-induced array of block-copolymer micelles. Small 3:1869–1872
72. Smith JT, Hang Q, Franklin AD et al (2008) Highly ordered diamond and hybrid triangle-diamond patterns in porous anodic alumina thin films. Appl Phys Lett 93:043108
73. Liu NW, Datta A, Liu CY et al (2003) High-speed focused-ion-beam patterning for guiding the growth of anodic alumina nanochannel arrays. Appl Phys Lett 82:1281–1283
74. Stasi V, Cattaneo G, Franz S et al (2007) Aluminium pre-patterning for highly ordered nanoporous anodized alumina. Photon Nanostruct 5:136–139
75. Kustandi TS, Loh WW, Gao H et al (2010) Wafer-scale near-perfect ordered porous alumina on substrates by step and flash imprint lithography. ACS Nano 4:2561–2568

76. Masuda H, Yamada H, Satoh M et al (1997) Highly ordered nanochannel-array architecture in anodic alumina. Appl Phys Lett 71:2770–2772
77. Masuda H, Yotsuya M, Asano M et al (2001) Self-repair of ordered pattern of nanometer dimensions based on self-compensation properties of anodic porous alumina. Appl Phys Lett 78:826–828
78. Masuda H, Yasui K, Sakamoto Y et al (2001) Ideally ordered anodic porous alumina mask prepared by imprinting of vacuum-evaporated Al on Si. Jpn J Appl Phys 40:L1267–L1269
79. Masuda H, Yasui K, Watanabe HM et al (2001) Fabrication of through-hole diamond membranes by plasma etching using anodic porous alumina mask. Electrochem Solid State Lett 4:G101–G103
80. Masuda H, Ohya M, Asoh H et al (1999) Photonic crystal using anodic porous alumina. Jpn J Appl Phys 38:L1403–L1405
81. Masuda H, Abe A, Nakao M et al (2003) Ordered mosaic nanocomposites in anodic porous alumina. Adv Mater 15:161–164
82. Asoh H, Nishio K, Nakao M et al (2001) Conditions for fabrication of ideally ordered anodic porous alumina using pretextured Al. J Electrochem Soc 148:B152–B156
83. Asoh H, Nishio K, Nakao M et al (2001) Fabrication of ideally ordered anodic porous alumina with 63 nm hole periodicity using sulfuric acid. J Vac Sci Technol B 19:569–572
84. Matsumoto F, Harada M, Koura N et al (2004) Fabrication and electrochemical behavior of nanodisk electrode arrays with controlled interval using ideally ordered porous alumina. Electrochem Solid-State Lett 7:E51–E53
85. Matsumoto F, Harada M, Nishio K et al (2005) Nanometer-scale patterning of DNA in controlled intervals on gold-disk array fabricated using ideally ordered anodic porous alumina. Adv Mater 17:1609–1612
86. Yasui K, Morikawa T, Nishio K et al (2005) Patterned magnetic recording media using anodic porous alumina with single domain hole configurations of 63 nm hole interval. Jpn J Appl Phys 44:L469–L471
87. Yasui K, Nishio K, Masuda H (2005) Fabrication of nanocomposites by filling nanoholes in highly ordered anodic porous alumina by vacuum deposition of metal. Jpn J Appl Phys 44: L1181–L1183
88. Nishio K, Nakao M, Yokoo A et al (2003) Ideally ordered metal hole arrays with high aspect ratios prepared from anodic porous alumina. Jpn J Appl Phys Part 2(42):L83–L85
89. Kondo T, Tanji M, Nishio K et al (2006) Cross-striped ordered arrays of Au nanoparticles in anodic porous alumina matrix. Electrochem Solid State Lett 9:C189–C191
90. Masuda H, Asoh H, Watanabe M et al (2001) Square and triangular nanohole array architectures in anodic alumina. Adv Mater 13:189–192
91. Yanagishita T, Nishio K, Nakao M et al (2002) Synthesis of diamond cylinders with triangular and square cross sections using anodic porous alumina templates. Chem Lett 2002:976–977
92. Asoh H, Ono S, Hirose T et al (2003) Growth of anodic porous alumina with square cells. Electrochim Acta 48:3171–3174
93. Asoh H, Ono S, Hirose T et al (2004) Detailed observation of cell junction in anodic porous alumina with square cells. Jpn J Appl Phys 43:6342–6346
94. Yanagishita T, Sasaki M, Nishio K et al (2004) Carbon nanotubes with a triangular cross-section, fabricated using anodic porous alumina as a template. Adv Mater 16:429–432
95. Choi J, Wehrspohn RB, Gösele U (2005) Mechanism of guided self-organization producing quasi-monodomain porous alumina. Electrochim Acta 50:2591–2595
96. Choi J, Schilling J, Nielsch K et al (2002) Large-area porous alumina photonic crystals via imprint method. Mat Res Soc Symp Proc 722:L5.2.1–L.5.2.6
97. Choi J, Sauer G, Göring P et al (2003) Hexagonally arranged monodisperse silver nanowires with adjustable diameter and high aspect ratio. Chem Mater 15:776–779
98. Choi J, Luo Y, Wehrspohn RB et al (2003) Perfect two-dimensional porous alumina photonic crystals with duplex oxide layers. J Appl Phys 94:4757–4762

99. Choi J, Nielsch K, Reiche M et al (2003) Fabrication of monodomain alumina pore arrays with an interpore distance smaller than the lattice constant of the imprint stamp. J Vac Sci Technol B 21:763–766
100. Choi J, Sauer G, Nielsch K et al (2003) Monodisperse metal nanowire arrays on Si by integration of template synthesis with silicon technology. J Mater Chem 13:1100–1103
101. Choi J, Wehrspohn RB, Gösele U (2003) Moire pattern formation on porous alumina arrays using nanoimprint litography. Adv Mater 15:1531–1534
102. Kwon N, Kim K, Heo J et al (2009) Fabrication of ordered anodic aluminum oxide with matrix arrays of pores using nanoimprint. J Vac Sci Technol A 27:803–807
103. Wolfrum B, Mourzina Y, Mayer D et al (2006) Fabrication of large-scale patterned gold-nanopillar arrays on a silicon substrate using imprinted porous alumina templates. Small 2:1256–1260
104. Li Y, Nagato K, Delaunay JJ et al (2014) Fabrication of highly ordered Ta2O5 and Ta3N5 nanorod arrays by nanoimprinting and through mask anodization. Nanotechnology 25:014013
105. Navas D, Sánchez O, Asenjo A et al (2007) Titanium nitride stamps replicating nanoporous anodic alumina films. Nanotechnology 18:165302
106. Matsui Y, Nishio K, Masuda H (2005) Highly ordered anodic porous alumina by imprinting using Ni molds prepared from ordered array of polystyrene particles. Jpn J Appl Phys 44:7726–7728
107. Yasui K, Nishio K, Nunokawa H et al (2005) Ideally ordered anodic porous alumina with sub-50 nm hole intervals based on imprinting using metal molds. J Vac Sci Technol B 23: L9–L12
108. Lee W, Ji R, Gösele U et al (2006) Fast fabrication of long-range ordered porous alumina membranes by hard anodization. Nature Mater 5:741–747
109. Lee W, Ji R, Ross CA et al (2006) Wafer-scale Ni imprint stamps for porous alumina membranes based on interference lithography. Small 2:978–982
110. Lee BK, Cha KJ, Kwon TH (2009) Fabrication of polymer micro/nano-hybrid lens array by microstructured anodic aluminum oxide (AAO) mold. Microelectron Eng 86:857–860
111. Harada M, Kondo T, Yanagishita T (2010) Anodic porous alumina masks with checkerboard pattern. Appl Phys Exp 3:015001
112. Choi J, Park YB, Scherer A (2005) Fabrication of a tungsten master stamp using self-ordered porous alumina. Nanotechnology 16:1655–1659
113. Mikulskas I, Juodkazis S, Tomašiunas R et al (2001) Aluminum oxide photonic crystals grown by a new hybrid method. Adv Mater 13:1574–1577
114. Fan Z, Razavi H, Do JW et al (2009) Three-dimensional nanopillar-array photovoltaics on low-cost and flexible substrates. Nat Mater 8:648–653
115. Masuda H, Matsui Y, Yotsuya M et al (2004) Fabrication of highly ordered anodic porous alumina using self-organized polystyrene particle array. Chem Lett 33:584–585
116. Lipson AL, Comstock DJ, Hersam MC (2009) Nanoporous templates and membranes formed by nanosphere lithography and aluminum anodization. Small 5:2807–2811
117. Fournier-Bidoz S, Kitaev V, Routkevitch D et al (2004) Highly ordered nanosphere imprinted nanochannel alumina (NINA). Adv Mater 16:2193–2196
118. Surawathanawises K, Cheng X (2014) Nanoporous anodic aluminum oxide with a long-range order and tunable cell sizes by phosphoric acid anodization on pre-patterned substrates. Electrochim Acta 117:498–503
119. Matsui Y, Nishio K, Masuda H (2006) Highly ordered anodic porous alumina with 13-nm hole intervals using a 2D array of monodisperse nanoparticles as a template. Small 2:522–525
120. Zhang L, Zhou Z, Cheng B et al (2006) Superhydrophobic behavior of a perfluoropolyether lotus-leaf-like topography. Langmuir 22:8576–8580
121. Fois G, Bolger CT, Holmes JD et al (2010) Novel pre-oxidation pattering on thin aluminum film generating ordered nanopores through anodization. Mater Res Soc Symp Proc 1258: Q05-0.1

122. Chen SH, Chan DS, Chen CK et al (2010) Nanoimprinting pre-patterned effects on anodic aluminum oxide. Jpn J Appl Phys 49:015201
123. Shingubara S, Maruo S, Yamashita T et al (2010) Reduction of pitch of nanohole array by self-organizing anodic oxidation after nanoimprinting. Microelectron Eng 87:1451–1454
124. Masuda H, Satoh M (1996) Fabrication of gold nanodot array using anodic porous alumina as an evaporation mask. Jpn J Appl Phys 35:L126–L129
125. Sulka GD (2008) Highly ordered anodic porous alumina formation by self-organised anodising. In: Eftekhari A (ed) Nanostructured materials in electrochemistry. Wiley-VCH, Weinheim, pp 1–116
126. Sulka GD, Stroobants S, Moshchalkov V et al (2002) Synthesis of well-ordered nanopores by anodising aluminium foils in sulphuric acid. J Electrochem Soc 149:D97–D103
127. Sulka GD, Stroobants S, Moshchalkov V et al (2004) Effect of tensile stress on growth of self-organised nanostructures on anodised aluminium. J Electrochem Soc 151:B260–B264
128. Sulka GD, Jaskuła M (2006) Defect analysis in self-ordered nanopore structures grown by anodization of aluminium at various temperatures. J Nanosci Nanotechnol 6:3803–3811
129. Sulka GD, Parkoła KG (2006) Anodizing potential influence on well-ordered nanostructures formed by anodization of aluminium in sulphuric acid. Thin Solid Films 515:338–345
130. Sulka GD, Parkoła KG (2007) Temperature influence on well-ordered nanopore structures grown by anodization of aluminium in sulphuric acid. Electrochim Acta 52:1880–1888
131. Sulka GD, Moshchalkov V, Borghs G et al (2007) Electrochemical impendance spectroscopic study of barrier layer thinning in nanostructured aluminium. J Appl Electrochem 37:789–797
132. Sulka GD, Stępniowski WJ (2009) Structural features of self-organized nanopore arrays formed by anodization of aluminum in oxalic acid at relatively high temperatures. Electrochim Acta 54:3683–3691
133. Sulka GD, Brzózka A, Zaraska L et al (2010) Through-hole membranes of nanoporous alumina as templates for fabricating silver and tin nanowire arrays. Electrochim Acta 55:4368–4376
134. Zaraska L, Stępniowski WJ, Ciepiela E et al (2013) The effect of anodizing temperature on structural features and hexagonal arrangement of nanopores in alumina synthesized by two-step anodizing in oxalic acid. Thin Solid Films 534:155–161
135. Zaraska L, Stępniowski WJ, Jaskuła M et al (2014) Analysis of nanopore arrangement of porous alumina layers formed by anodizing in oxalic acid at relatively high temperatures. Appl Surf Sci 305:650–657
136. Zaraska L, Stępniowski WJ, Sulka GD et al (2014) Analysis of nanopore arrangement and structural features of anodic alumina layers formed by two-step anodizing in oxalic acid using the dedicated executable software. Appl Phys A 114:571–577
137. Zaraska L, Sulka GD, Jaskuła M (2009) Properties of nanostructures obtained by anodization of aluminum in phosphoric acid at moderate potentials. J Phys Conf Ser 146:012020
138. Zaraska L, Sulka GD, Jaskuła M (2010) The effect of n-alcohols on porous anodic alumin formed by self-organized two-step anodizing of aluminum in phosphoric acid. Surf Coat Technol 204:1729–1737
139. Zaraska L, Sulka GD, Szeremeta J et al (2010) Porous anodic alumina formed by anodization of aluminum alloy (AA1050) and high purity aluminum. Electrochim Acta 55:4377–4386
140. Zaraska L, Sulka GD, Jaskuła M (2010) Porous anodic alumina membranes formed by anodization of AA1050 alloy as templates for fabrication of metallic nanowire arrays. Surf Coat Technol 205:2432–2437
141. Chu SZ, Wada K, Inoue S et al (2005) Fabrication of ideally ordered nanoporous alumina films and integrated alumina nanotubule arrays by high-field anodization. Adv Mater 17:2115–2119

142. Chu SZ, Wada K, Inoue S et al (2006) Large-scale fabrication of ordered nanoporous alumina films with arbitrary pore intervals by critical-potential anodization. J Electrochem Soc 153:B384–B391
143. Li Y, Zheng M, Ma L et al (2006) Fabrication of highly ordered nanoporous alumina films by stable high-field anodization. Nanotechnology 17:5101–5105
144. Li Y, Ling ZY, Chen SS et al (2008) Fabrication of novel porous anodic alumina membranes by two-step hard anodization. Nanotechnology 19:225604
145. Li D, Jiang C, Jiang J et al (2008) Investigation on highly ordered porous anodic alumina membranes formed by high electric field anodization. Mater Chem Phys 111:168–171
146. Schwirn K, Lee W, Hillebrand R et al (2008) Self-ordered anodic aluminum oxide formed by H_2SO_4 hard anodization. ASC Nano 2:302–310
147. Nagaura T, Takeuchi F, Inoue S (2008) Fabrication and structural control of anodic alumina films with inverted cone porous structure using multi-step anodizing. Electrochim Acta 53:2109–2114
148. Nagaura T, Takeuchi F, Yamauchi Y et al (2008) Fabrication of ordered Ni nanocones using a porous anodic alumina template. Electrochem Commun 10:681–685
149. Yamauchi Y, Nagura T, Ishikawa A et al (2008) Evolution of standing mesochannels on porous anodic alumina substrates with designed conical holes. J Am Chem Soc 130: 10165–10170
150. Yamauchi Y, Nagaura T, Takai K et al (2009) Generation of electron moire fringes on designed nanoporous anodic alumina films and their replicated Ni cone arrays: exploration of domain sizes and nanopore arrangements. J Phys Chem C 113:9632–9637
151. Yamauchi Y, Wang L, Ataee-Esfahani H et al (2010) Electrochemical design of two-dimensional Au nanocone arrays using porous anodic alumina membranes with conical holes. J Nanosci Nanotechnol 10:4384–4387
152. LeClere DJ, Thompson GE, Derby B (2009) Conical tungsten stamps for the replication of pore arrays in anodic aluminium oxide films. Nanotechnology 20:245304
153. Yanagishita T, Kondo T, Nishio K et al (2008) Optimization of antireflection structures of polymer based on nanoimprinting using anodic porous alumina. J Vac Sci Technol B 26:1856–1859
154. Nagaura T, Wada K, Inoue S (2010) Hexagonally ordered Ni nanocone array; controlling the aspect ratio. Mater Trans 51:1237–1241
155. Choi K, Park SH, Song YM et al (2010) Nano-tailoring the surface structure for the monolithic high-performance antireflection polymer film. Adv Mater 22:3713–3718
156. Li J, Li C, Gao X (2011) Structural evolution of self-ordered alumina tapered nanopores with 100 nm interpore distance. Appl Surf Sci 257:10390–10394
157. Li J, Li C, Chen C et al (2012) Facile method for modulating the profiles and periods of self-ordered three-dimensional alumina taper-nanopores. ACS Appl Mater Interfaces 4:5678–5683
158. Li J, Zhou C, Jin X et al (2012) Controlled nanoscale diffusion-limited chemical etching for releasing polystyrene nanocones from recyclable alumina templates. Chem Commun 48:11322–11324
159. Li C, Li J, Chen C et al (2012) Tailoring ordered taper-nanopore arrays by combined nanosphere self-assembling, imprinting, anodizing and etching. Chem Commun 48:5100–5102
160. Li J, Hu L, Li C et al (2013) Tailoring hexagonally packed metal hollow-nanocones and taper-nanotubes by template-induced preferential electrodeposition. ACS Appl Mater Interfaces 5:10376–10380
161. Lin Q, Leung SF, Tsui KH et al (2013) Programmable nanoengineering templates for fabrication of three-dimensional nanophotonic structures. Nanoscale Res Lett 8:268
162. Brzózka A, Brudzisz A, Hnida K et al (2015) Chemical and structural modifications of nanoporous alumina and its optical properties. In: Losic D, Santos A (eds) Electrochemically engineered nanoporous materials—methods, properties and applications. Springer International Publishing, Switzerland, pp 219–288

163. Brzózka A, Szeliga D, Kurowska-Tabor E et al (2016) Synthesis of copper nanocone array electrodes and its electrocatalytic properties toward hydrogen peroxide reduction. Mater Lett 174:66–70
164. Cui J, Wu Y, Wang Y et al (2012) A facile and efficient approach for pore-opening detection of anodic aluminum oxide membranes. Appl Surf Sci 258:5305–5311
165. Tian YT, Meng GM, Wang GZ et al (2006) Step-shaped bismuth nanowires with metal–semiconductor junction characteristics. Nanotechnology 17:1041–1045
166. Yang D, Meng G, Zhu C et al (2010) Synthesis and thermal expansion of copper nanotubes and nanowires with Y- and step-shaped topologies. Small 6:381–385
167. Zhang LD, Fang XS (2008) Controlled growth and characterization methods of semiconductor nanomaterials. J Nanosci Nanotechnol 8:149–201
168. Jeong KO, Choi YC, Kim J et al (2007) Porous alumina templates with various shaped nanochannels. J Korean Phys Soc 51:S105–S110
169. Lee W, Park BG, Kim DH et al (2010) Nanostructure-dependent water-droplet adhesiveness change in superhydrophobic anodic aluminum oxide surfaces: from highly adhesive to self-cleanable. Langmuir 26:1412–1415
170. Im WS, Cho YS, Choi GS et al (2004) Stepped carbon nanotubes synthesized in anodic aluminum oxide templates. Diamond Relat Mater 13:1214–1217
171. Zhao Z, Meng G, Han F et al (2013) Nanocontainers made of various materials with tunable shape and size. Sci Rep 3:2238
172. Macias G, Hernández-Eguía LP, Ferré-Borrull J et al (2013) Gold-coated ordered nanoporous anodic alumina bilayers for future label-free interferometric biosensors. ACS Appl Mater Interfaces 5:8093–8098
173. Salem MS, Sergelius P, Corona RM (2013) Magnetic properties of cylindrical diameter modulated $Ni_{80}Fe_{20}$ nanowires: interaction and coercive fields. Nanoscale 5:3941–3947
174. Ji R, Lee W, Scholz R et al (2006) Templated fabrication of nanowire and nanoring arrays based on interference lithography and electrochemical deposition. Adv Mater 18:2593–2596
175. Santos A, Formentín P, Pallarès J et al (2011) Structural engineering of nanoporous anodic alumina funnels with high aspect ratio. J Electroanal Chem 655:73–78
176. Krishnan R, Thompson CV (2007) Monodomain high-aspect-ratio 2D and 3D ordered porous alumina structures with independently controlled pore spacing and diameter. Adv Mater 19:988–992
177. Zaraska L, Jaskuła M, Sulka GD (2016) Porous anodic alumina layers with modulated pore diameters formed by sequential anodizing in different electrolytes. Mater Lett 171:315–318
178. Liu S, Tang S, Zhou H et al (2013) Fabrication of AAO films with controllable nanopore size by changing electrolytes and electrolytic parameters. J Solid State Electrochem 17:1931–1938
179. Zhu X, Song Y, Yu D et al (2013) A novel nanostructure fabricated by an improved two-step anodizing technology. Electrochem Commun 29:71–74
180. Liu CY, Lai MY, Tsai KT et al (2013) A novel vertical fan-out platform based on an array of curved anodic alumina nanochannels. Nanotechnology 24:055306
181. Lim JH, Rotaru A, Min SG et al (2010) Synthesis of mild–hard AAO templates for studying magnetic interactions between metal nanowires. J Mater Chem 20:9246–9252
182. Vega V, Böhnert T, Martens S et al (2012) Tuning the magnetic anisotropy of Co–Ni nanowires: comparison between single nanowires and nanowire arrays in hard-anodic aluminum oxide membranes. Nanotechnology 23:465709
183. Chen B, Xu Q, Zhao X et al (2010) Branched silicon nanotubes and metal nanowires via AAO-template-assistant approach. Adv Funct Mater 20:3791–3796
184. Kong LB, Meng ML, Li K et al (2003) Branched silver nanowires obtained in porous anodic aluminum oxide template. J Mater Sci Lett 22:701–702
185. Kong LB (2005) Synthesis of Y-junction carbon nanotubes within porous anodic aluminum oxide template. Solid State Commun 133:527–529

186. Lee KP, Mattia D (2013) Monolithic nanoporous alumina membranes for ultrafiltration applications: characterization, selectivity–permeability analysis and fouling studies. J Membrane Sci 435:52–61
187. Li J, Papadopoulos C, Xu JM (1999) Growing Y-junction carbon nanotubes. Nature 402:253–254
188. Lukatskaya MR, Trusov LA, Eliseev AA et al (2011) Controlled way to prepare quasi-1D nanostructures with complex chemical composition in porous anodic alumina. Chem Commun 47:2396–2398
189. Meng G, Jung YJ, Cao A et al (2005) Controlled fabrication of hierarchically branched nanopores, nanotubes, and nanowires. Proc Natl Acad Sci USA 102:7074–7078
190. Meng G, Han F, Zhao X et al (2009) A general synthetic approach to interconnected nanowire/nanotube and nanotube/nanowire/nanotube heterojunctions with branched topology. Angew Chem Int Ed 48:7166–7170
191. Petukhov DI, Napolskii KS, Eliseev AA (2012) Permeability of anodic alumina membranes with branched channels. Nanotechnology 20:335601
192. Subhramannia M, Ramaiyan K, Aslam M et al (2009) Y-junction nanostructures of palladium: Enhanced electrocatalytic properties for fuel cell reactions. J Electroanal Chem 627:58–62
193. Tian YT, Meng GW, Gao T et al (2004) Alumina nanowire arrays standing on a porous anodic alumina membrane. Nanotechnology 15:189–191
194. Tian Y, Meng G, Biswas SK et al (2004) Y-branched Bi nanowires with metal–semiconductor junction behavior. Appl Phys Lett 85:967–969
195. Xu L, Yuan Z, Zhang X (2006) Fabrication of multi-level branched metal nanowires by AAO template electro-deposition. Chin Sci Bull 51:2055–2058
196. Xu Q, Meng G, Han F et al (2009) Controlled fabrication of gold and polypyrrole nanowires with straight and branched morphologies via porous alumina template-assisted approach. Mater Lett 63:1431–1434
197. Zaraska L, Kurowska E, Sulka GD et al (2012) Porous alumina membranes with branched nanopores as templates for fabrication of Y-shaped nanowire arrays. J Solid State Electrochem 16:3611–3619
198. Gao T, Meng G, Zhang J et al (2002) Template synthesis of Y-junction metal nanowires. Appl Phys A 74:403–406
199. Shuoshuo C, Zhiyuan L, Xing H et al (2010) Competitive growth of branched channels inside AAO membranes. J Mater Chem 20:1794–1798
200. Rauf A, Mehmood M, Yuan ZH et al (2009) The effects of ordering on the morphology of two-layer alumite-forming a patterned interface. Mater Lett 63:1601–1604
201. Mahima S, Kannan R, Komath I et al (2008) Synthesis of platinum Y-junction nanostructures using hierarchically designed alumina templates and their enhanced electrocatalytic activity for fuel-cell applications. Chem Mater 20:601–603
202. Yang Y, Hu Z, Wu Q et al (2003) Template-confined growth and structural characterization of amorphous carbon nanotubes. Chem Phys Lett 373:580–585
203. Kwon JT, Shin HG, Seo YH et al (2009) Simple fabrication method of hierarchical nano-pillars using aluminum anodizing processes. Curr Appl Phys 9:e81–e85
204. Shuoshuo C, Zhiyuan L, Xing E et al (2009) Controlled growth of branched channels by a factor of $1/\sqrt{n}$ anodizing voltage? J Mater Chem 19:5717–5719
205. Ho AYY, Gao H, Lam YC et al (2008) Controlled fabrication of multitiered three-dimensional nanostructures in porous alumina. Adv Funct Mater 18:2057–2063
206. Cheng W, Steinhart M, Gösele U et al (2007) Tree-like alumina nanopores generated in a non-steady-state anodization. J Mater Chem 17:3493–3495
207. Ghaffari M, Ramazani A, Kashi MA (2013) Improvement in the microstructure and magnetic properties in arrays of dc pulse electrodeposited Co nanowires induced by Cu pre-plating. J Phys D Appl Phys 46:295002

208. Marquardt B, Eude L, Gowtham M et al (2008) Density control of electrodeposited Ni nanoparticles/nanowires inside porous anodic alumina templates by an exponential anodization voltage decrease. Nanotechnology 19:405607
209. Ramazani A, Kashi MA, Seyedi G, Crystallinity and magnetic properties of electrodeposited Co nanowires in porous alumina. J Magn Magn Mater 324:1826–1831
210. Sousa CT, Leitao DC, Ventura J et al (2012) A versatile synthesis method of dendrites-free segmented nanowires with a precise size control. Nanoscale Res Lett 7:168
211. Sousa CT, Apolinario A, Leitao DC et al (2012) Precise control of the filling stages in branched nanopores. J Mater Chem 22:3110–3116
212. Sui YC, Acosta DR, González-León JA et al (2001) Structure, thermal stability, and deformation of multibranched carbon nanotubes synthesized by CVD in the AAO template. J Phys Chem B 105:1523–1527
213. Sui YC, González-León JA, Bermúdez A et al (2001) Synthesis of multi branched carbon nanotubes in porous anodic aluminum oxide template. Carbon 39:1709–1715
214. Zhu X, Liu L, Song Y et al (2008) Oxygen bubble mould effect: serrated nanopore formation and porous alumina growth. Monatsh Chem 139:999–1003
215. Li D, Jiang C, Jiang J et al (2009) Self-assembly of periodic serrated nanostructures. Chem Mater 21:253–258
216. Li D, Zhao L, Jiang C et al (2010) Formation of anodic aluminum oxide with serrated nanochannels. Nano Lett 10:2766–2771
217. Zhu XF, Song Y, Liu L et al (2009) Electronic currents and the formation of nanopores in porous anodic alumina. Nanotechnology 20:475303
218. Zaraska L, Brudzisz A, Wierzbicka E et al (2016) The effect of electrolyte change on the morphology and degree of nanopore order of porous alumina formed by two-step anodization. Electrochim Acta 198:259–267
219. Wang B, Fei GT, Wang M et al (2007) Preparation of photonic crystals made of air pores in anodic alumina. Nanotechnology 18:365601
220. Wang B, Fei GT, Wu B et al (2006) Preparation of three-dimensional netlike mesoporous alumina membrane. Chem Lett 35:1336–1337
221. Hu X, Ling ZY, Chen SS et al (2008) Influence of light scattering on transmission spectra of photonic crystals of anodized alumina. Chin Phys Lett 25:3284–3287
222. Zheng WJ, Fei GT, Wang B et al (2009) Modulation of transmission spectra of anodized alumina membrane distributed Bragg reflector by controlling anodization temperature. Nanoscale Res Lett 4:665–667
223. Zheng WJ, Fei GT, Wang B et al (2009) Distributed Bragg reflector made of anodic alumina membrane. Mater Lett 63:706–708
224. Rahman MM, Marsal LF, Pallarès J et al (2013) Tuning the photonic stop bands of nanoporous anodic alumina-based distributed Bragg reflectors by pore widening. ACS Appl Mater Interfaces 5:13375–13381
225. Ling ZY, Chen SS, Hu X et al (2009) Optical transmission spectra of anodic aluminum oxide membranes with a dual layer-by-layer structure. Chin Phys Lett 5:054213
226. Hu X, Pu YJ, Ling ZY et al (2009) Coloring of aluminum using photonic crystals of porous alumina with electrodeposited Ag. Opt Mater 32:382–386
227. Su Y, Fei GT, Zhang Y et al (2011) Controllable preparation of the ordered pore arrays anodic alumina with high-quality photonic band gaps. Mater Lett 65:2693–2695
228. Xing H, Zhiyuan L, Kai W et al (2011) Fabrication of three dimensional interconnected porous carbons from branched anodic aluminum oxide template. Electrochem Commun 13:1082–1085
229. Shang GL, Fei GT, Zhang Y et al (2013) Preparation of narrow photonic bandgaps located in the near infrared region and their applications in ethanol gas sensing. J Mater Chem C 1:5285–5291
230. Shang GL, Fei GT, Zhang Y et al (2014) Fano resonance in anodic aluminum oxide based photonic crystals. Sci Rep 4:3601

231. Ferré-Borrull J, Rahman MM, Pallarès J et al (2014) Tuning nanoporous anodic alumina distributed-Bragg reflectors with the number of anodization cycles and the anodization temperature. Nanoscale Res Lett 9:416
232. Guo DL, Fan LX, Wang FH et al (2008) Porous anodic aluminum oxide Bragg stacks as chemical sensors. J Phys Chem C 112:17952–17956
233. Beck G (2012) Alternating nanopore diameters in anodic alumina grown within aluminium tubes. Mater Chem Phys 137:5–11
234. Lee W, Schwirn K, Steinhart M et al (2008) Structural engineering of nanoporous anodic aluminium oxide by pulse anodization of aluminium. Nat Nanotechnol 3:234–239
235. Pitzschel K, Montero Moreno JM, Escrig J et al (2009) Controlled introduction of diameter modulations in arrayed magnetic iron oxide nanotubes. ACS Nano 3:3463–3468
236. Chung CK, Zhou RX, Liu TY et al (2009) Hybrid pulse anodization for the fabrication of porous anodic alumina films from commercial purity (99%) aluminum at room temperature. Nanotechnology 20:055301
237. Lee W, Kim JC (2010) Highly ordered porous alumina with tailor-made pore structures fabricated by pulse anodization. Nanotechnology 21:485304
238. Sulka GD, Brzózka A, Liu L (2011) Fabrication of diameter-modulated and ultrathin porous nanowires in anodic aluminum oxide templates. Electrochim Acta 56:4972–4979
239. Sulka GD, Hnida K (2012) Distributed Bragg reflector based on porous anodic alumina fabricated by pulse anodization. Nanotechnology 23:075303
240. Lee W, Scholz R, Gösele U (2008) A continuous process for structurally well-defined Al_2O_3 nanotubes based on pulse anodization of aluminum. Nano Lett 8:2155–2160
241. Esmaeily AS, Venkatesan M, Razavian AS et al (2013) Diameter-modulated ferromagnetic CoFe nanowires. J Appl Phys 113:17A327
242. Moradi M, Noormohammadi M, Behzadi F (2011) Three-dimensional structural engineering of nanoporous alumina by controlled sprinkling of an electrolyte on a porous anodic alumina (PAA) template. J Phys D Appl Phys 44:045301
243. Noormohammadi M, Moradi M (2012) Structural engineering of nanoporous alumina by direct cooling the barrier layer during the aluminum hard anodization. Mater Chem Phys 135:1089–1095
244. Raoufi M, Schönherr H (2013) Improved synthesis of anodized aluminum oxide with modulated pore diameters for the fabrication of polymeric nanotubes. RSC Adv 3:13429–13436
245. Lee W, Kim JC, Gösele U (2010) Spontaneous current oscillations during hard anodization of aluminum under potentiostatic conditions. Adv Funct Mater 20:21–27
246. Noormohammadi M, Moradi M, Kashi MA et al (2013) Structural engineering of nanoporous alumina by controlling the anodization voltage during the spontaneous current oscillation in hard anodization. Surf Coat Technol 223:104–109
247. Santos A, Vojkuvka L, Alba M et al (2012) Understanding and morphology control of pore modulations in nanoporous anodic alumina by discontinuous anodization. Phys Status Solidi A 209:2045–2048
248. Losic D, Lillo M, Losic D Jr (2009) Porous alumina with shaped pore geometries and complex pore architectures fabricated by cyclic anodization. Small 5:1392–1397
249. Losic D, Losic D Jr (2009) Preparation of porous anodic alumina with periodically perforated pores. Langmuir 25:5426–5431
250. Pitzschel K, Bachmann J, Martens S et al (2011) Magnetic reversal of cylindrical nickel nanowires with modulated diameters. J Appl Phys 109:033907
251. Calvo ME, Colodrero S, Hidalgo N et al (2011) Porous one dimensional photonic crystals: novel multifunctional materials for environmental and energy applications. Energy Environ Sci 4:4800–4812
252. Crouse D, Lo YH, Miller AE et al (2000) Self-ordered pore structure of anodized aluminum on silicon and pattern transfer. Appl Phys Lett 76:49–51
253. Sander MS, Tan LS (2003) Nanoparticle arrays on surfaces fabricated using anodic alumina films as templates. Adv Funct Mater 13:393–397

254. Rabin O, Herz PR, Lin YM et al (2013) Formation of thick porous anodic alumina films and nanowire arrays on silicon wafers and glass. Adv Funct Mater 13:631–638
255. Shimizu T, Xie T, Nishikawa J et al (2007) Synthesis of vertical high-density epitaxial Si (100) nanowire arrays on a Si(100) substrate using an anodic aluminum oxide template. Adv Mater 19:917–920
256. Shimizu T, Senz S, Shingubara S et al (2007) Synthesis of epitaxial Si(100) nanowires on Si (100) substrate using vapor–liquid–solid growth in anodic aluminum oxide nanopore arrays. Appl Phys A 87:607–610
257. Kokonou M, Nassiopoulou AG, Giannakopoulos KP et al (2006) Growth and characterization of high density stoichiometric SiO_2 dot arrays on Si through an anodic porous alumina template. Nanotechnology 17:2146–2151
258. Kokonou M, Nassiopoulou AG (2007) Nanostructuring Si surface and Si/SiO_2 interface using porous-alumina-on-Si template technology. Electrical characterization of Si/SiO_2 interface. Phys E 38:1–5
259. Kokonou M, Rebholz C, Giannakopoulos KP et al (2007) Fabrication of nanorods by metal evaporation inside the pores of ultra-thin porous alumina templates. Nanotechnology 18:495604
260. Zhou HY, Qu SC, Wang ZG et al (2006) Preparation and AFM characterization of self-ordered porous alumina films on semi-insulated GaAs substrate. Mater Sci Semicond Process 9:337–340
261. Yasui K, Sakamoto Y, Nishio K et al (2005) Preparation of anodic porous alumina mask with ideally arranged holes on InP single crystals. Chem Lett 34:342–343
262. He LP, Mai YW, Chen ZZ (2004) Effects of anodization voltage on CaP/Al_2O_3–Ti nanometre biocomposites. Nanotechnology 15:1465–1471
263. Chu SZ, Wada K, Inoue S et al (2002) Formation and microstructures of anodic alumina films from aluminum sputtered on glass substrate. J Electrochem Soc 149:B321–B327
264. Chu SZ, Wada K, Inoue S et al (2002) Fabrication and characteristics of ordered Ni nanostructures on glass by anodization and direct current electrodeposition. Chem Mater 14:4595–4602
265. Chu SZ, Wada K, Inoue S et al (2003) Fabrication and characteristics of nanostructures on glass by Al anodization and electrodeposition. Electrochim Acta 48:3147–3153
266. Chu SZ, Inoue S, Wada K et al (2004) Fabrication and structural characteristics of nanocrystalline Fe-Pt thin films and Fe-Pt nanowire arrays embedded in alumina films on ITO/Glass. J Phys Chem B 108:5582–5587
267. Chu SZ, Inoue S, Wada K et al (2005) Fabrication of integrated arrays of ultrahigh density magnetic nanowires on glass by anodization and electrodeposition. Electrochim Acta 51:820–826
268. Foong TRB, Sellinger A, Hu X (2008) Origin of the bottlenecks in preparing anodized aluminum oxide (AAO) templates on ITO glass. ACS Nano 2:2250–2256
269. Schierhorn M, Boettcher SW, Kraemer S et al (2009) Photoelectrochemical performance of CdSe nanorod arrays grown on a transparent conducting substrate. Nano Lett 9:3262–3267
270. Musselman KP, Mulholland GJ, Robinson AP et al (2008) Low-temperature synthesis of large-area, free-standing nanorod arrays on ITO/glass and other conducting substrates. Adv Mater 20:4470–4475
271. Peitao G, Zhilin X, Yiyu X et al (2011) Morphology and transmittance of porous alumina on glass substrate. Appl Surf Sci 257:3307–3312
272. Mozalev A, Sakairi M, Saeki I et al (2003) Nucleation and growth of the nanostructured anodic oxides on tantalum and niobium under the porous alumina film. Electrochim Acta 48:3155–3170
273. Mozalev A, Sakairi M, Takahashi H (2004) Structure, morphology, and dielectric properties of nanocomposite oxide films formed by anodizing of sputter-deposited Ta-Al bilayers. J Electrochem Soc 151:F257–F268
274. Mozalev A, Khatko V, Bittencourt C et al (2008) Nanostructured columnlike tungsten oxide film by anodizing Al/W/Ti layers on Si. Chem Mater 20:6482–6493

275. Mozalev A, Smith AJ, Borodin S et al (2009) Growth of multioxide planar film with the nanoscale inner structure via anodizing Al/Ta layers on Si. Electrochim Acta 54:935–945
276. Park IH, Lee JW, Jeong SH et al (2005) Shape change of self-organized NbOx nanopillar arrays by high density plasma etching. Electrochem Solid-State Lett 8:C117–C120
277. Oh J, Thompson CV (2008) Selective barrier perforation in porous alumina anodized on substrates. Adv Mater 20:1368–1372
278. Kisner A, Heggen M, Fischer W et al (2012) In situ fabrication of ultrathin porous alumina and its application for nanopatterning Au nanocrystals on the surface of ion-sensitive field-effect transistors. Nanotechnology 23:485301
279. Kokonou M, Giannakopoulos KP, Nassiopoulou AG (2007) Few nanometer thick anodic porous alumina films on silicon with high density of vertical pores. Thin Solid Films 515:3602–3606
280. Yan B, Pham HTM, Ma Y et al (2007) Fabrication of in situ ultrathin anodic aluminum oxide layers for nanostructuring on silicon substrate. Appl Phys Lett 91:053117
281. Jin X, Hu Y, Wang Y et al (2012) Template-based synthesis of Ni nanorods on silicon substrate. Appl Surf Sci 258:2977–2981
282. Xu CL, Li H, Xue T et al (2006) Fabrication of CoPd alloy nanowire arrays on an anodic aluminum oxide/Ti/Si substrate and their enhanced magnetic properties. Scripta Mater 54:1605–1609
283. Jung YW, Byun JS, Woo DH et al (2009) Ellipsometric analysis of porous anodized aluminum oxide films. Thin Solid Films 517:3726–3730
284. Jee SW, Jung JY, Seo HS et al (2009) Successive nucleation and a rodlike growth of SiOx nanoparticles into the pore bottoms of an anodic alumina template. Mater Chem Phys 117:478–481
285. Myung NV, Lim J, Fleurial JP et al (2004) Alumina nanotemplate fabrication on silicon substrate. Nanotechnology 15:33–838
286. Seo HS, Jung YG, Jee SW et al (2007) Compositionally bilayered feature of interfacial voids in a porous anodic alumina template directly formed on Si. Scripta Mater 57:968–971
287. Yang KS, Kim HJ, Ahn JK et al (2009) Microfluidic chip with porous anodic alumina integrated with PDMS/glass substrate for immuno-diagnosis. Curr Appl Phys 9:e60–e65
288. Gao H, Gosvami NN, Deng J et al (2006) Template-assisted patterning of nanoscale self-assembled monolayer arrays on surfaces. Langmuir 22:8078–8082
289. Park M, Yu G, Shin K (2012) Alumina templates on silicon wafers with hexagonally or tetragonally ordered nanopore arrays via soft lithography. Bull Korean Chem Soc 33:83–89
290. Nasir ME, Allsopp DWE, Bowen CR et al (2010) The fabrication of mono-domain highly ordered nanoporous alumina on a wafer scale by a guided electric field. Nanotechnology 21:105303
291. Banerjee A, Halder N (2010) Electrochemical growth of ordered nickel nano-rods within a composite structure of anodic-alumina-membrane/metal/silicon substrate. J Nanosci Nanotechnol 10:1–7
292. Evans P, Hendren WR, Atkinson R et al (2006) Growth and properties of gold and nickel nanorods in thin film alumina. Nanotechnology 17:5746–5753
293. Habouti S, Mátéfi-Tempfli M, Solterbeck CH et al (2011) On-substrate, self-standing Au-nanorod arrays showing morphology controlled properties. Nano Today 6:12–19
294. Zhao X, Lee UJ, Seo SK et al (2009) The nanoporous structure of anodic aluminum oxide fabricated on the Au/Nb/Si substrate. Mater Sci Eng, C 29:1156–1160
295. Holubowitch N, Nagle LC, Rohan JF (2012) Porous alumina thin films on conductive substrates for templated 1-dimensional nanostructuring. Solid State Ionics 216:110–113
296. Mei YF, Huang GS, Li ZM et al (2004) Formation of Si-based nano-island array on porous anodic alumina. Acta Mater 52:5633–5637
297. Feil AF, Migowski P, Dupont J et al (2011) Nanoporous aluminum oxide thin films on Si substrate: Structural changes as a function of interfacial stress. J Phys Chem C 115:7621–7627

298. Liu P, Singh VP, Rajaputra S (2010) Barrier layer non-uniformity effects in anodized aluminum oxide nanopores on ITO substrates. Nanotechnology 21:115303
299. Sjöström T, Fox N, Su B (2010) A study on the formation of titania nanopillars during porous anodic alumina through-mask anodization of Ti substrates. Electrochim Acta 56:203–210
300. Asoh H, Fujihara K, Ono S (2013) Sub-100-nm ordered silicon hole arrays by metal-assisted chemical etching. Nanoscale Res Lett 8:410
301. Kwon N, Kim N, Yeon J et al (2011) Fabrication of ordered Au nanodot arrays utilizing anodic aluminum oxide templates formed on Si substrate. J Vac Sci Technol, B 29:031805
302. Wu MT, Leu IC, Yen JH et al (2004) Preparation of Ni nanodot and nanowire arrays using porous alumina on silicon as a template without a conductive interlayer. Electrochem Solid-State Lett 7:C61–C63
303. Zhang ZW, Lam H, Li J et al (2009) Growth of copper-indium nanorods on Si substrate using porous anodic alumina as template. J Phys Conf Ser 188:012028
304. Franklin AD, Maschmann MR, DaSilva M et al (2007) In-place fabrication of nanowire electrode arrays for vertical nanoelectronics on Si substrates. J Vac Sci Technol B 25: 343–347
305. Tasaltın N, Öztürk S, Kılınç N et al (2009) Simple fabrication of hexagonally well-ordered AAO template on silicon substrate in two dimensions. Appl Phys A 95:781–787
306. Trezza M, Cirillo C, Vorobjeva AI et al (2013) Vortex matching effects in Nb thin films due to Ni nanopillars embedded in anodic aluminum oxide substrates. Supercond Sci Technol 26:035001
307. Asoh H, Matsuo M, Yoshihama M et al (2003) Transfer of nanoporous pattern of anodic porous alumina into Si substrate. Appl Phys Lett 83:4408–4410
308. Chen LC, Liu BH (2010) Porous silicon layer patterned from anodic aluminum oxide and application in ZnPc hybrid solar cell. Electrochem Solid-State Lett 13:H108–H111
309. Zacharatos F, Gianneta V, Nassiopoulou AG (2008) Highly ordered hexagonally arranged nanostructures on silicon through a self-assembled silicon-integrated porous anodic alumina masking layer. Nanotechnology 19:495306
310. Zacharatos F, Gianneta V, Nassiopoulou AG (2009) Highly ordered hexagonally arranged sub-200 nm diameter vertical cylindrical pores on p-type Si using non-lithographic pre-patterning of the Si substrate. Phys Stat Sol A 206:1286–1289
311. Takizawa C, Kato S, Goto A et al (2013) Fabrication of a porous alumina mask on the large surface area of a semi-insulating semiconductor substrate. J Ceram Soc Japan 121:516–519
312. Gianneta V, Nassiopoulou AG, Krontiras CA et al (2008) Porous anodic alumina thin films on Si: interface characterization. Phys Stat Sol (c) 5:3686–3689
313. Gong J, Butler WH, Zangari G (2008) Electrochemical/chemical synthesis of nanostructured arrays of oxide cones or rings. J Mater Chem 18:1741–1746
314. Wu CT, Lin CH, Cheng C et al (2010) Design of artificial hollow moth-eye structures using anodic nanocones for high-performance optics. Chem Mater 22:6583–6589
315. Park SH, Kim YH, Lee TG et al (2012) Synthesis and electrochemical capacitance of long tungsten oxide nanorod arrays grown vertically on substrate. Mater Res Bull 47:3612–3618
316. Sjöström T, Fox N, Su B (2009) Through-mask anodization of titania dot-and pillar-like nanostructures on bulk Ti substrates using a nanoporous anodic alumina mask. Nanotechnology 20:135305
317. Zaraska L, Kurowska E, Sulka GD et al (2014) The effect of anode surface area on nanoporous oxide formation during anodizing of low purity aluminum (AA1050) alloy. J Solid State Electrochem 18:361–368
318. Tian M, Xu S, Wang J et al (2005) Penetrating the oxide barrier in situ and separating freestanding porous anodic alumina films in one step. Nano Lett 5:697–703
319. Sharma G, Kripesh V, Sim MC et al (2007) Synthesis and characterization of patterned and nonpatterned copper and nickel nanowire arrays on silicon substrate. Sensor Actuat A-Phys 139:272–280

320. Chahrour KM, Ahmed NM, Hashim MR et al (2014) Controllable fabrication of highly ordered thin AAO template on Si substrate for electrodeposition of nanostructures. Appl Phys A 116:1389–1393
321. Sharma G, Chong CS, Ebin L et al (2007) Patterned micropads made of copper nanowires on silicon substrate for application as chip to substrate interconnects. Nanotechnology 18:305306
322. Shen L, Ali M, Gu Z et al (2013) Preparation of anodic aluminum oxide (AAO) nano-template on silicon and its application to one-dimensional copper nano-pillar array formation. Korean J Chem Eng 30:221–227
323. Liu CM, Tseng YC, Chen C et al (2009) Superparamagnetic and ferromagnetic Ni nanorod arrays fabricated on Si substrates using electroless deposition. Nanotechnology 20:415703
324. Sharma G, Chong SC, Ebin L et al (2007) Fabrication of patterned and non-patterned metallic nanowire arrays on silicon substrate. Thin Solid Films 515:3315–3322
325. Kim KT, Cho SM (2006) A simple method for formation of metal nanowires on flexible polymer film. Mater Lett 60:352–355
326. Matefi-Tempfli S, Matefi-Tempfli M (2009) Vertically aligned nanowires on flexible silicone using a supported alumina template prepared by pulsed anodization. Adv Mater 21:4005–4010
327. Bernardin T, Dupré L, Burnier L et al (2011) Organized porous alumina membranes for high density silicon nanowires growth. Microelectron Eng 88:2844–2847
328. Gorisse T, Dupré L, Gentile P et al (2013) Highly organised and dense vertical silicon nanowire arrays grown in porous alumina template on <100> silicon wafers. Nanoscale Res Lett 8:287
329. Leinartas K, Miečinskas P, Selskis A et al (2011) Formation of gold-capped silicon nanocolumns on silicon substrate. J Solid State Electrochem 15:2419–2425
330. Yuldashev SU, Choi SW, Kang TW (2003) Growth of ZnO nanowires by electrochemical deposition into porous alumina on silicon substrates. J Korean Phys Soc 42:S216–S218
331. Masuda H, Nishio K, Baba N (1993) Fabrication of a one-dimensional microhole array by anodic oxidation of aluminum. Appl Phys Lett 63:3155–3157
332. Cojocaru CS, Padovani JM, Wade T et al (2005) Conformal anodic oxidation of aluminum thin films. Nano Lett 5:675–680
333. Zhang H, Chen Z, Li T et al (2005) Fabrication of a one-dimensional array of nanopores horizontally aligned on a Si substrate. J Nanosci Nanotechnol 5:1745–1748
334. Xiang Y, Lee W, Nielsch K et al (2008) Microstructured horizontal alumina pore arrays as growth templates for large area few and single nanowire devices. Phys Stat Sol (RRL) 2:59–61
335. Gowtham M, Eude L, Cojocaru CS et al (2008) Controlled fabrication of patterned lateral porous alumina membranes. Nanotechnology 19:035303
336. Lee MH, Jung SH, Jeong SH (2009) Carbon nanotubes based on laterally formed anodic aluminum oxide template. Jpn J Appl Phys 48:040204
337. Xiang Y, Keilbach A, Moreno Codinachs L et al (2010) Multiple nanowire species synthesized on a single chip by selectively addressable horizontal nanochannels. Nano Lett 10:1341–1346
338. Bolger CT, Fois G, Petkov N et al (2012) Mechanical constraint and release generates long, ordered horizontal pores in anodic alumina templates. Nanotechnology 23:175602

Chapter 6
Porous Thin Films from Sol-Gel

M. Ramirez-del-Solar and E. Blanco

6.1 Introduction

The manufacture of porous materials in the form of thin films allows, first, to have a high ratio (surface/volume) of these by increasing their ability to interact with the medium using small amounts of material, which is of great interest for many applications of technological interest.

Additionally, this type of conformation allows the creation of coatings on existing devices increasing added value. It is therefore appropriate to have preparation methods, without high requirements, allow the manufacture of porous thin films that can be adapted to different geometries.

The sol-gel method offers a way of preparing porous materials at low temperature and low cost. It also allows the manufacture of porous thin films of very different compositions from the corresponding substrate wet with suitable precursor sol. Brinker [14, 15] reviewed underlying physics and chemistry of sol-gel thin-film formation using conventional coating techniques such as dip, spin or spray coating. Sol-gel gained very much interest for thin-film processing after the work of Dislich [26], because it provides a very simple and low-cost path which gives us the ability to tailor the microstructure of the deposited film by the control of diverse reaction and deposition parameters. One significant advantage of sol-gel films processing is the control of stoichiometry and homogeneity, based on the knowledge of precursor chemistry, in multicomponent systems, organic–inorganic composites or inclusion of a dopant. Another point of interest is the conservation of the pore network typical

M. Ramirez-del-Solar · E. Blanco (✉)
Dep. Física de la Materia Condensada and Instituto de Microscopía Electrónica y Materiales (IMEYMAT), Universidad de Cadiz, 11510 Puerto Real, Cádiz, Spain
e-mail: eduardo.blanco@uca.es

M. Ramirez-del-Solar
e-mail: milagrosa.ramirez@uca.es

P. Bettotti (ed.), *Submicron Porous Materials*,
DOI 10.1007/978-3-319-53035-2_6

of dried gels that can also be modified and organized according to the requirements of each application.

This chapter is intended as a review of recent activity in the preparation, characterization and applications of thin porous films produced by sol-gel. Derived from the experience of the authors in this review becomes more relevant porous titanium dioxide films for solar applications and their characterization by ellipsometry.

6.2 Films Preparation

6.2.1 From Sol to Film

Fabrication of thin films and coating by sol-gel based technology is accomplished in the common of cases by one of these three procedures: dipping, spinning or spraying.

In dip coating process, a clean substrate immersed in a sol bath is withdrawn normally at a precise constant speed, under controlled condition of environment and temperature. In the moving substrate, the solvent is evaporating and draining and a well-defined drying line appears (Fig. 6.1, left). Considering the colloidal nature of the bath liquid, there are entrained inorganic species that concentrate with solvent evaporation during dip coating, as they are transported from bath to drying line in the film. As a consequence the viscosity increases abruptly leading to aggregation, gelation and ageing. The film thickness is determined by the competition between six forces that determines the position stream line dividing upward and downward moving layers on extraction. Several authors evaluated the effect on the final film of different parameters related to the sol-gel transition (precursor, solvent, temperature, catalysis, etc.) or the dip coating technique (withdrawal speed, viscosity, capillary pressure, substrate, environment, etc.). By controlling these factors it is possible to

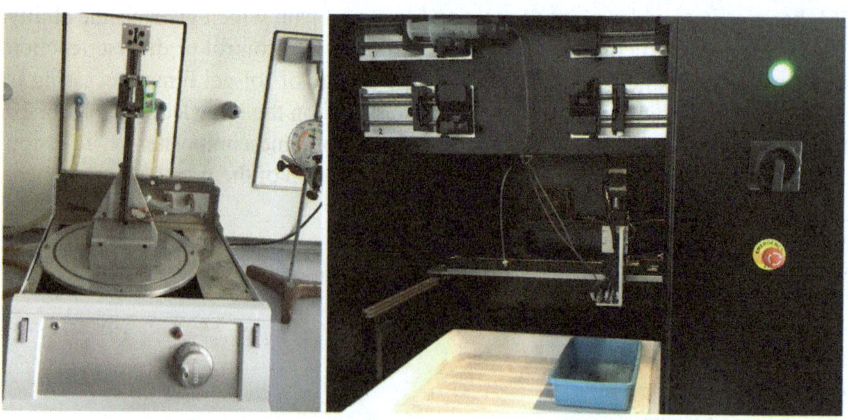

Fig. 6.1 Dip coater (*left*) and spry coater (*right*) devices

tailor the film thickness, resistance, pore size, surface area, porous volume and refractive index [14]. The most determining factors for the film thickness are the liquid viscosity and the withdrawal speed. In general, thickness is proportional to the power law of both parameters with a variable exponent, depending on the viscosity and speed values, that for Newtonian fluids ranges from 0.50 to 0.66. Detailed evaluation of the film thickness dependence on substrate withdrawal speed can be found in [60]. Anyway, for a given technological process, the proportionality index and power can be determined empirically. Global timescale of the deposition process, determined by thickness and the evaporation rate, is typically of several seconds, shortening considerably aggregation and gelation processes compared to bulk gels. Dip coating processing is well adjusted for the flat or curved substrates and other geometries like bulb, tubes and bottles.

Silica- and Titania-based film have been widely studied and used for a variety of applications. We have reported [10] about dip coating of sol-gel TiO_2 thin films onto glass substrates and the effects on processing parameters like annealing temperature and withdrawal speed on their morphological and optical features. On the basis of a multilayer air-film-substrate-film-air optical scheme, UV-visible absorption behaviour was described on the basis of the Tauc–Lorentz dispersion model, allowing for the evaluation of thickness, optical constants, absorption edges and optical band gaps of the dip-coated TiO_2 films. Thicknesses derived from this model were in a good linear correlation with both the number of layers deposited and the withdrawal speed of the dip coating process (Fig. 6.2).

High-resolution film thickness profiles, determined from ellipsometric measurements revealed their relatively flat profiles with a characteristic dip-coated film shape.

Raman Spectroscopy proved that as-deposited films were amorphous but, after annealing at temperatures above 300 °C, crystallization of anatase phase occurs, increasing the number and size of crystals with annealing temperature, as followed from the position and width of E_g anatase band occurring at 144 cm^{-1}. Film

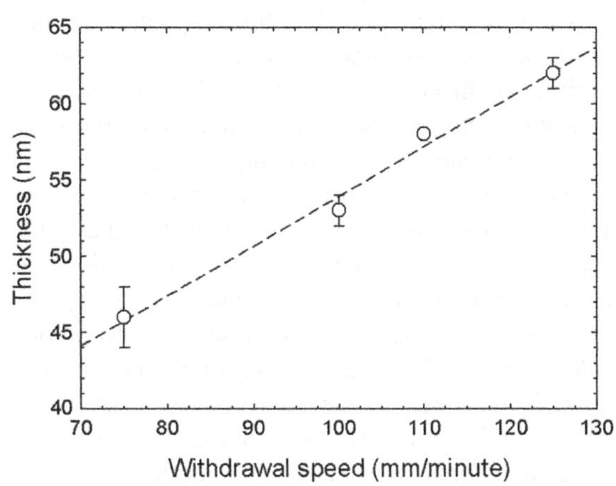

Fig. 6.2 Film thickness as a function of withdrawal speed. Reproduced with permission from [10]

microstructure evolution on heating was evaluated from refractive indexes, using Lorentz–Lorenz expression and considering the effective medium approximation. Higher annealing temperatures lead to greater refractive indexes and reduced porous fraction. Samples treated at 500 °C were selected for photocatalytic application because they are residue free and present small anatase crystals and highly porous network.

Spin coating processing was developed for spin-on glasses in microelectronics, based on spinning the substrate around a perpendicular vertical axis. The film results from evaporation and centrifugal draining of the liquid in four steps: deposition on the substrate of a liquid excess, spin-up when it flows radially driven by the centrifugal force, spin-off of the excess of liquid leaving a gradually thinner film and evaporation of the volatile components of the liquid. In this case, the balance between the centrifugal force outward and viscous force inward, whose balance results in a uniform thickness of the film [6]. In the case of aggregating sol, a lower viscosity associated to the lower shear rate in the centre can originate certain radial thickness gradient. In general, the quality of the coating depends on the rheological parameters of the coating liquid and on the rotating velocity that can alter the air friction. Spinning creates forced convection that increases the evaporation rate and significantly reduces the gelation time in comparison with bulk gels.

Several physical and chemical properties of sol-gel TiO_2 film prepared by spin coating have been compared with commercial CVD films [52]. In general no great differences were found in surface composition and roughness, crystal phase and size and UV-Vis absorption. Authors reported that sol-gel films exhibit comparable or in some cases better photocatalytic performance that is increased after film reaction in model (neutral, acidic and basic) solutions.

ZnO is an n-type semiconductor appropriate as chemical sensor, surface acoustic wave device or photoanode of solar cells because of its transparency in the visible and high infrared reflectivity, acoustic characteristics, high electrochemical stability and excellently electronic properties. Liu et al. [45] prepared ZnO porous thin films by sol-gel and related the chemical and physical parameters of the processing with the characteristic of the resulting films.

A dip/spin coating method for the preparation of titanium oxide films has been proposed which combined sol immersion and rotation. The resulting TiO_2 films had small and uniform grains with a unique structure, resulting in an increased photocatalytic activity when tested for degradation of methyl orange under UV irradiation, compared to the films prepared by dip or spin conventional method [68].

Spray coating (Fig. 6.1, right) is a more flexible technique for coating glasses with very irregular shapes when high thickness uniformity is not required. This technique, used in other industrial applications, has higher deposition speed, requires smaller stock solution that suffers reduced ageing and contamination. Therefore, it is adequate for short pot life sols, produces lower amount of waste sol and is well suited to an in-line industrial production. In this case, the desired flow of liquid is delivered to the spray nozzle from an infusion pump combining the solution and a compressed carrier gas. It is well known that coating solution must be suitable for homogeneous film formation on spraying [2]. Thus, low boiling point solvents, commonly used for dip coating in order to guarantee a fast evaporation during withdrawal, are not adequate for spray because they already evaporate during flight phase and an almost powder hits the substrate without enough mobility to level the film. Evaporation rate can be modulated by experimental conditions but, in general, glycol and glycol ethers are used as additives for increasing the boiling point and improving substrate wetting.

Other film processing methods, like flow coating, capillary coating [28], electrophoretic coating, roll coating, have been proposed but, in some cases, correspond to modified strategies from the former ones, exploiting the advantages of each of them for a particular application.

Brinker et al. proposed the Evaporation induced Self-Assembly (EISA) method for the synthesis of highly ordered mesoporous silica and hybrid materials via dip- or spin coating procedure [47–49]. This method has been applied in order to synthesize a highly aligned mesoporous silica film at low temperature with honeycomb-packed mesochannels controlled and formed over the entire thickness of the film with a very narrow distribution [53]. Boissiere et al. published a comprehensive revision of the major advances in the field of design of complex hierarchical porous structures by aerosol or spray processing of sols [12]. They reported about the integration of chemical strategies, like sol-gel chemistry and diverse templating (nanocasting, biotemplating, self-assembling, etc.), with aerosol process for the construction of bioinspired materials with complex hierarchical structures with multiscale texturation.

6.2.2 Pores Formation

One of the most outstanding advantages of sol-gel materials is that porosity can be tailored by means of the synthesis parameters. The presence of pores of given size, volume fraction and spatial arrangement is needed to improve sensing performance, through an increase of surface area available for interaction or binding, to convey a molecular sieve behaviour by size constrains, or to provide a matrix where catalyst particles and reactants can conveniently be trapped for localized processing, between others. We have mentioned that shorter times are involved in coating gelation process, which leads to a microstructure different from a bulk in the same conditions. Thus a specific analysis of the synthesis parameters on the film porosity is required. Thus the effect of the catalyst used on the morphology of porous silica films, prepared by dip coating, have been evaluated by Vincent et al. [67]. They showed that acid catalysis resulted in smoother surface, less porous and non-particulate network structure with a lower optical transmission. On the contrary, base-catalyzed coatings exhibited a coarse and porous surface, larger surface area and pore volume, superior optical transmission, higher surface roughness and lower refractive indices. Base catalysis provided a greater flexibility in controlling the particle and pore size, refractive index, and hence optical coatings suitable for antireflective applications. However, the mechanical strength of the acid-catalyzed coatings was found to be superior to that of base-catalyzed coatings because of the stronger particle–particle binding. Then, by controlling the porosity and particle size, it is possible to attain almost 99% transmission in the visible range, which can be useful for preparing sol-gel based porous antireflective optical coating. Many other authors reported about the control of the catalytic conditions (basic, base/acid two-step and acid) of the starting sol in order to adjust refractive index of nano-porous silica films. Wu et al. [72] reported that the reflectance and the refractive index of dip-coated silica films in the solar spectrum range changes continuously with the type of catalysis and other hydrolysis conditions.

Another usual porogen path considers the use of hybrid precursors containing Si–C bonds from which the inorganic matrix is built up around the organic moiety through the hydrolytic polycondensation [19]. The use of Si–C bonds provides a stable linkage between the organic unit and the oxide matrix that decompose when heating. Thus organic–inorganic hybrid materials are often used for the preparation of systems with controlled porosity. One group of precursors are bridged silsesquioxanes, where the bridging group acts as a molecular spacer capable of distancing silicon atoms in the silica network and, with appropriate synthesis conditions, providing it of specific textural properties. This approach has been followed to produce porous inorganic thin films from phenyl-bridged silsesquioxanes by Brigo et al. [13]. They reported that thermal treatments in the range of 60–800 °C leaded to a progressive formation of the crosslinked silica network that, after template elimination and film densification, result in completely inorganic porous thin films of low refractive index. Thus, the controlled and uniform dispersion of the benzene porogen, through covalent bonds, within the three-dimensional hybrid network determined the small pore dimension and regular arrangement obtained. Widespread developed mesoporous thin films with large pores are produced by the extraction of self-assembled polymeric units as discussed below.

Porous structure of the film is often modified by adding an organic phase that decomposes on thermal treatment leaving additional porosity. Thus, polyethylene glycol **(PEG)** is often used as template for preparing superhydrophilic sol-gel TiO_2 coating films [76]. The superhydrophilic character is interesting because it allows water to spread completely across the surface, rather than remain as droplets, thus making the surface anti-fogging and easy to wash. The contact angle for water on the film depends on the chemistry and microstructure of the resultant surface [75]. The larger the amount and molecular weight of PEG, the larger the size and numbers of pores produced in the resultant coating films after PEG decomposition on heat treatment. However, the final morphologies of porous films are sensitive to experimental parameters such as concentration and type of PEG templates, amount of water, complexing agents and solvents [16, 17]. This trend has been also showed for ZnO porous thin films prepared with PEG as organic template, Zn $(CH_3COO)_2 \cdot 2H_2O$ as precursor, ethanol as solvent and $NH(C_2H_2OH)_2$ as chelating agent [45]. Authors interpreted the mechanism for porous network formation from organic template on the basis of phase separation.

Comparison of the use of surfactant templating (soft) or dual surfactant/nanospheres templating (soft/hard) for hierarchical porous TiO_2 thin-film processing has been undertaken by Henrist et al. [36]. Soft templating route, using PEG and block copolymer surfactant, lead to very thin films with high specific surface and bimodal porosity (10 and 54 nm). Dual templating route, using the surfactant and polystyrene (PS) nanospheres, leads to thicker titania films, with a bimodal pore size distribution (8 nm and 165–200 nm). Thus, PS nanospheres induce an additional macroporosity with pore size 20–30% smaller than the original beads.

Finally, different etching strategies are also used in order to modify the porous network, increase the substrate roughness for improving the film adhesion, increase the external surface presented by the film, in catalytic applications for instance, and for the generation of ordered structures in nanopattering processes. For instance, electrochemical etching has been used to create a nanopatterned porous structure in a semiconductor substrate (GaAS) before depositing TiO_2 thin film. This approach tried to reduce extended defect density and the stress in the over deposited layer, improving the lattice mismatch between both materials [1]. It was shown a direct effect of etching time with the grain size and thickness of titania films, where some decrease of the gap energy was also found.

Titania nanotubes array, for photocatalysis and/or solar energy applications, can be directly synthesized on porous alumina by the combination of anodization of sputtered aluminium and the sol-gel process [24]. Highly pure aluminium RF sputtered film, deposited on a glass substrate with an ITO layer, was anodized in a phosphoric acid solution. The porous alumina structure was coated with a TiO_2 film after dipping, drying and curing. Then, the titania nanotubes array was obtained taking out the surface titania layer, by mechanical abrading, and selectively removing the anodic alumina template, by chemical etching (Fig. 6.3). The resultant TiO_2 is 4–20 nm polycrystalline of anatase structure with (101) preferential orientation, with a high transmittance in visible light and a strong absorbance within ultraviolet range. Moreover, addition of appropriate SiO_2 and TeO_2 to TiO_2 enhanced the adhesion to the substrate and the mechanical strength of the nanotubes. In fact, the composite TiO_2-SiO_2-TeO_2 filled the interface zones of anodic alumina and joined the separate nanotubules into an integrated and sturdy "network" structure on glass [23].

Fabrication of 2D structures of highly ordered TiO_2 nanocones arrays with controlled surface roughness has been also reported by Van et al. [66]. In this case, a mask of self-assembled monolayers of SiO_2 spheres fabricated by Langmuir–Blodgett technique was used and nanopatterning performed by dry etching. The roughness changes caused by nanopatterning affect the resulting contact angle. Thus, modified-TiO_2 layers exhibited tunability for hydrophilicity or hydrophobicity, pointing out a relationship between the wettability and the structure of TiO_2 arrays. So, increasing etching time causes a decrease of the water contact angle due to the higher roughness induced, according to Wenzel equation from which water

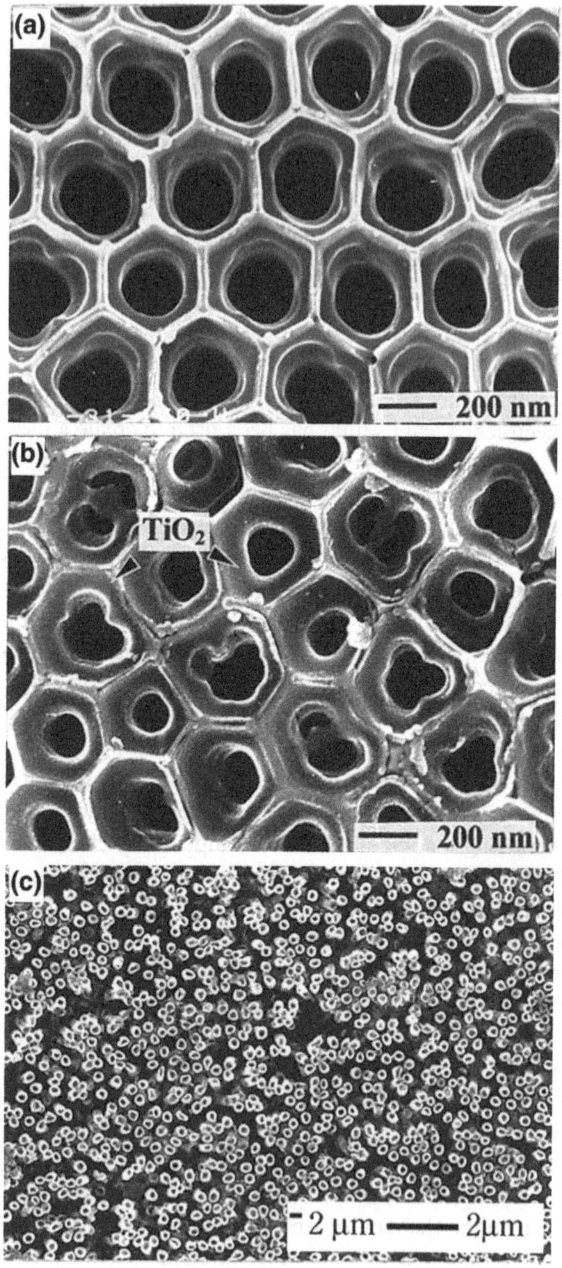

Fig. 6.3 Field emission scanning electron microscopy images of **a** a bottom view of anodic alumina film after pore widening and before sol-gel coating, **b** a bottom view of an $Al2O_3/TiO_2$ composite layer after dip coating and heating at 400 °C and **c** TiO_2 (*top view*) of nanotubules arrays standing on a glass substrate after removal of anodic alumina by selective chemical etching. Reprinted and adapted with permission from [24]. Copyright 2002, American Chemical Society

prefers to spread on rougher surfaces [70]. Conversely, super-hydrophobicity can be conferred to these nanostructures by certain coating for surface silanization but, in this case, they found a more complex behaviour with etching time. Thus, modified-TiO_2 layers exhibited tunability for hydrophilicity or hydrophobicity, pointing out a relationship between the wettability and the structure of TiO_2 arrays. In consequence, these nanostructures can be used for multifunctional transparent layers to hinder microbial adhesion and dirt accumulation and concurrently enable photocatalytic degradation of surface-bound organics.

A more complex processing was followed in the preparation of Perovskite (Ba,Sr) TiO_3 (BST) thin films on SiO_2/Si(1 0 0) substrate by a modified sol-gel method [59]. To realize highly integrated devices involving (BST) thin film, it is crucial to develop etching processes providing simultaneously near vertical sidewall angle, fence-free profile, good selectivity over resist, no sidewall residues and high etch rates. For this purpose, the films were etched using CHF_3/Ar plasma following an etching mechanism based on the cooperation of diverse effects: ion bombardment, ions assist chemical reaction and reaction etching effects. For ZnO etching in a BCl_3/Ar plasma, combination of ion bombardment in addition to physical sputtering allowed high etch rates, in spite of the relatively high volatility of the by-products formed [71]. In fact, etching rate showed a non-monotonic behaviour with an increasing BCl_3 fraction in the BCl_3/Ar plasma, the RF power and gas pressure.

6.3 Films Porosity Characterization

6.3.1 Ellipsometry

The most widespread methods for characterization of thin films are optical. Thus, the spectrophotometric methods using reflection and/or transmission of light through a film-substrate system can be used to determine both their optical constants and its geometric characteristics.

Beside them emerges ellipsometry, a technique capable of performing a complete optical characterization of substrates and film-substrate systems. With the development of modern and precise equipment, among advances in computer technology that allowed the automation of ellipsometry instruments as well as ellipsometry data analysis, spectroscopic ellipsometry has become a standard technique for characterization of thin films [65]. Essentially, reflection ellipsometry is an optical technique that measures the change in polarized light upon light reflection on a sample. It allows a high precision evaluation of both optical constants and film thickness because ellipsometry measures relative light intensities modulated by optical elements instead of absolute light intensities, as it is the case in reflection or transmission spectroscopies.

Reflection ellipsometry on an ambient-film-substrate system is based on the change of polarization that takes place on reflection due to the different amplitude attenuation and phase shift experienced by the parallel, p, and perpendicular, s, components of the incident light. Figure 6.4 represents a light wave travelling in a medium of refractive index $\tilde{n}_0$ that impinges on a dielectric layer ($\tilde{n}_1$) on top of a substrate ($\tilde{n}_2$). The light will partially be reflected, characterized by the reflection coefficient r_{01}, and the remaining fraction will be transmitted, quoted by the

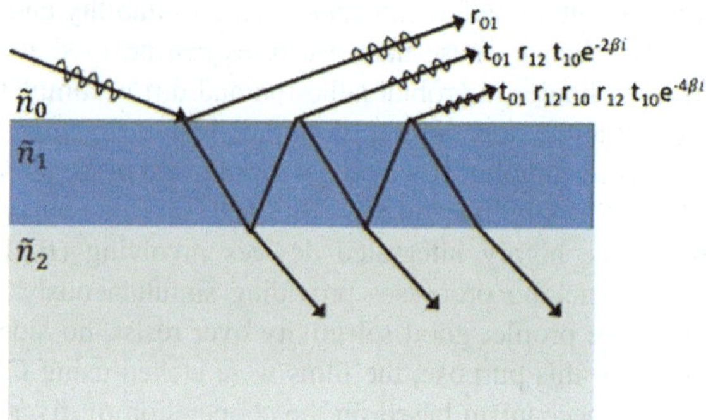

Fig. 6.4 Optical model of a film-substrate system for reflexion ellipsometry

transmission coefficient t_{01}. The transmitted fraction will penetrate through the dielectric layer to the interface between layer and substrate. There, the light will again be partially reflected and transmitted, represented by the coefficients r_{12} and t_{12}, respectively. This process will be repeated again and again, thus multiple beams will leave the sample and contribute to the measured result. Determination of the layer thickness or refractive index from the reflected light intensities involves a huge amount of calculations and, therefore, it is done by specific ellipsometer software.

The actual quantity measured by a common ellipsometer is the time-dependent reflected beam intensity at the detector, which is proportional to the magnitude of the squared electric field. By multiplying the electric field of the input beam by Jones matrices appropriate for each element in the optical system, the electric field at the detector can be expressed as follows [7]:

$$E_D = [Analyzer\ matrix][Sample\ matrix][Polarizer\ matrix][Input\ beam]$$

E_D is a function of the ratio, $\tilde{\rho}$, of the complex parallel, $\widetilde{R}_p$, and perpendicular, $\widetilde{R}_s$ Fresnel reflection coefficients and it is commonly expressed in terms of the two ellipsometric parameters measured, Ψ and Δ as follows:

$$\tilde{\rho} = \frac{\widetilde{R}_p}{\widetilde{R}_s} \equiv tan\Psi\, e^{i\Delta} \tag{1}$$

In principle, ellipsometry measures the two values (Δ, Ψ) that represent the amplitude ratio, Ψ, and phase difference, Δ, between light waves p- and s-polarized. In spectroscopic ellipsometry, (Δ, Ψ) spectra are measured by changing the wavelength of light in the ultraviolet (UV)/visible/NIR region or infrared region.

Ellipsometry allows the direct calculation of the refractive index *n* and extinction coefficient *k*, which is also referred to as optical constants. From the two values (*n*, *k*), the complex refractive index defined by $\widetilde{N} \equiv n - ik(i = -1)$ is determined. The complex dielectric constant $\tilde{\varepsilon} \equiv \varepsilon_1 - \varepsilon_{i2}$ and absorption coefficient α can also be obtained from $\widetilde{N}$.

Then, Ψ and Δ are given explicitly as functions of the angle of incidence, the light wavelength in air, the optical constants of the film and the substrate, as well as the thickness of the film. The separation of Eq. (1) into its real and imaginary parts yields one equation for Δ and one for Ψ. The experimental ellipsometrics angles Ψ and Δ can be correlated with the thickness and optical constants of a film by

$$tan\Psi e^{i\Delta} = \mathrm{K} = \frac{r_{01}^{p} + r_{12}^{p} e^{-2i\beta}}{1 + r_{01}^{p} r_{12}^{p} e^{-2i\beta}} \frac{1 + r_{01}^{s} r_{12}^{s} e^{-2i\beta}}{r_{01}^{s} + r_{12}^{s} e^{-2i\beta}}, \tag{2}$$

where $\beta = \frac{360}{\lambda} d(n_1^2 - sin^2\phi)$ is known as *film phase thickness* and it is expressed in degrees.

The solution of the equation for K is quite complicated and can be solved satisfactory by computers because of its nonlinear transcendental nature and their explicitly dependence of nine real arguments. There are two general restrictions on the ellipsometry measurement: (1) surface roughness of samples has to be rather small and (2) the measurement can be performed at oblique incidence. When light scattering in surface roughness reduces severely the reflected light intensity, the ellipsometry measurement becomes difficult, given that ellipsometry determines a polarization state from its light intensity. Sensitivity of the measurement is maximized for incidence angles near the Brewster angle. On the contrary, at normal incidence the ellipsometry measurement becomes impossible since p- and s-polarization cannot be distinguished anymore at this angle [30].

For a simple characterization, ellipsometry measurement at a single wavelength using a He–Ne laser (632.8 nm) has been performed widely. We have reported single wavelength ellipsometry characterization [10] of TiO_2 sol-gel thin film dip-coated onto a glass substrate. In this case the parameters used to calculate (Δ, Ψ) are reduced to $tan\Psi e^{i\Delta} = \mathrm{K}(\widetilde{N}_0, \widetilde{N}_1, \widetilde{N}_2, d, \phi)$. As usual, values of $\widetilde{N}_2$ and ϕ were known and, $\widetilde{N}_0 = 1$, so only three parameters (n_1, k_1 and d) were unknown. In the case of TiO_2, the thin layer does not show any absorption at 632.8 nm ($k_1 = 0$), therefore n_1 and d could be determined from two (Δ, Ψ) values. Figure 6.5 shows a surface profile, obtained from single wavelength ellipsometry, corresponding to a four dip-coated films at 100 mm/min, cured at 150 °C and, finally, annealed at 500 °C. The raster scan of thickness profile obtained from Ψ and Δ shows a variation of 14 nm between the highest and the lowest points along the film surface. In the transversal section, film shows a parabolic profile with a higher thickness in the central region and the thinner part close to the borders. At the same time, there is a smoothly increase of the thickness along the wetting direction, induced by gravitational forces during the dipping process, as expected for dip-coated films.

On the other hand, spectroscopic ellipsometry (SE) gives us the (Δ, Ψ) spectra in the measured region (see Fig. 6.6). From the analysis of these spectra, additional information about the thin film can be obtained, including the complex dielectric constant and band gap or other geometrical parameters as surface roughness or thickness inhomogeneity (wedging). In order to evaluate the optical constants and thickness of samples from SE, it is necessary to perform ellipsometry data analysis

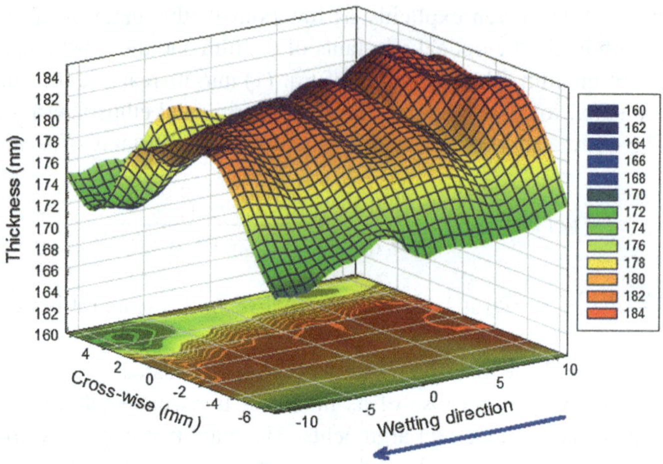

Fig. 6.5 Raster Scan Ellipsometry of a dip-coated TiO_2 thin films on a glass substrate, four layers at 100 mm/min, cured at 150 °C and annealed at 500 °C

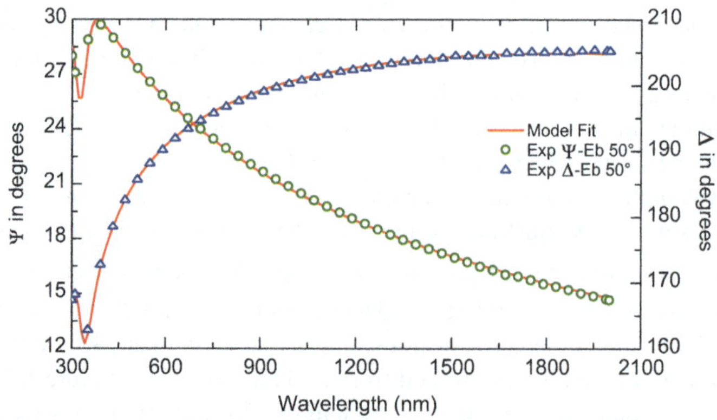

Fig. 6.6 Ellipsometric angles spectra for a TiO_2 sol-gel thin film dip-coated onto a glass substrate and heat treated at 550 C during 1 h, modelled (*red line*) with a Tauc–Lorentz dielectric function

that consists of three major parts, i.e., the construction of an optical model (Fig. 6.4), dielectric function modelling, and fitting to (Δ, Ψ) measured spectra [74].

The modelling of the dielectric function of the film-substrate system is the most important step in the SE results interpretation. Since a material absorbs light in a specific spectral region, we can find a transparency spectral region for almost any sample. For this region, with no light absorption, $\varepsilon_2 = 0$ and $\varepsilon_1 = n^2$, the substrate-film system can be modelled by a Cauchy function:

$$n(\lambda) = A_n + \frac{B_n}{\lambda^2} + \frac{C_n}{\lambda^4}$$

The Cauchy function can describe dispersion for a material that is essentially non-absorbing over the applied wavelength range and it provides a means to curve-fit dispersion. In the Cauchy equation, the A_n term describes the long-wavelength asymptotic index value, while the B_n and C_n are the dispersion terms that add upward slope to the index curve as wavelengths become shorter (B_n and $C_n \geq 0$). The Cauchy dispersion relation was designed to model the normal dispersion realized in transparent materials. This curve fit can give us also the film thickness.

The absorption region can be fitted using forced oscillators models by light. Thus, Gauss, Drude, Lorentz or Tauc–Lorentz oscillators are commonly used for modelling ε_2 in these regions.

For some materials, it is desirable to have an oscillator which is not symmetric, as Tauc–Lorentz (TL), where the ε_2 function drops more rapidly in one side of the band than in the other. That is the case of amorphous and polycrystalline materials, as a-Si or TiO_2, or transparent conductive oxides thin films. The TL oscillator considers a four-parameter function, where ε_2 is modelled from the product of a single band gap of amorphous materials (Tauc's gap) and a Lorentz oscillator:

$$\varepsilon_2 = \frac{AE_{n_0}C(E_n - E_g)^2}{\left(E_n - E_{n_0}^2\right)^2 + C^2E_n^2} \frac{1}{E_n} (E_n > E_g) \text{ and } \varepsilon_2 = 0\ (E_n \leq E_g)$$

being A the amplitude, E_{n_0} the peak position, C is the broadening parameter and E_gAmong these, Lorentz the Tauc–Lorentz gap. The equation for ε_1 can be derived using the Kramers–kronig relations because ε_1 and ε_2 (n and k) have to be Kramers–Kronig consistent. Figure 6.6 shows a Tauc–Lorentz fit of a (Δ, Ψ) spectra of a TiO_2/glass thin film annealed at 550 C. The fit provides the optical constants, the thickness and the gap energy, 3.37 eV in this case.

Figure 6.7 shows the dispersion curves for the same system submitted at different annealed treatments between 400 and 800 C. At this wavelength range, n presents a normal dispersion behaviour with a sharper reduction at lower wavelengths. Comparison of the dispersion curves set reflects the densification process taking place when the thin film is annealed at increasing temperature. The increase of the refractive index with the annealing temperature observed over all the wavelength range is in agreement with a porosity reduction occurring that would change the film microstructure. The inset shows the evolution with the temperature of the refractive index at 632 nm.

Ellipsometry is quite sensitive to surface and interface structures. Then, if we are able to incorporate in some way these structures to the optical model, additional information about them can be inferred, as roughness or thickness non-uniformity.

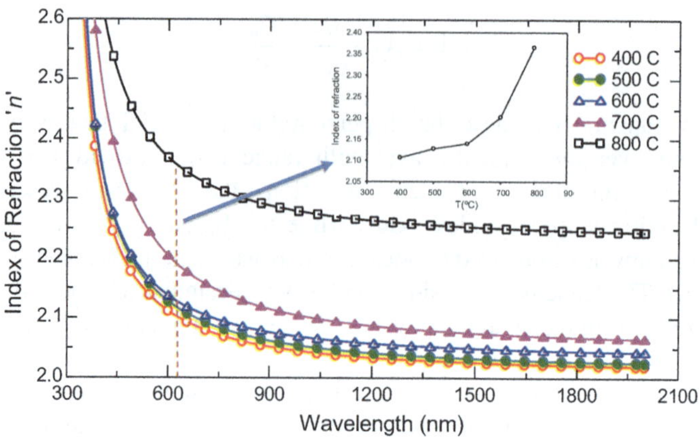

Fig. 6.7 TiO_2 thin film on a silica glass substrate refractive index dispersion annealed at temperatures in the range 400–800 C. The inset shows the refractive index temperature evolution at 632 nm

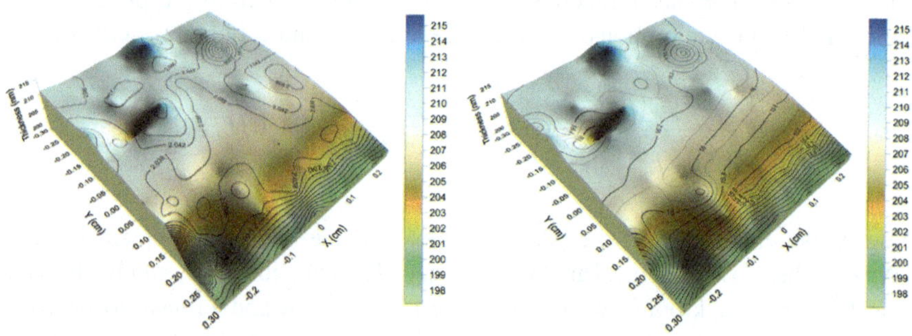

Fig. 6.8 Thickness map of a TiO_2 thin film dip-coated onto a glass substrate annealed at 600 C. Over the map, Cauchy parameter A_n (*left*) and local (point) porosity obtained from MG-EMA model (*right*) are plotted as contour lines

Figure 6.8 displays the result of an SE scan experiment performed with a focus probe with a spot size of 200 μm, over a 6 mm × 5 mm region of a TiO_2 sol-gel thin film annealed at 600 °C. At each point, a SE experiment was performed in a transparent region, between 500 and 800 nm, and the resulting data were fitted to a Cauchy model. Figure 6.8 (left) shows the thin-film thickness distribution along the studied area with a maximum peak to peak difference of 20 nm but with a very regular profile over the bulk surface analyzed showing thicknesses between 204 and 210 nm. The same figure presents a contour plot of the Cauchy parameter, A_n, along the studied surface. This parameter is directly related to the refractive index and can give additional microstructural information of the sample. In this case, the profile of A_n, with most of the values between 2.02 and 2.06, is consistent with a very homogeneous microstructure. The main differences between A_n values are found in the edge region of the film, where a larger variation of the thickness also occurs. In this area, A_n increases with decreasing thickness, indicating that the microstructure becomes more compact. However, in the upper right area, a singularity with a relatively sharp decrease in A_n is observed in a region with no remarkable thickness.

Thus, thin-films microstructure can be characterized from ellipsometric results, regarding the refractive index as an effective optical property. If we consider the Effective Medium Approximation, EMA [4], we can characterize each volume fraction of the mesoporous thin film modelled as a composite pore-matrix material. Thus, EMA of nanocomposite and mesoporous thin films results in the calculation of relative volume fractions of their components. For an ideal mixing of a dispersed pore volume (refractive index n_P) in a dense film (refractive index n_D) the effective refractive index n_{eff} of the porous film is given by

$$f(n_{eff}) = (1-P)f(n_D) + Pf(n_p),$$

where $f(n_i)$ is the specific refractive index of a certain model, P is the volume fraction of pores (porosity) and (1 − P) is the volume fraction of pore-free material. Diverse mixture models have been reported to treat a heterogeneous thin film as it was homogeneous with an effective refractive index. Among these, Lorentz–Lorenz, Maxwell–Garnett, and Bruggeman are most frequently employed [51]. The use of these principal mixture rules results in either a linear refractive index-porosity dependence or in a negative or positive deviation from a linear additivity of n_{eff} versus P.

Lorentz–Lorenz expression is suited for treating heterogeneous materials which are mixtures of constituents of different polarizabilities, α_a and α_b, and its expression in terms of the dielectric functions of the pure phases a and b, ε_a, ε_b is [4]:

$$\frac{\varepsilon-1}{\varepsilon+2} = f_a\frac{\varepsilon_a-1}{\varepsilon_a+2} + f_b\frac{\varepsilon_b-1}{\varepsilon_b+2},$$

where f_i represents the volume fraction of the i_{th} phase.

A more general expression is derived when the heterogeneous material consists of regions that are small compared to the wavelength of light but still large to possess their own dielectric identity and they are not mixed on an atomic scale. Then the assumption of vacuum ($\varepsilon = 1$) as the host medium is not good, and we consider a host dielectric function ε_h. Under these assumption, the Lorentz–Lorenz expression changes to

$$\frac{\varepsilon-\varepsilon_h}{\varepsilon+2\varepsilon_h} = f_a\frac{\varepsilon_a-\varepsilon_h}{\varepsilon_a+2\varepsilon_h} + f_b\frac{\varepsilon_b-\varepsilon_h}{\varepsilon_b+2\varepsilon_h}$$

Thus if b represents the dilute phase, $f_a > f_b$, and $\varepsilon_h = \varepsilon_a$:

$$\frac{\varepsilon - \varepsilon_a}{\varepsilon + 2\varepsilon_a} = f_b \frac{\varepsilon_b - \varepsilon_h}{\varepsilon_b + 2\varepsilon_h}$$

The alternative equation obtained with $\varepsilon_h = \varepsilon_b$ when $f_b > f_a$, is the Maxwell-Garnet EMA (MG-EMA) expressions [5]. The Maxwell-Garnet expressions have the same general form than Lorentz–Lorenz.

Bruggeman proposed an expression when there is no clear majority phase in the heterogeneous material but the inclusions of a and b are considered embedded in the effective medium itself. This is equivalent to consider $\varepsilon_h = \varepsilon$ in the MG-EMA model. Thus, the Bruggeman EMA model expression is

$$0 = f_a \frac{\varepsilon_a - \varepsilon}{\varepsilon_a + 2\varepsilon} + f_b \frac{\varepsilon_b - \varepsilon}{\varepsilon_b + 2\varepsilon}$$

It is clear that although superficially the above expressions may seem very different, actually are related to each other, differing only in the choice of host dielectric ε_h.

Below is presented a study of the porosity evolution in a porous TiO_2 sol-gel thin film during annealing. The MG-EMA model has been considered, where the host dielectric is a TiO_2 matrix and phase b is the the air-filled pores embedded in it ($\varepsilon_b = 1$). Thus, the ellipsometric (Δ, Ψ) can be fitted to a Maxwell-Garnet function, where f_{por}, in this case, would give the pore volume fraction. Figure 6.9 shows, as an example, the ellipsometric spectra, corresponding to a 500 °C annealed TiO_2 thin-film dip-coated on the two faces of a glass substrate, fitted to the MG-EMA model. The fitting with a mean square error (MSE) of 4.7 results in a 210.6 nm thickness with a pore volume fraction of 18.3%.

MG-EMA can be applied for a film submitted to increasing annealing treatment temperatures in order to follow the changes induced in the porosity. Figure 6.10

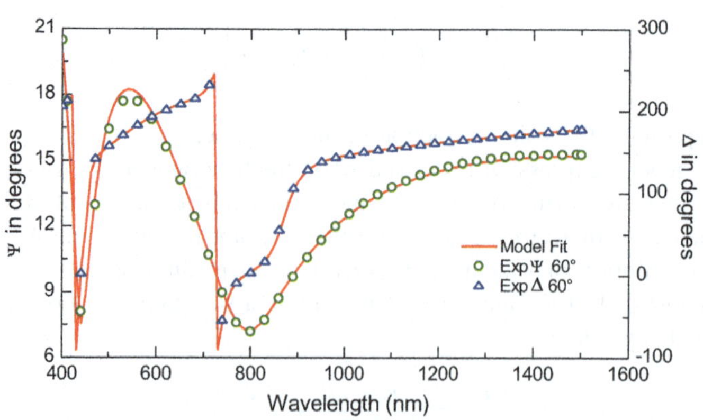

Fig. 6.9 Ellipsometric (Δ, Ψ) spectra for a double-sided TiO_2 thin film onto silica glass annealed at 500 °C. Both spectra have been fitted according to a MG-EMA model (*red line*) giving rise to a porosity of 18.3%

shows the results obtained by applying MG-EMA model to the spectra corresponding to TiO_2 films annealed at different temperatures. Results indicate that the porosity evolution with temperature is not linear and it can be fitted to a decreasing polynomial function.

Likewise, MG-EMA can be also applied to the ellipsometric spectra (Δ, Ψ) obtained at each point of the surface scan experiment presented before. Figure 6.8 (right) shows the surface scan of a sample of TiO_2 annealed at 600°C, with a contour plot indicating the variation of the porosity in percentage, obtained from MG-EMA model at each point, using a focus probe with a spot size of 200 μm. The combination of these two results allows obtaining a map of the porosity presented by the film at different points of its surface.

Antireflective thin films can be designed by changing the porosity in such a way that a refractive index gradient along the film thickness is induced. In these cases, it is possible to adjust the ellipsometric angle spectra to a MG-EMA model combined with an index graded function, by fitting two parameters: percentage variation range of refractive index and an exponent.

Figure 6.11 (left) shows (Δ,Ψ) spectra of a sol-gel TiO_2 film performed by three dip-coated layers on a glass substrate. Layers were annealed at 650, 550 and 400 °C, respectively, from bottom to top, preserving a different porosity degree, as predicted by Fig. 6.10, and consequently different refractive index.

(Δ, Ψ) spectra were fitted to MG-EMA model in which the percentage of refractive index variation decreases as we move away from the substrate. The best fit obtained gives a percentage of variation of −8.85% and an exponent of 1.66. The refraction index profile of the best fitting is shown in Fig. 6.11 (right) where a variation from 2.1, in the external part, to 2.3, in its innermost part, is observed.

Another application of the MG-EMA in the characterization of thin-film porosity is the *molecular probing ellipsometry* technique [29]. This technique is based on the possibility of using a liquid with a known molecular size and refractive index as adsorbate and measuring the change in the refractive index of the film/probe

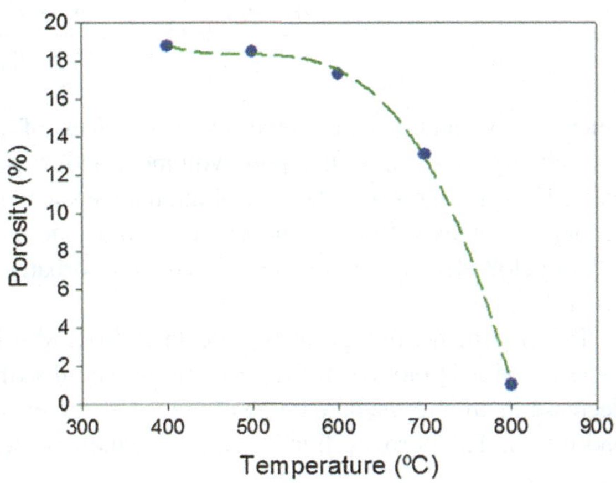

Fig. 6.10 Porosity evolution of a TiO_2 thin film dip-coated onto a silica glass annealed at temperatures in the range of 400–800 C

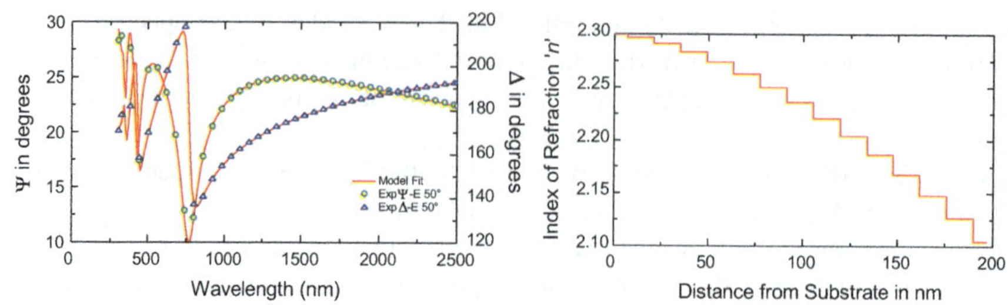

Fig. 6.11 (Δ, Ψ) spectra of a three dip-coated sol-gel TiO_2 layers on a glass substrate, fitted to a MG-EMA function with a graded refractive index from top to bottom (*left*). Refractive index profile (*right*)

molecule/matrix system by reflection ellipsometry. A pore volume distribution can be obtained according to the pore which each molecule can fill using different sized molecules. Thus, Silica-Alumina mullite like films were deposited by dip coating onto AISI-304 stainless steel and submitted to different heat treatments. For the experiments, samples were placed in a hemispherical sealed chamber located on the sample stage of the ellipsometer. First dry N_2 gas was passed through the chamber and pumped out by a vacuum pump, in order to empty the pores of any condensed adsorbate, until Ψ and Δ values remain steady. Then, N_2 gas, previously bubbled through the liquid adsorbate to drag it, was passed over the sample to fill the pores; in each case the refractive index was measured when Ψ and Δ reached a steady value. In the first stage, the refractive index grows because the pores are being filled with a liquid with $n > 1$, while the thickness remains constant. When all the accessible pores are completely full, an external layer is formed and the refractive index decreases. If it is assumed that all the accessible pores are completely emptied, in dry N_2, or totally filled with adsorbate, when the atmosphere is saturated, the pore volume and the solid skeleton index can be determined from the MG-EMA:

$$\frac{n_f^2 - n_p^2}{n_f^2 - 2n_p^2} = (1 - v_p)\frac{n_s^2 - n_p^2}{n_s^2 - 2n_p^2},$$

where n_s, n_f and n_p are the refractive indices of the matrix, film and pores, respectively, and v_p is the pore volume fraction filled by the adsorbate. This expression accounts for the contribution of each component to the composite's refractive index. Water, methanol, carbon tetrachloride, cyclohexane and 1,5,9-cyclododecatriene have been used as adsorbates, which size range from 3 to 10 Å.

The volume fraction porosity, calculated from MG-EMA equation, versus probe molecule size is plotted in Fig. 6.12 (left). The distributions present an important decrease in an intermediate range of the measurements interval mostly between 4.5 and 6.5 Å. The porosity fraction values remaining nearly constant are out of this

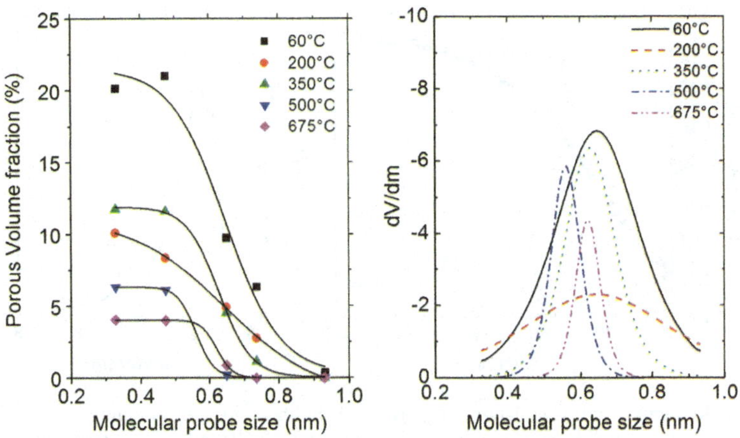

Fig. 6.12 Volume fraction porosity obtained by MPE at different temperatures (*left*). The curves are obtained by a sigmoidal fitting of experimental data and, after differentiation, pore distribution (*right*) is obtained. The sizes are referred in all cases at each probe molecule to van der Waals equivalent sphere diameter. Reproduced with permission from [29]

range. This general behaviour suggests that pore volume fraction follows a sigmoidal law as a function of pore diameter. The maximum porosity appears at 60 °C. For higher temperatures, a progressive porosity reduction is found due to the pore collapse caused by the film densification.

Figure 6.12 (right) shows the pore size distribution obtained differentiating the sigmoidal curves of Fig. 6.12 (left). Peak position, width and area are related to the mean pore size, root mean square deviation and total pore volume, respectively. After all the organics removal, between 350 and 675 °C the peak area decreases being the final pore distribution very narrow. This behaviour was expected due to the film sintering. It is possible to detect some remaining porosity in samples heated at the highest temperature. Molecular probe ellipsometry on films deposited over stainless steel is an adequate method to evaluate film microporosity, as well as a convenient alternative to other techniques.

Environmental Ellipsometry Porosimetry (EEP) is obtained through capillary condensation of water into film porosity using an atmospheric control chamber. The technique consists of plotting a gas adsorption–desorption isotherm from the variations of film refractive index and thickness induced by the change of relative pressure of solvent, which is water in this case. The porous volume was evaluated using the Bruggemann effective medium approximation model. When the studied thin film does not adsorb light in the considered wavelength range, the dielectric constants are taken to be the square of refractive index values (real part of the dielectric constants). Pore size distribution of micro- and mesopores of the film can also be directly obtained via porosimetry models, usually based on Kelvin equation, and modified for the case of water [13].

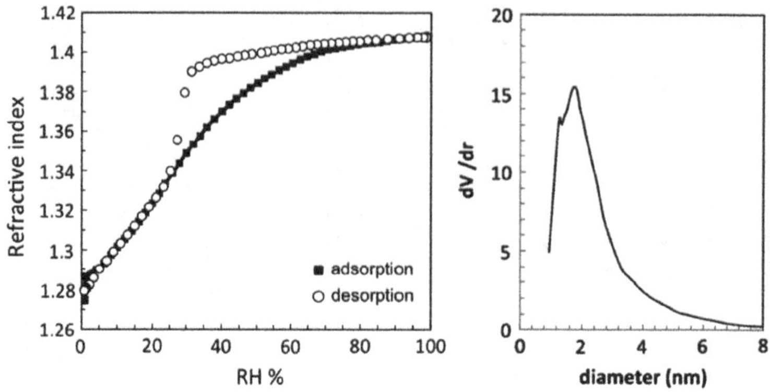

Fig. 6.13 Refractive index variation of porous phenyl-bridged polysilsequioxane thin films treated at 700 °C, for increasing and decreasing Relative Humidity at fixed wavelength ($\lambda = 700$ nm) (*left*). Pore size distribution as determined by the Kelvin equation modified for the case of water (*right*). Reproduced with permission from [13]

Figure 6.13 (left) displays refractive index variation of porous phenyl-bridged polysilsequioxane thin films treated at 700 °C, for increasing and decreasing Relative Humidity (RH) at fixed wavelength ($\lambda = 700$ nm). The constant increase of n at lower RH can be attributed to water condensation into the micro- and mesoporosity. Figure 6.13 (right) shows the pore size distribution as determined by the Kelvin equation modified for water.

6.3.2 Other Characterization Techniques

Another important group of techniques for thin-film porosity characterization are gas adsorption techniques. Thus, in [21] to obtain detailed information on the texture of sol-gel dip-coated TiO_2 films onto stainless steel, N_2 adsorption analysis was performed. The results show that all samples correspond to type IV (BDDT classification) isotherms. The pore size distributions (PSD) for the TiO_2 film, annealed at different temperatures, are determined from the isotherms adsorption branch. The average pore size and the pore size distribution show a clear dependence with the calcination temperature. The specific surface area decreases when the calcination temperature increases from 400 to 600 °C due to pore collapse and shrinkage.

For the most frequently used adsorptives, namely nitrogen and argon, saturation pressures are very high, which leads to a relatively large number of molecules being trapped within the void volume of the sample cell. Because of the very small total pore volumes and surface areas of thin porous films, pressure changes due to adsorption cannot be measured with sufficient precision. As an alternative, an adsorptive with a substantially lower saturation vapour pressure can be used, such as krypton at the boiling point of liquid nitrogen.

As the analysis of adsorption isotherms of Kr at 77 K is far from being straightforward and a well-established procedure has not been implemented yet, a simple method based on comparative plots can be used to determine the mean pore size and the pore size distribution.

Using the above-described semi-empirical method [79] have measured thin-films-specific surface area as low as 350 cm^2/g and pore volumes of $0.052.10^{-3}$ cm^3/g.

Krylova et al. [41] have employed adsorption–desorption isotherms of hexane vapour recorded at 293 K using an apparatus equipped with a McBain–Bakr quartz microbalance (sensitivity 2.3 mg/mm) and a cathetometer, to estimate the specific surface area, the pore volume, and the pore size distribution of silica and titania films.

6.4 Applications

6.4.1 Sensors

Semiconducting metal oxides are frequently used as gas-sensing materials. For sensing applications, mesoporous films with large surface/volume ratios but also with uniform pore structures are desired. Well-defined porosity confers to the films additional performance with respect to selectivity, self-diagnosis and long-term stability. A comprehensive review about the improvement of gas-sensing performance by the control of the textural properties of metal oxide film can be found in Ref. [64].

Porous silicon (PS) is used as an alternative simple material to metal oxide semiconductors for **hydrogen sensor** because of its low operating temperature, very high surface to volume ratio and the compatibility with the silicon technology. However, PS lacks of the required thermal and long-term stability because its fragile structure, easy oxidation and rapid degradation. These problems were overcome by coating PS with a thin ZnO layer, after surface passivation with nanocrystalline Pd, that stabilized it up to 150 °C [39]. The performance of this sensor was improved in respect of the operating temperature, the response time and stability but the slow gas desorption kinetics and related long recovery times owing to the use of the inert carrier gas for operation. [27] evaluate the use as H_2 sensors of TiO_2 thin films spin-coated onto oxidized silicon and doped with Pt. After heat treatment at 500 °C, the film were able to withstand a full microelectronic processing sequence, including dry etching for gas sensors sensitive area definition, Ti/Pt contact formation, and heater processing on the backside of the sensor substrates. H_2 gas-sensing tests evidenced that the anatase phase was much more sensitive than the rutile one, but the presence of Pt further enhanced the gas-sensing properties, lowering the optimum sensor operation temperature to near 330 °C and allowing for the detection of a minimum H_2 concentration of about 1000 ppm.

WO_3–Cr_2O_3 semiconducting thin films, prepared by sol-gel method, have been tested as acetone gas sensors for diabetes diagnosis based on human breath analysis. As-prepared films exhibit excellent sensing properties with a fast response and recovery time. However, the sensing properties were found to be strongly influenced by the porous structure, the content of Cr_2O_3, the sintering temperature and the cooling path of the films [31].

Zhao et al. have developed an ethanol-sensitive field-effect transistor combining sol-gel processing with the NMOS integrated circuit technology, using a $LaFeO_3$ nanocrystalline (perovskite-type) thin film as a gate electrode of an ordinary n-channel MOSFET [78]. Films obtained by sol-gel coating and patterned by photo-etching technique, had excellent selectivity, good sensitivity to ethanol and a good stability.

Sol-gel nanocrystalline porous tin oxide thin film have been proposed as gasoline [46] and CO sensor [38]. Gas sensor films exhibited particles and pore size of 7–15 nm and 1.6–9 nm, respectively, but their surface morphology, electrical conductance and CO sensitivity depend on calcination temperature (500–800 °C). The tin oxide film calcined at 700 °C deposited onto interdigitated electrodes showed high CO sensitivity with relatively fast response (4 min) and recovery time (1 min), proving the convenience of nanocrystalline materials for sensor development.

6.4.2 *Photocatalysis and Solar Applications*

Many human activities, like paint and adhesive industry or combustion processes, release volatile organic compounds that are recognized as dangerous pollutant compounds. These organic pollutants are carcinogen agents, contribute to ozone production in the troposphere and, unfortunately, present high resistance to physical, chemical or biological treatments. Photocatalysis is an efficient, attractive and clean technology for pollutant reduction either in aqueous media or in gas phase. On the other hand, one of the most prevalent and universal problems is the access to clean and safe water. Conventional methods for water decontamination and disinfection are effective but not environmental friendly [11]. Using photocatalytic technology to decompose pollutants by advanced oxidation processes is an economical method, with lower energy consuming and minimizes the environmental impact. In essence, heterogeneous photocatalysis detoxification consists in the near-ultraviolet photoexcitation (<400 nm) of a wide band semiconductor catalyst in contact with water and in the presence of oxygen which produces hydroxyl radicals (•OH) in the water, due to the photo-generated holes, with a high oxidizing power. Almost any toxic pollutant present in the water will be oxidized and mineralized yielding to CO_2 and inorganic ions in water [56].

The electronic structure of semiconductors (filled valence band an empty conduction band) allows them to act as sensitizers for this light-induced redox process by the electron/hole pair formed after absorption of a photon with an energy greater than band gap. In general this energy will be dissipated unless recombination is prevented and trapped hole can induce the oxidation. TiO_2 is one of the most active photocatalyst with strong chemical and photocorrosion resistance that, at the same time, is low cost and non-toxic. Anatase phase of TiO_2 has a relative high-energy band gap (~ 3.2 eV) that can only be excited efficiently by high-energy UV irradiation that constrains the practical usage of the TiO_2. Other semiconductors absorb larger fraction of the solar spectrum and can be used as photocatalyst but often they are chemically degraded after a number of catalytic cycles and produce final toxic products. For that reason, many efforts have been made in order to improve its photocatalytic efficiency or to extent the energy absorption range of TiO_2.

The first strategy involves increasing the number of attainable surface activation sites either using nanocrystalline TiO_2 or inducing highly porous surface structure that offers a much larger number of catalytic sites than a dense surface thus improving the photocatalytic performance. In general, larger porous structure, hydrophilic properties and higher specific surface area make it easier for the solution to enter, spread and further migrate into films where a larger number of accessible reactive sites are found. For instance, it has been shown that porous TiO_2-anatase films prepared by sol-gel route exhibit a higher photocatalytic activity (conversion 20% higher) for degradation of trichloroethylene (TCE) in air than dense titania [3]. Incorporation of polyethylene glycol (PEG) as pore generating agent also reduced the time for TiO_2-anatase phase conversion, which is complete after 3 h at 350 °C. The decomposition of PEG during high temperature treating is considered to be responsible for the generation of nanoporous structure in the films in such a way that a larger surface area and porosity is obtained that increase the contact surface exposed to TCE, enhancing crystallinity and the degradation rate. Although the presence of anatase is necessary, the crystal fraction and crystal size seem not to play a critical role in photocatalyst behaviour. The influence of the particular pore generator used (polyethylene glycol, pluronic F-127 and polyethylene glycol hexadecyl ether) was studied by Suarez et al. [62] for the same degradation process. They also consider two different chelating agents for the titania alkoxide (acetic acid or acetyl acetone) because they can also affect the particular structural properties. Results at different gas flows and water vapour contents indicated that the photocatalytic activity mainly depends on the nature of the pore generating agent which influences both the TiO_2 crystallite size and the textural properties of the final material. The samples obtained using acetic acid, as the modifier ligand, and either Brij®58 or PEG, as the pore generating agent, exhibited the best photocatalytic behaviour even in the presence of high quantities of water vapour. The direct trichloroethylene oxidation pathway to CO_2 is favoured at low water vapour contents, but $COCl_2$ is also generated even at low water partial pressures.

The influence of the PEG molecular weight effect on the photocatalytic activity has been evaluated for TiO_2 thin films, spin-coated on glass substrates, to the decomposition of methyl orange in water [33]. After PEG decomposition during heat treatment, hydrophilic, transparent and crack-free films resulted with improved thermal stability. As it can be expected, the higher the PEG molecular weight, the larger the pore size and specific surface area of the films. It was reported that the photocatalytic activities of the TiO_2 films is promoted with the use and the molecular weight of PEG and the irradiation time, as revealed by the decrease of 460 nm absorbance peak of treated dye solutions.

Four different experimental parameters of the TiO_2 sol-gel thin-film processing have been studied and optimized by the Taguchi experimental design method [44]: acetylacetone (AcAc) molar ratio, addition of HCl, addition of PEG and its molecular weight. Among them, AcAc ratio showed the greatest impact on the photocatalytic degradation under UV irradiation of dyes like methylene blue, methyl orange, and indigo in water, being the optimized AcAc: Alkoxide ratio 2:1.

The second strategy for improving titania photocatalytic activity often involves the addition of a second active element. In some cases, red shift of TiO_2 band gap is caused for the formation of new energy levels between valence and conduction bands. In other cases, they act as electrons traps and promote interfacial charge transfer, thus decreasing the electron–hole pair recombination rate. Peláez et al. have published a comprehensive review of different strategies to develop visible light active titania photocatalyst, including non-metal and/or metal doping, dye sensitization and coupling semiconductors [56].

Chang et al. proposed the use of Ag as electron trap but also because it facilitates electron excitation, by creating a local electric field [20]. Thus, nanocrystalline sol-gel TiO_2/Ag thin films were spin-coated, using polystyrene spheres of three different sizes (50, 200 and 400 nm) as pore generators after calcination at 500 °C. Photocatalytic efficiency of these porous thin films for the UV degradation of methylene blue was improved both by increasing porosity and by adding active Ag elements. The better photocatalytic performance was found for films doped with 1 mol% Ag and prepared with polystyrene spheres of 200 nm.

Highly porous TiO_2–4%SiO_2–1%TeO_2 sol-gel nanostructures, directly fixed on glass substrates with AlO_3/TiO_2 sputtered and anodized layers, have also been proposed for photocatalytic applications [23]. As mentioned in the etching paragraph of the preparation section, anodization results in a porous alumina structure that is used as support for titania nanotubes array formation. The addition of SiO_2 and TeO_2 improved the wetting ability of TiO_2 sol and the affinity of the calcined layer to the anodic alumina walls, leading to a hollow composite structure with open apertures. Large surface areas (~10000 m^2/substrate m^2), high porosities (~40%) and transparency were shown to enhance photocatalytic performance in decomposing acetaldehyde gas under UV illumination, in comparison to that of commercial P-25.

A different approximation consists in the use of a macroporous silica film as mechanical support to immobilize TiO_2 nanoparticles doped with erbium [18]. Macroporous network of the spin coating film was induced, in this case, by adding Rhodamine 6G to the sol that is photodegraded under illumination with a fluorescent lamp. The photocatalytic activity of trapped nanoparticles was evaluated by measuring the degradation rate of methylene blue under visible light, which followed a first-order kinetics. The results confirm that the addition of erbium ions induces a red shift in the TiO_2 band gap, sensitizing it to visible light.

In the same way, Gregori et al. [32] evaluated the photocatalytic efficiency of commercial anatase nanoparticles dispersed in hybrid silica film matrices in order to address the structure/reactivity relationship. Under UV exposure, the structure of the coatings is evolving for a limited period of time due to a partial elimination of the organic groups brought by the organosilane precursors. Thus, a cleaning preliminary step is done to obtain an active and stable photocatalytic material, where pending organic groups are removed and particular porosity induced open efficiently the interaction between the catalysts and the adsorbed pollutants. The effect of optimized coating in the degradation of formic acid and methylene blue in different conditions demonstrated that these materials can be efficiently used as self-cleaning materials and for water depollution using solar light.

Photocatalytic activity of many other semiconductors has been also checked. For instance, porous sol-gel ZnO thin films of spongy particle aggregates with uniform pores at the surface have been prepared on glass substrate from Zinc acetate [55]. These films decompose, with high rate constants and efficiently, aqueous solutions of phenol, chlorophenol, naphthalene and anthracene to CO_2 but detection of hydroquinone, benzoquinone and hydroxylated carbonyl compounds as the major intermediate photoproducts during the degradation processes were reported.

Anti-reflection coatings (ARCs) are widely used in energy-related applications including solar thermal collectors, windows of high-rise buildings and covers of solar cells. Enhancement in the conversion efficiency of solar cells, improvement of the photocatalytic activity and reduction of reflection losses in photovoltaics are some benefits associated to their use. Essentially, anti-reflection is achieved for coating with a thin film with quarter wavelength thickness and a refractive index matched to the geometric mean of those of the substrate and the environment which leads to destructive interference of the reflected light from upper and lower interfaces. For instance, for a silica glass substrate a single material with ~1.23 refractive index, like MgF_2 [58], a porous silica thin film [50, 63] or a graded index coating can be used. However, most technological applications demand multi-functional coatings combining anti-reflection capacity with other such as anti-fogging, anti-abrasion and self-cleaning. Additionally, they should be resistant to heat, water, heat-transfer solvents and corrosive environments.

One approach consists in addition of a surface active agent to the precursor sol which considerably reduces the surface tension. For instance, Triton X-100™ has been used to prepare highly porous silica xerogel films [40] and its content and the heating temperature was fitted to control the refractive index and thickness of the films for antireflective applications. As expected, film porosity ($\sim 50\%$) caused a decrease of the refractive index ($n = 1.2$–1.3), a change with depth of the effective refractive index was also detected probably due to a certain normal porous gradient. This optical inhomogeneity should be taken into account for theoretical predictions as antireflective coating, particularly for porous SiO_2 layer with thickness less than 100 nm or more than 1000 nm.

Commonly materials for solar applications are required to show a high absorption in the solar wavelength range (0.3–2.5 μm) and low thermal infrared emittance. A selective solar absorber has been proposed [22] based on carbon nanoparticles dispersed in a titania sol-gel film in which a macroporous structure was generated, without templating, by phase separation induced by photopolymerization. The spectral selective absorption performance was controlled by varying the HNO_3 concentration in the solution in order to regulate the surface morphology of the thin films, such as the pore size and the degree of roughness. The remaining large number of amorphous carbon dispersed in the titanium dioxide network enhances the structural stability of the composite films, and simultaneously, the exposure of the well-defined interconnected pores effectively enhances the scattering effect as a second phase. So, by reducing HNO_3 concentration in the solution, the mean size of pores, carbon content, and the thickness of the nanocomposite films decrease, leading to an optimized single layer film 300 nm thick with a solar absorption coefficient $\alpha = 0.95$ and thermal emittance $\varepsilon = 0.12$.

Cu_2ZnSnS_4 (CZTS) is a promising semiconductor material for thin-film solar cells. Thin-film CZTS have been processed by non-aqueous thiourea–metal–oxygen sol-gel processing with extrinsic dopants and chemical etching [61]. For photovoltaic technologies, high-efficiency and low-cost films are required with high-quality, phase-control, homogeneous elemental distribution and low impurity content. Based on the modified sol-gel method, solar cells with a structure of Ni:Al/ZAO/i-ZnO/CdS/CZTS/Mo/glass have been fabricated with a power conversion efficiency of 5.10%, indicating its potential for photovoltaic devices.

6.4.3 Bio-applications

High chemical homogeneity, fine grain structure and low crystallization temperature of the coating as well as low cost and technical simplicity are features of sol-gel technique that make it adequate for coating surfaces of implant materials dedicated for hard tissue replacements [9]. Thus, calcium phosphate coatings are usually prepared by dip coating the substrate in calcium salt and phosphorus gels followed by sintering at different temperatures to improve their adhesion strength and density. Weng et al. reported the preparation of hydroxyapatite (HAP) thin film by dip coating from a mixed ethanol solution of $Ca(NO_3)_2 \cdot 4H_2O$ and P_2O_5 in a Ca/P ratio of 1.67. These coatings presented good HAP crystallinity and adhesive strength with a porosity that can be controlled by adjustment of the preparation conditions [69].

Fluoridated hydroxyapatite coatings with different fluoride contents prepared by sol-gel methods have shown an ambivalent behaviour: either to improve the cell viability or, at high fluoride contents, to decrease calcium release in culture medium either due to reduced solubility. For clinical applications, it is suggested that a moderate content of fluorides may be suitable as a compromise among cell attachment, proliferation, apatite deposition and dissolution resistance. Sol-gel technology is also used for tailored porous bioactive and osteoconductive Si-substituted HA, which is a highly promising material in the sense of bioactivity improvement [34].

Sol-gel hydroxyapatite (HAP) coating was successfully deposited on the external surface and within the pores of a porous NiTi alloy, displaying the inner surface of porous NiTi alloy an increased apatite deposition rate due to the rough surface. The coating shows good stability in Tris solution and much reduced rate of Ni ion release in simulated body fluid. Authors' points out a remarkable increase of apatite forming ability in calcification solution and the larger apatite deposition on the inner surface of porous NiTi alloy due to the rough surface [77]. Macroporous Ti scaffolds with highly porous and interconnected structure, prepared by polymer impregnating method, had been also coated by sol-gel Ag/Hidroxiapatite composite in order to achieve both bacterial inhibition and enhancement of osteoblast function [57].

A mesoporous hydroxyapatite (HA) coating was prepared on a β-tricalcium phosphate (β-TCP) porous scaffold by a sol-gel dip coating method using the block copolymer Pluronic F127 ($EO_{106}PO_{70}EO_{106}$) as the template [73]. Cell growth and bone-related protein expression in MC3T3-E1 cells were investigated on the mesoporous HA coated β-TCP porous scaffold.

Nanoporous sol-gel Silica Coating loaded with Bioactive (sol-gel) Glass nanoparticles (nBG/NSC) have been deposited on titanium dental implant. The coating nanotopography in conjunction with the chemical bioactivity of the nanoparticles enhances the in vitro bone-like apatite formation and stimulates the osteogenic differentiation of stem cells in the absence of osteogenic supplements [25]. These properties allow improving the osseointegration responses in vivo, accelerating the formation of bone tissue in intimate contact with the implant surface within a short time of implantation.

Ti–6Al–4 V implants formed with a sintered porous surface for implant fixation by bone ingrowth were prepared with or without the addition of a thin surface layer of sol-gel calcium phosphate (Ca–P). The implants were placed transversely across the tibiae of 17 rabbits. The formation of poorly crystallized hydroxyapatite layer over the sintered porous surface zone of Ti–6Al–4 V implants significantly increases the rate of bone ingrowth into the porous region and, thereby, produced a faster implant fixation and osseointegration. Furthermore, the Ca–P film promotes more complete bone fill of the pore regions [54].

WO_3–Cr_2O_3 semiconducting thin films, prepared by sol-gel method, have been tested as acetone vapour sensor for diabetes diagnosis based on human breath analysis [31]. Sensing properties of these films are strongly influenced by the porous structure, Cr_2O_3 content, sintering temperature and the cooling path. Authors postulated that dissymmetrical monoclinic structure obtained after quenching improved selectivity to acetone. This is related to the low symmetry and the large dipole moment, which lead to a stronger interaction between the WO_3 surface and acetone molecules. Thus these sensors exhibit a fast response, high sensitivity and good selectivity to low concentrations of acetone.

6.4.4 Other Applications

Sol-gel processed porous metal oxide coatings based on alumina, silica and titanium oxide have been developed by dip coating on a Fe-based alloy, as support for catalytically active components, to enhance the inner surface of microchannel reactors. By fitting the sol composition and viscosity and the thermal treatment procedure a surface enhancement factor reached 800 m^2/substrate m^2 [35].

Double-layered structures of $LiMn_2O_4$ thin-film cathodes coated with Mn_2O_3 layer, for **Li-ion microbattery tests,** have been prepared by the sol-gel spin coating method. The electrochemical properties of $Mn_2O_3/LiMn_2O_4$ cathodes characterized by charge–discharge (C–D) cycling performance, cyclic voltammetry, and rate capability were compared with those of the pristine $LiMn_2O_4$ cathodes. The improved electrochemical performance is ascribed to the porous Mn_2O_3 over layer that protects the underlying $LiMn_2O_4$ film and also provides efficient Li^+ diffusion path during the C–D cycling [43].

Porous orthorhombic tungsten oxide (o-WO_3) thin films, stabilized by nanocrystalline anatase, are obtained by sol-gel, using organic porosity template, dip coating and subsequent annealing at 600 C. An asymmetric electrochromic device was constructed based on this porous o-WO_3 layer where the intercalation of Li ion was analyzed as a function of applied coloration/bleaching voltages. Additionally, a photochromic device was constructed, by sensitizing the o-WO_3 layer with a ruthenium-based dye, where the presence of nanocrystalline anatase TiO_2 enhanced to 51% photochromic optical transmittance contrast in the NIR region. The combination of these photochromic and electrochromic properties of the o-WO_3/TiO_2 films opens up new vista for its application in energy saving smart windows [8].

Nanoporous thin film of alumina (γ-Al_2O_3), prepared by sol-gel method, with high surface to volume ratio and high distribution of micropores have been used to develop a low-cost moisture sensor. For that purpose, porous film of 3–6 μm thickness was dip-coated in between two parallel gold electrodes to form a capacitive sensor with commercial applications as trace moisture sensor for a 2.5–25 ppm range [37].

Piezoelectric acoustic sensors based on partially unclamped PZT-coated cantilever and bridge have been fabricated by silicon micromachining. High sensitivity at low frequencies (5–100 Hz) has been achieved by patterning very narrow slits (3–5 μm) around the structures. Stress compensation, dry etching and integration of high performance piezoelectric thin films were the key issues. PZT/Pt/SiO_2 stacks have been patterned by reactive ion etching and PZT film's tensile stress was compensated by adjusting the thickness of a thermal SiO_2 layer. The integration of sol-gel PZT films with a transverse piezoelectric coefficient $e_{31}f$ of -12.8 C m^{-2} has been realized without any degradation of the properties. The microphones were successfully integrated into a miniature photoacoustic detector and tested for CO_2 detection. [42].

The applications of sol-gel porous films are numerous and diverse, as can be seen by reference to specialized databases, having registered more than 7,500 patents in the last 10 years. At present, it continues to develop new lines of research in this field, while presented here is a set of the most recent contributions where the great diversity in them is shown.

References

1. Abdellaoui T, Bennaceur J, Chtourou R (2010) Effect of nanostructured porous n plus GaAs substrates on the energy band gap of nanocrystalline TiO_2 thin films analyzed by spectroscopic ellipsometry. Phys E-Low-Dimension Syst Nanostruct 43:239–247
2. Aegerter M, Puetz J, Gasparro G, Al-Dahoudi N (2004) Versatile wet deposition techniques for functional oxide coatings. Opt Mater 26:155–162
3. Arconada N, Duran A, Suarez S, Portela R, Coronado JM, Sanchez B, Castro Y (2009) Synthesis and photocatalytic properties of dense and porous TiO_2-anatase thin films prepared by sol-gel. Appl Catal B-Environ 86:1–7
4. Aspnes D (1982) Local-field effects and effective-medium theory—a microscopic perspective. Am J Phys 50:704–709
5. Aspnes D (1982) Optical-properties of thin-films. Thin Solid Films 89:249–262
6. Attia S, Wang J, Wu G, Shen J, Ma J (2002) Review on sol-gel derived coatings: Process, techniques and optical applications. J Mater Sci Technol 18:211–218
7. Azzam RMA, Bashara NM (1987) Ellipsometry and polarized light. Amsterdam etc., North-Holland, XVII, 539 p. p
8. Balaji S, Djaoued Y, Albert A-S, Bruening R, Beaudoin N, Robichaud J (2011) Porous orthorhombic tungsten oxide thin films: synthesis, characterization, and application in electrochromic and photochromic devices. J Mater Chem 21:3940–3948
9. Bauer S, Schmuki P, von der Mark K, Park J (2013) Engineering biocompatible implant surfaces part I: materials and surfaces. Prog Mater Sci 58:261–326
10. Blanco E, Gonzalez-Leal JM, Ramirez-del Solar M (2015) Photocatalytic TiO_2 sol-gel thin films: optical and morphological characterization. Sol Energy 122:11–23
11. Blanco J, Malato S, Fernandez-Ibanez P, Alarcon D, Gernjak W, Maldonado M (2009) Review of feasible solar energy applications to water processes. Renew Sustain Energy Rev 13:1437–1445
12. Boissiere C, Grosso D, Chaumonnot A, Nicole L, Sanchez C (2011) Aerosol route to functional nanostructured inorganic and hybrid porous materials. Adv Mater 23:599–623
13. Brigo L, Faustini M, Pistore A, Kang HK, Ferraris C, Schutzmann S, Brusatin G (2016) Porous inorganic thin films from bridged silsesquioxane sol-gel precursors. J Non-Cryst Solids 432:399–405
14. Brinker C, Frye G, Hurd A, Ashley C (1991) Fundamentals of sol-gel dip coating. Thin Solid Films 201:97–108
15. Brinker C, Hurd A, Schunk P, Frye G, Ashley C (1992) Review of sol-gel thin-film formation. J Non-Cryst Solids 147:424–436
16. Bu SJ, Jin ZG, Liu XX, Yang LR, Cheng ZJ (2005) Synthesis of TiO_2 porous thin films by polyethylene glycol templating and chemistry of the process. J Eur Ceram Soc 25:673–679
17. Calderon-Moreno JM, Preda S, Predoana L, Zaharescu M, Anastasescu M, Nicolescu M, Stoica M, Stroescu H, Gartner M, Buiu O, Mihaila M, Serban B (2014) Effect of polyethylene glycol on porous transparent TiO_2 films prepared by sol-gel method. Ceram Int 40:2209–2220
18. Castaneda-Contreras J, Maranon-Ruiz VF, Chiu-Zarate R, Perez-Ladron de Guevara H, Rodriguez R, Michel-Uribe C (2012) Photocatalytic activity of erbium-doped TiO_2 nanoparticles immobilized in macro-porous silica films. Mater Res Bull 47:290–295

19. Cerveau G, Corriu R, Framery E (2001) Nanostructured organic-inorganic hybrid materials: kinetic control of the texture. Chem Mater 13:3373–3388
20. Chang C-C, Chen J-Y, Hsu T-L, Lin C-K, Chan C-C (2008) Photocatalytic properties of porous TiO_2/Ag thin films. Thin Solid Films 516:1743–1747
21. Chen Y, Dionysiou D (2006) Correlation of structural properties and film thickness to photocatalytic activity of thick TiO_2 films coated on stainless steel. Appl Catal B-Environ 69:24–33
22. Cheng B, Wang KK, Wang KP, Jiang W, Cong BJ, Song CL, Jia SH, Han GR, Liu Y (2015) Porous carbon-titania nanocomposite films for spectrally solar selective absorbers. Sol Energy Mater Sol Cells 133:126–132
23. Chu SZ, Inoue S, Wada K, Li D, Haneda H, Awatsu S (2003) Highly porous (TiO_2-SiO_2-TeO_2)/$Al2O_3$/TiO_2 composite nanostructures on glass with enhanced photocatalysis fabricated by anodization and sol-gel process. J Phys Chem B 107:6586–6589
24. Chu S, Wada K, Inoue S, Todoroki S (2002) Synthesis and characterization of titania nanostructures on glass by Al anodization and sol-gel process. Chem Mater 14:266–272
25. Covarrubias C, Mattmann M, Von Marttens A, Caviedes P, Arriagada C, Valenzuela F, Pablo Rodriguez J, Corral C (2016) Osseointegration properties of titanium dental implants modified with a nanostructured coating based on ordered porous silica and bioactive glass nanoparticles. Appl Surf Sci 363:286–295
26. Dislich H (1971) New routes to multicomponent oxide glasses: Angewandte Chemie-International Edition, vol 10, p 363
27. Epifani M, Helwig A, Arbiol J, Diaz R, Francioso L, Siciliano P, Mueller G, Morante J (2008) TiO_2 thin films from titanium butoxide: synthesis, Pt addition, structural stability, microelectronic processing and gas-sensing properties. Sens Actuators B-Chem 130:599–608
28. Floch H, Belleville P (1994) A scratch-resistant single-layer antireflective coating by a low temperature sol-gel route. J Sol-Gel Sci Technol 1:293–304
29. FuentesGallego J, Blanco E, Esquivias L (1997) Ellipsometric characterization of an AISI 304 stainless steel protective coating. Thin Solid Films 301:12–16
30. Fujiwara H (2007) Spectroscopic ellipsometry: principles and applications. Wiley, Chichester
31. Gao P, Ji H, Zhou Y, Li X (2012) Selective acetone gas sensors using porous WO_3-Cr_2O_3 thin films prepared by sol-gel method. Thin Solid Films 520:3100–3106
32. Gregori D, Guillard C, Benchenaa I, Leonard D, Parola S (2015) Hybrid sol-gel porous nanocomposites as efficient photocatalytic coatings: Insights in the structure/reactivity relationships. Appl Catal B-Environ 176:472–479
33. Guo B, Liu ZL, Hong L, Jiang HX (2005) Sol gel derived photocatalytic porous TiO_2 thin films. Surf Coat Technol 198:24–29
34. Gupta R, Kumar A (2008) Bioactive materials for biomedical applications using sol-gel technology. Biomed Mater 3
35. Haas-Santo K, Fichtner M, Schubert K (2001) Preparation of microstructure compatible porous supports by sol-gel synthesis for catalyst coatings. Appl Catal A-Gen 220:79–92
36. Henrist C, Dewalque J, Cloots R, Vertruyen B, Jonlet J, Colson P (2013) Hierarchical porous TiO_2 thin films by soft and dual templating a quantitative approach of specific surface and porosity. Thin Solid Films 539:188–193
37. Islam T, Kumar L, Khan SA (2012) A novel sol-gel thin film porous alumina based capacitive sensor for measuring trace moisture in the range of 2.5–25 ppm. Sens Actuators B-Chem 173:377–384
38. Jin ZH, Zhou HJ, Jin ZL, Savinell RF, Liu CC (1998) Application of nano-crystalline porous tin oxide thin film for CO sensing. Sens Actuators B-Chem 52:188–194
39. Kanungo J, Saha H, Basu S (2010) Pd sensitized porous silicon hydrogen sensor-Influence of ZnO thin film. Sens Actuators B-Chem 147:128–136
40. Karasinski P, Jaglarz J, Reben M, Skoczek E, Mazur J (2011) Porous silica xerogel films as antireflective coatings—fabrication and characterization. Opt Mater 33:1989–1994

41. Krylova G, Gnatyuk Y, Smirnova N, Eremenko A, Gun'ko V (2009) Ag nanoparticles deposited onto silica, titania, and zirconia mesoporous films synthesized by sol-gel template method. J Sol-Gel Sci Technol 50:216–228
42. Ledermann N, Muralt P, Baborowski J, Forster M, Pellaux J (2004) Piezoelectric Pb(Zr-x, Ti1-x)O-3 thin film cantilever and bridge acoustic sensors for miniaturized photoacoustic gas detectors. J Micromech Microeng 14:1650–1658
43. Lee JH, Kim KJ (2013) Superior electrochemical properties of porous Mn2O3-coated $LiMn_2O_4$ thin-film cathodes for Li-ion microbatteries. Electrochim Acta 102:196–201
44. Lin W-C, Yang W-D, Jheng S-Y (2012) Photocatalytic degradation of dyes in water using porous nanocrystalline titanium dioxide. J Taiwan Inst Chem Eng 43:269–274
45. Liu ZF, Jin ZG, Li W, Qiu JJ (2005) Preparation of ZnO porous thin films by sol-gel method using PEG template. Mater Lett 59:3620–3625
46. Liu G, Zhao Z, Jiao X, Chen D (2014) Preparation of porous SnO_2 thin film with high gasoline sensing performance. Mater Technol 29:167–171
47. Lu Y, Fan H, Stump A, Ward T, Rieker T, Brinker C (1999) Aerosol-assisted self-assembly of mesostructured spherical nanoparticles. Nature 398:223–226
48. Lu Y, Ganguli R, Drewien C, Anderson M, Brinker C, Gong W, Guo Y, Soyez H, Dunn B, Huang M, Zink J (1997) Continuous formation of supported cubic and hexagonal mesoporous films by sol gel dip-coating. Nature 389:364–368
49. Lu Y, Yang Y, Sellinger A, Lu M, Huang J, Fan H, Haddad R, Lopez G, Burns A, Sasaki D, Shelnutt J, Brinker C (2001) Self-assembly of mesoscopically ordered chromatic polydiacetylene/silica nanocomposites. Nature 410:913–917
50. Mahadik DB, Lakshmi RV, Barshilia HC (2015) High performance single layer nano-porous antireflection coatings on glass by sol-gel process for solar energy applications. Sol Energy Mater Sol Cells 140:61–68
51. Matthias A, Raicevic N, Tchana R, Kip D, Deubener J (2014) Density dependence of refractive index of nanoparticle-derived titania films on glass. Thin Solid Films 558:86–92
52. Mellott N, Durucan C, Pantano C, Guglielmi M (2006) Commercial and laboratory prepared titanium dioxide thin films for self-cleaning glasses: Photocatalytic performance and chemical durability. Thin Solid Films 502:112–120
53. Miyata H, Kawashima Y, Itoh M, Watanabe M (2005) Preparation of a mesoporous silica film with a strictly aligned porous structure through a sol-gel process. Chem Mater 17:5323–5327
54. Nguyen HQ, Deporter DA, Pilliar RM, Valiquette N, Yakubovich R (2004) The effect of sol-gel-formed calcium phosphate coatings on bone ingrowth and osteoconductivity of porous-surfaced Ti alloy implants. Biomaterials 25:865–876
55. Pal B, Sharon M (2002) Enhanced photocatalytic activity of highly porous ZnO thin films prepared by sol-gel process. Mater Chem Phys 76:82–87
56. Pelaez M, Nolan N, Pillai S, Seery M, Falaras P, Kontos A, Dunlop P, Hamilton J, Byrne J, O'Shea K, Entezari M, Dionysiou D (2012) A review on the visible light active titanium dioxide photocatalysts for environmental applications. Appl Catal B-Environ 125:331–349
57. Qu J, Lu X, Li D, Ding Y, Leng Y, Weng J, Qu S, Feng B, Watari F (2011) Silver/hydroxyapatite composite coatings on porous titanium surfaces by sol-gel method. J Biomed Mater Res Part B-Appl Biomater 97B:40–48
58. Raut HK, Dinachali SS, Ansah-Antwi KK, Ganesh VA, Ramakrishna S (2013) Fabrication of highly uniform and porous MgF2 anti-reflective coatings by polymer-based sol-gel processing on large-area glass substrates. Nanotechnology 24
59. Shi P, Yao X, Zhang L (2004) Reactive ion etching of sol-gel-derived BST thin film. Ceram Int 30:1513–1516
60. Strawbridge I, James P (1986) The factors affecting the thickness of sol-gel derived silica coatings prepared by dipping. J Non-Cryst Solids 86:381–393
61. Su Z, Sun K, Han Z, Cui H, Liu F, Lai Y, Li J, Hao X, Liu Y, Green M (2014) Fabrication of Cu2ZnSnS4 solar cells with 5.1% efficiency via thermal decomposition and reaction using a non-toxic sol-gel route. J Mater Chem A 2:500–509

62. Suarez S, Arconada N, Castro Y, Coronado JM, Portela R, Duran A, Sanchez B (2011) Photocatalytic degradation of TCE in dry and wet air conditions with TiO_2 porous thin films. Appl Catal B-Environ 108:14–21
63. Thomas IM (1992) Method for the preparation of porous silica antireflection coatings varying in refractive-index from 1.22 to 1.44. Appl Opt 31:6145–6149
64. Tiemann M (2007) Porous metal oxides as gas sensors. Chem-A Eur J 13:8376–8388
65. Tompkins HG, Irene EA (2005) Handbook of ellipsometry. William Andrew Springer, Norwich Heidelberg, xvi, 870
66. Van T, Lee Y, Lee J, Park J (2013) Tuning hydrophobicity of TiO_2 layers with silanization and self-assembled nanopatterning. Langmuir 29:3054–3060
67. Vincent A, Babu S, Brinley E, Karakoti A, Deshpande S, Seal S (2007) Role of catalyst on refractive index tunability of porous silica antireflective coatings by sol-gel technique. J Phys Chem C 111:8291–8298
68. Wang X, Shi F, Gao X, Fan C, Huang W, Feng X (2013) A sol-gel dip/spin coating method to prepare titanium oxide films. Thin Solid Films 548:34–39
69. Weng WJ, Baptista JL (1998) Sol gel derived porous hydroxyapatite coatings. J Mater Sci Mater Med 9:159–163
70. Wenzel R (1936) Resistance of solid surfaces to wetting by water. Ind Eng Chem 28:988–994
71. Woo J, Um D, Kim C (2010) The dry etching of a sol-gel deposited ZnO thin film in a high density BCl_3/Ar plasma. Thin Solid Films 518:2905–2909
72. Wu GM, Wang J, Shen J, Yang TH, Zhang QY, Zhou B, Deng ZS, Fan B, Zhou DP, Zhang FS (2000) A novel route to control refractive index of sol-gel derived nano-porous silica films used as broadband antireflective coatings. Mater Sci Eng B-Solid State Mater Adv Technol 78:135–139
73. Ye X, Cai S, Xu G, Dou Y, Hu H, Ye X (2013) Preparation and in vitro evaluation of mesoporous hydroxyapatite coated beta-TCP porous scaffolds. Mater Sci Eng C-Mater Biol Appl 33:5001–5007
74. Yoshizawa TR (2009) Handbook of optical metrology: principles and applications. CRC, Boca Raton, Fla., London
75. Yu JC, Yu JG, Tang HY, Zhang LZ (2002) Effect of surface microstructure on the photoinduced hydrophilicity of porous TiO_2 thin films. J Mater Chem 12:81–85
76. Yu JG, Zhao XJ, Zhao QN, Wang G (2001) Preparation and characterization of super-hydrophilic porous TiO_2 coating films. Mater Chem Phys 68:253–259
77. Zhang JX, Guan RF, Zhang XP (2011) Synthesis and characterization of sol-gel hydroxyapatite coatings deposited on porous NiTi alloys. J Alloy Compd 509:4643–4648
78. Zhao S, Sin J, Xu B, Zhao M, Peng Z, Cai H (2000) A high performance ethanol sensor based on field-effect transistor using a $LaFeO_3$ nano-crystalline thin-film as a gate electrode. Sens Actuators B-Chem 64:83–87
79. Zita J, Krysa J, Cernigoj U, Lavrencic-Stangar U, Jirkovsky J, Rathousky J (2011) Photocatalytic properties of different TiO_2 thin films of various porosity and titania loading. Catal Today 161:29–34

Chapter 7
Synthesis Strategies and Emerging Catalytic Applications of Siliceous Materials with Hierarchically Ordered Porosity

Jean-Philippe Dacquin, Carmen Ciotonea and Sébastien Royer

7.1 Introduction

The quest for controlling materials morphology and hierarchical porosity to meet the demands of ever efficient-cost industrial processes has led many scientists to explore rather intuitive methodologies for building complex hierarchical architectures. Since the first example of the bioinspired hierarchical structure obtained by Ozin et al. in 1995 [1], many developments have been undertaken to engineer pore size, shape, connectivity, and multiscale porosity following templating chemistry. During the 1990s, we assisted to the successive development of the mesostructuring route by the teams of Kuroda et al. [2] and Beck et al. [3] and macrostructuring route by the teams of Velev et al. [4] and Holland et al. [5]. In those pioneer works, the pore size and shape depends on the structuring agent nature, while the pore volume and connectivity depend on the volume fraction of the structuring agent. Nowadays, thanks to the wide availability of templates and recent original synthesis routes, the field of materials chemistry is still experiencing an explosive growth of hierarchical morphologies that mimic natural architectures (diatoms) found in the living world. The ultimate objective of such tremendous efforts is to get access to the controlled preparation of hierarchically and spatially structured 3D or 2D materials with sharp pore size distributions at all scales. These multiple properties combined with high surface area and chemical functionalization would be highly valuable for relevant applications in the domain of catalysis amongst others. In this line, porous silica (and alumina) materials are widely used as classical supports for heterogeneous catalysis. Solid catalysts with bi-modal porosity would be very useful for relevant industrial applications since size or shape-selectivity involving molecular species can be solved by the presence of micropores/mesopores and easy transport/access to the

J.-P. Dacquin (✉) · C. Ciotonea · S. Royer
Univ. Lille, CNRS, Centrale Lille, ENSCL, Univ. Artois, UMR 8181 - UCCS - Unité de Catalyse et Chimie du Solide, 59000 Lille, France
e-mail: jean-philippe.dacquin@univ-lille1.fr

P. Bettotti (ed.), *Submicron Porous Materials*,
DOI 10.1007/978-3-319-53035-2_7

active sites would be allowed by the macropore network. However, such exciting area is still looking forward the bottom-up approach used by natural microorganisms (diatoms and radiolarian) to achieve complex hierarchical architectures under mild conditions (ambient pressure and temperature). This chapter will give an overview of the tools designed to the continuous access from the mesoporous to the macroporous structures as well as hierarchical macro-mesoporous structures. Emphasis will be given to the surfactant templating approaches involving species self-assemblies which determine the final microstructural features of monopore-sized and hierarchically structured porous materials. Only phenomena involving weak physical interactions during inorganic–organic assemblies formation will be reported here as well as influences of the experimental conditions that determine the final physico-chemical properties of the solid. A specific focus will be given on the catalytic applications of hierarchical porous materials.

7.2 Synthesis Strategies

7.2.1 Structure-Directed Materials Syntheses: From Mesoporous to Macroporous Architectures Design

7.2.1.1 Surfactant Templating Chemistry

The development of porous silica showing large and accessible pores has gathered considerable attention because of their potential applications toward adsorption, separation and heterogeneous catalysis which could involve bulky/viscous molecules (i.e., proteins, triacylglycerides, long-chain hydrocarbons…) or mass transfer limitations. The preparation of porous siliceous (and non-siliceous) materials is mainly based on molecular or supramolecular templating chemistry. The control of this method, which seems obvious at first sight, requires the knowledge of both sol-gel chemistry and assembly processes of organic and inorganic species, from molecular to colloidal sizes. These complex phenomena result in the geometry transfer of the sacrificial template into the matrix pore architecture. The first example of molecular templating to control pore size and shape of a silicate material can be found in the work of Dickey [6]. Using a sol-gel process, he developed a method which consists in “forming the structure of the adsorbent in the presence of the methyl orange compound for which it is desired to prepare a specific adsorbent”. Weak interactions were suggested responsible of the silicate species attraction around the methyl orange molecules. Since then a true race for materials pore size control began and an explosive growth of publications in the field has comforted the use of structure directing agents (SDA) for the preparation of texturally controlled architectures. Among numerous SDA found now in the literature, surfactant templates are the most used to direct the material pore properties.

Surfactant templates are typically amphiphilic molecules having a hydrophilic and a hydrophobic part covalently bound to each other.

On one hand, when using surfactant templating chemistry (Fig. 7.1), surfactants are present in aqueous solution and can self-associate in order to form supramolecular entities such as micelles, vesicles, and liquid-crystalline phase with regard to the concentration/temperature/pH/surfactant geometry used during the synthesis. On the other hand, inorganic species can be stabilized under a cationic, anionic or neutral form depending on the pH fixed regarding the oxide point of zero charge. Hence, one of the critical step during the material's formation is related to the charge density matching at the surfactant-inorganic species interfaces (Monnier et al. [7]) with the presence or not of small ions (Fig. 7.1c). Subsequent condensation step of the inorganic precursor on the external corona of micelles occurs and will govern the assembly process via non-covalent interactions (Van der Waals, hydrogen bonding, coulombic forces). Hybrid mesophase with a well-determined geometry is obtained following intermicellar condensation (Fig. 7.1). Finally, aging treatment of the reaction mixture will consolidate the mesostructured hybrid material through the complete silica polycondensation. The elimination of the "sacrificial template" can proceed either by calcination or solvent extraction. The first report on the synthesis of ordered mesoporous silica (i.e., MCM-41) was performed in basic medium using hexadecyltrimethylammonium bromide (CTAB). In such system, material's structural organization strictly depends on the direct interaction S^+I^- between anionic silicate species and the positively charged quaternary ammonium surfactant (CTAB) (Fig. 7.1c). As predicted by Monnier et al. [7], Huo et al. [8] found a generalized synthesis route from basic to acidic pH to extend the range of preparation of mesoporous molecular sieves. Pinnaivaia et al. [9] opened a new synthetic pathway where the surfactant–silicate interfaces can be obtained by nonionic surfactant in acidic medium as illustrated by the Hexagonal Mesoporous Silica (HMS) materials family. According to Pinnavaia [10], cooperative self-assemblies involving neutral routes ($S°I°$), like in the synthesis of HMS, led to mesoporous molecular sieves with larger pore size control and thermal stability with respect to mechanisms involving direct electrostatic bonds between the SDA and the inorganic species. By anticipating the interactions between the surfactant head and the inorganic species (Fig. 7.1c), Stucky et al. developed numerous SBA-n materials exhibiting well-determined pore geometry with 2D (SBA-3, SBA-8, SBA-15) or 3D mesostructures (SBA-6, SBA-12, and SBA-16) (Fig. 7.2). These materials are named Santa Barbara Amorphous (SBA-n where n is corresponding to a well-determined pore structure). More importantly, Stucky et al. [11] paved the way to the use of block copolymers of larger molecular weight (compared to alkylammonium surfactants) for the preparation of ordered mesoporous silica with tunable pore properties. Nowadays, SBA-15 is still one of the most used hexagonal mesoporous model material in the literature and is prepared using pluronic P123 poloxamers as nonionic surfactant (i.e., $(EO)_{20}(PO)_{70}(EO)_{20}$ with EO = (CH_2CH_2O) and PO = $(CH_2CH(CH_3)O)$). Block copolymers are ideal to change pore geometry and

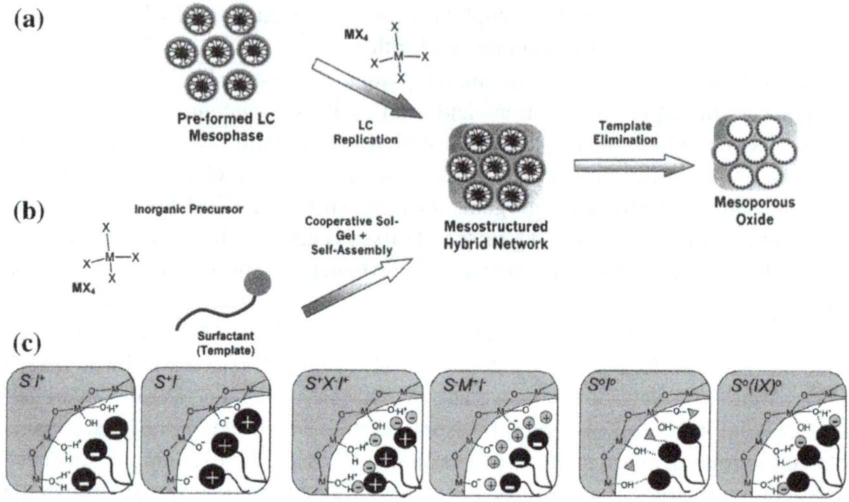

Fig. 7.1 Schematic views of the **a** true liquid crystal and **b** cooperative templating approaches to obtain a mesostructured siliceous oxide and **c** the different interfacial interactions between silica and the surfactant. Surfactant is named "S", the inorganic species "I", and the corresponding counterions "M^{+}" and "X". Dashed lines correspond to H-bonding interactions. Adapted with permission from Ref. [101]. Copyright (2002) American Chemical Society

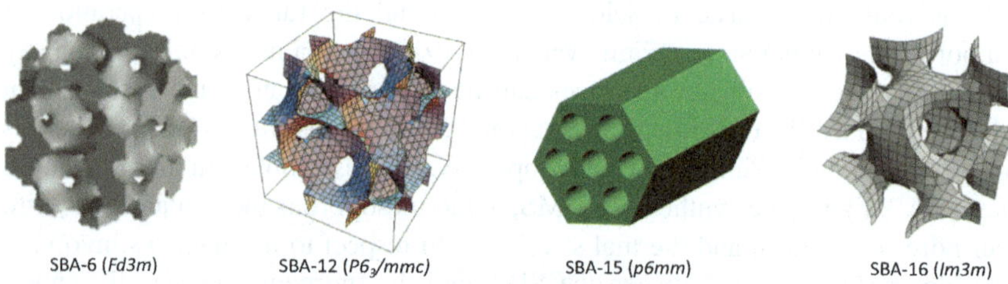

Fig. 7.2 Pore structure geometries of SBA-6, SBA-12, SBA-15 and SBA-16. Adapted with permission from Refs. [12, 102]. Copyright (2002) and (2007) American Chemical Society

connectivity while keeping a highly organized pore network. For instance, by selecting a block copolymer having increased EO/PO ratio such as pluronic F127 ($E0_{106}PO_{70}EO_{106}$), a body centered cubic SBA-16 can be prepared using similar synthetic conditions with respect to SBA-15 (Fig. 7.2). Surfactant templated synthesis has expanded continuously and parallel development of other families of mesoporous silicas appeared such as MSU, FDU [12], KIT [13] and AMS [14] where the materials are synthesized with variations in synthesis conditions and surfactants.

7.2.1.2 Mesopore Expansion in Micelle-Templated Synthesis

Following the pioneer report on the preparation of SBA-15, considerable efforts were dedicated to the rationalization of the SBA-15 synthesis to anticipate the material's formation. As reported in the literature physicochemical properties (morphology, porosity, surface area…) strongly depend on numerous synthesis parameters such as temperature of the reaction solution, acid nature and concentration, P123/TEOS ratio, stirring speed [15–19]. By a fine tuning of synthesis parameters, SBA-15 materials can reach up to 1000 m^2/g, pore mean diameters are usually spanning size around 5 nm with a sharp pore size distribution. However, if one calculates the specific surface area following a rough theoretical porous model derived from SBA-15 (Gurwitsch relation), we would obtain lower values compared to the experience (200–300 m^2g^{-1} against 600–900 m^2g^{-1}). Such differences have been explained in the literature [16, 20–23] by the additional presence of microporosity and secondary mesoporosity whose coexistence in the material depends primarily on the temperature of the hydrothermal aging (Fig. 7.3). Hence, SBA-15 materials should normally be considered as, at least, bi-modal porous materials.

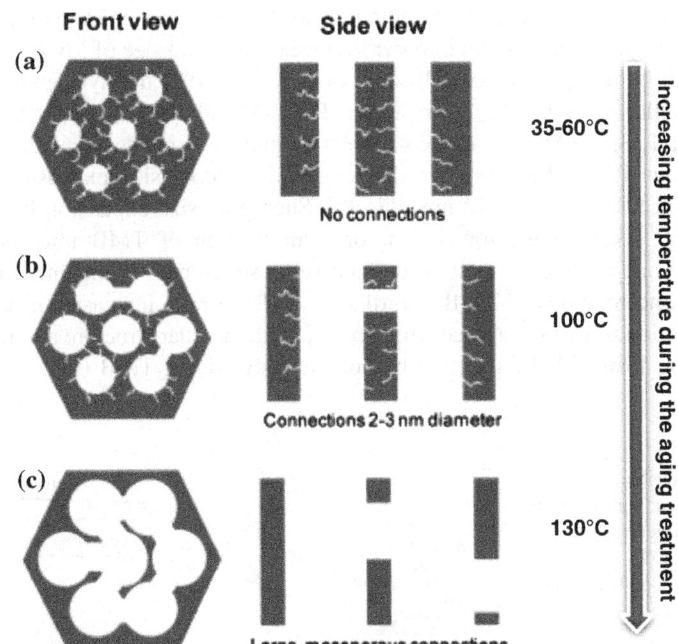

Fig. 7.3 Schematic view of the SBA-15 pore structure modification upon increasing the hydrothermal aging temperature: **a** 35–60 °C aging temperature; **b** 100 °C aging temperature and **c** 130 °C aging temperature. Adapted with permission from Ref. [103]. Copyright (2010) American Chemical Society

As illustrated in Fig. 7.3, the pore system of the SBA-15 is highly dependent on the temperature of the hydrothermal treatment. For low temperatures (Fig. 7.3a), the main mesoporous structure (5 nm) comprises walls filled by micropores which are the result of the imprinting left in the siliceous framework by PEO chains. When the aging temperature is reached to 100 °C (Fig. 7.3b), there is the coexistence of micropores and small mesopores (1.5–4 nm) that each form a junction between two adjacent mesopores (5 nm). Finally at 130 °C, structural mesopores are expanded up to 9.5 nm and the micropores disappear from the siliceous framework in favor of the multiplication of mesoporous junctions between structural SBA-15 mesopores (Fig. 7.3c). The absence of microporosity is justified by the weakening of interaction forces between the surfactant and silicate species since PEO chains are becoming more hydrophobic at high temperature. Besides, the silica walls become thinner upon temperature increase. Similar behaviors have been observed with other ordered mesoporous silica like SBA-16 [24], MCM-41 [25] or KIT-6 [14] materials. Hence, hydrothermal treatment of the reaction solution can expand by itself the size of the final mesopores up to ~10 nm but careful attention must be taken on modifications of the pore system. Albeit hydrothermal treatment resulted in slight pore expansion in the mesoporous domain because of the interaction modification between the surfactant head and silicate species, organic molecules and block copolymers are inevitably physically limited by their polymer chain dimensions. In 1992, Beck et al. introduced "swelling agents" (i.e. 1, 3, 5-Trimethylbenzene (TMB)) to the reaction solution in order to increase the pore size of MCM-41 (from 4 to 10 nm) by preferential solubilization of 1, 3, 5 TMB in the hydrophobic parts of the micellar assembly [26]. Regarding PEO-PPO-PEO templated system, the addition of 1, 3, 5 TMB organic co-solvent as micelle expander generate as well homogeneous P123/TMB droplets that could allow pore size expansion of the well-ordered SBA-15 up to 12 nm [27, 28]. Such pore size expansion limitations are directly linked to the low extent of solubilization of TMB into the P123 micelles. While the mesoporous hexagonal silica structure is maintained only for very low concentrations of TMB, a further TMB/P123 ratio increase will induce a phase transition from a hexagonal structure to a multilamellar structure and turn into mesocellular foam (MCF) structure by continuously adding TMB (Fig. 7.4).

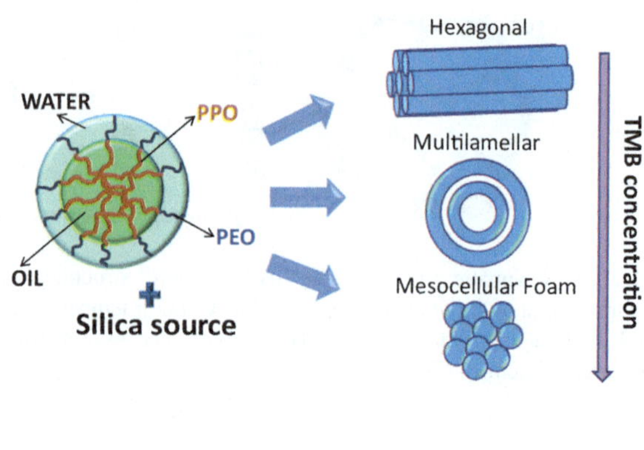

Fig. 7.4 Plausible mechanism leading to various silica pore structures using TEOS-P123-TMB precursor solution. Adapted from Ref. [104]

Despite the loose of meso-ordering, the MCF derived from SBA-15 still exhibits interconnected pores spanning size between 22 and 42 nm [26, 29]. Hence, stable microemulsions can also be prepared, which played the part of a template of mesostructured cellular foams [30]. Regarding other swelling agents like triisopropylbenzene (TIPB) or aliphatic hydrocarbons…, their extensive addition led as well in the lowering of the structural degree ordering or phase transition of the siliceous materials under the classical temperature synthesis conditions (i.e., 308 K for SBA-15 [17]).

In 2005, Su et al. [31] proposed an alternative alkane-assisted protocol with additives such as inorganic salts (i.e., NH_4F) in order to develop SBA-15 having large mesopore size. By properly adjusting the reaction synthesis temperature (290 K < T < 308 K), the alkane source, and the ionic strength, they were able to enlarge continuously the pore size of SBA-15 while maintaining meso-ordering up to 15 nm. On one hand, the doping of PPO blocks with short-chain alkanes (hexane, heptane, and nonane) allowed to direct the micellization at lower temperatures while swelling hydrophobic cores. On the other hand, at the same time, the inorganic salt (i.e., NH_4F) enabled the hydrolysis-condensation of silicate species at low temperature. From this smart work, low temperature synthesis of ultralarge pore SBA-15 were developed and conducted with other swelling agents, such as TIPB [31–33]. The use of triisopropylbenzene (TIPB) swelling agent and fine tuning of the synthesis parameters (temperature synthesis, TIPB amount, hydrothermal treatment and time) are shown to promote the formation of ultralarge pore silicas with cylindrical shape up to 26 nm. Authors demonstrated that low temperature synthesis (between 11 and 20 °C) is required for the true templating by TIPB-swollen micelles with respect to TMB-P123 systems where microemulsions are formed and led to less ordered structures and mesocellular foams. Such synthesis route has been now applied to other 2D and 3D mesostructures [34]. It has been shown that the micelle expander must be selected from the surfactant nature (mainly based on the hydrophobic/hydrophilic ratio for pluronic derived surfactant), in order to reach a controlled swelling effect within the surfactant micelles without inducing pore structure disorder. However, regarding parallel studies for catalytic applications employing micelle expanders with low temperature synthesis [35, 36] only poor textural properties of expanded mesoporous SBA-15 materials have been obtained, suggesting that the use of the low temperature synthesis method still remains sensitive to the swelling agent behavior toward solubilization and can easily lead to microemulsion formation or phase separation. Another alternative for increasing the pore diameter of ordered mesoporous silica materials went in the use of lab-made di-block copolymer surfactants with larger molar masses. Compared to the pluronic family, these alternative polymeric SDAs have long hydrophobic chains and a high carbon content. Despite these physical properties make them very interesting for the synthesis of large mesoporous materials, their immiscibility with typically used alcohol and water solvents (used in fact for polymer precipitation) imply the use of a co-solvent such as tetrahydrofuran (THF). The ability of the latter to dissolve and hydrogen bond is appropriate for many systems involving high molar mass polymer blocks. First demonstrated by Wiesner et al. [37] amphiphilic diblock copolymers regained in attention recently as illustrated by the works of Deng et al. [38–40], Wei et al. [39] and Bloch et al. [40]. Hydrophobic lab-made copolymers were prepared like polyisoprene-block-polyethylene oxide (PI-b-PEO), polystyrene-block-polyethylene oxide (PS-b-PEO), and polyethylene oxide-block-polymethyl-methacrylate (PEO-b-PMMA) that can expand the mesopore size up to 37 nm. Following these works, a "persistent micelle" templating strategy based on the use of polyethylene oxide-block-polyhexylacrylate (PEO-b-PHA) allowed a continuous access from mesoporous to macroporous silica structures with pore size up to 75 nm [41]. Specifically, isomorphic meso- and macrostructures have been successfully obtained by controlling the SDA molar mass and solution conditions for maintaining persistent micelle sizes during the co-assembly process. This very recent example clearly shows that creative polymer chemistry associated with sol-gel chemistry remain a key for the future development of ordered pore materials in the macroporous domain.

7.2.1.3 Colloidal Templating Approach

Numerous techniques can be found in the literature to generate macroporosity within a solid framework. Such techniques involve gas bubbles (foam), liquid drops (emulsion), or solid particles as template (replica and sacrificial hard templating agent). Herein, we will focus on the properties of polymer spheres as hard templating agents for the generation of inverse porous siliceous replicas. Polymer spheres offer one of the most convenient pathway for the preparation of macrostructured materials with long range pore ordering and macropores spanning size from 50 nm to several micrometers with respect to other techniques [42, 43]. Colloidal crystal templating described here is different from soft templating since the structuring agent possesses a well-defined shape and a rigid structure in the solution synthesis. Typically, the interstitial voids of the polymer spheres network are filled with a silicate precursor having the ability to interact and condensate onto the polymer surface. The "sacrificial" organic template is then decomposed by a thermal activation under air or chemically extracted to obtain a macroporous inverse replica (see Fig. 7.5). Among the different families of polymer spheres, polystyrene (PS), and polymethyl-methacrylate (PMMA) are widely used when applying the colloidal templating approach. The preparation of such polymer spheres with the desired size and morphology is quite straightforward and is grounded on the emulsion polymerization technique with or without the use of surfactant. Specifically, the use of an emulsion for the polymerisation of spherical particles allows to control the droplets size of the monomer and the subsequent final sphere size. It results in nearly monodisperse beads spanning size between 50 nm and 1 μm [44]. Besides, surface charge of the polymeric species (cationic, anionic, or nonionic) can be easily obtained from this method by selecting the appropriate radical initiator. This is of critical importance since charge matching between the silicate precursor and the polymer surface is mandatory for the development of the ordered composite material.

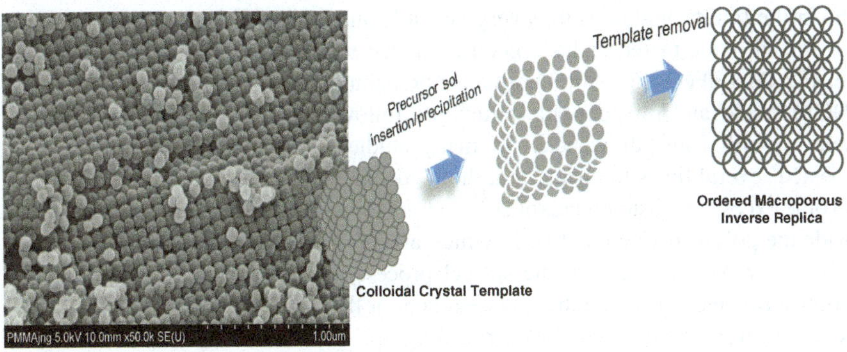

Fig. 7.5 Schematic view of the colloidal crystal templating approach leading to 3D ordered macroporous oxide materials

7.2.1.4 Precursor Infiltration Technique

This synthetic procedure is based on the hard template prearrangement before insertion of the sol precursors within the interstitial spaces of the macrotemplate. Numerous investigations have been done on spheres deposition in order to elaborate photonic crystals [45, 46]. Deposition methods are well documented and are nicely resumed in the following reviews [47–49]. To our knowledge, sedimentation, centrifugation, filtration are the most used techniques to self-assemble spheres into a 3D colloidal crystal having either face centered cubic or hexagonal ordered arrays. In 1997, Velev et al. conducted the first deposition of monodisperse PS spheres on a membrane surface by filtration [4, 50]. Subsequent formation of ordered layers of PS lead to a sphere packing having 10 mm thickness (which can be then penetrated by a cationic surfactant (hexadecyltrimethyl ammonium bromide (HTAB) and a silica solution. When the infiltrated silica solution had gelled inside the interstitial voids of the polymer network, an excess solution was removed from the system. The PS/silica hybrid material was finally dried and calcined at 450 °C in order to form the first 3D Ordered Macroporous (3-DOM) material from polymer spheres imprinting. In 1999, another synthesis route was demonstrated by Holland et al. [5] using an infiltration technique. The synthesis consists of compacting the polymer beads by centrifugation and deposit a layer of a few millimeters on a filter paper in a Buchner funnel via the assistance of a vacuum. Then a dilute solution of metal alkoxide (molecular precursor) in ethanol is deposited drop by drop on the beads. As for Velev and coworkers, the composite formed from the beads coated by the precursor is dried under vacuum to remove excess liquid. Calcination in air at 575 °C allows to obtain a 3D porous network organized with fully exposed macropores interconnected by windows throughout the silica framework. The windows produced between the empty spheres originate from the interpenetrated spheres imprinting. Size of the windows can be tuned by sintering the spheres before infiltration at mild temperature (i.e., 60 °C $< t <$ 100 °C). This method can be applied with molecular precursors, sols, or nanoparticle suspensions [51–53].

The infiltration technique is thus very versatile and a wide variety of 3-DOM oxide and mixed oxide materials have been produced such as TiO_2, Al_2O_3, ZrO_2, Y_2O_3, $Ce_{1-x}Zr_xO_2$, $LaFeO_3$ [54–56]. Despite the preparation of smart and highly organized 3DOM oxide and mixed oxide structures following these synthetic procedures, there are still some drawbacks. In many studies, precursor infiltration must be repeated several times to ensure that the voids are sufficiently filled, often resulting in some excess of waste precursor solution. In some cases, the weak filling fractions inside the polymer sphere network, which arise from the interstitial voids occupied by the solvent necessary for the sol-gel process, and the volume loss due to the partial shrinkage during the thermal treatment or the chemical extraction can restrict the application of such materials for scaling-up.

7.2.1.5 Sedimentation/Aggregation Technique

With respect to the infiltration technique, sedimentation–aggregation is an alternative process to produce macrostructured materials without the need of polymer spheres prearrangement before starting the synthesis [57]. Typically, a colloidal dispersion of polymer spheres is added directly into the silica precursor solution (i.e., TEOS) and left for several hours under stirring for impregnation. For this synthesis, cross-linked polystyrene spheres were allowed to gravity sediment in an ethanolic solution containing the molecular precursor. Classically, sedimentation is a very slow process (several days, depending on the polymer sphere size/density) but in this procedure, sedimentation occurs more rapidly during the evaporation of the supernatant solution. Vaudreuil et al. suggested that TEOS reacted with the residual water in the ethanol solution, giving rise to a layer of silica precursor on the microsphere surface, allowing the composite beads to agglomerate and sedimentate with a higher rate at the bottom of the reactor. However, after calcination, the polymer free silica exhibits spherical but disordered interconnected macropores which possess size smaller than the parent PS bead diameter (30–40% shrinkage) [58]. Nevertheless, the constant wall thickness throughout the sample indicates that deposition of the silica precursor on the beads occurs through a homogeneous wetting process. The process has been further optimized by Ihm and coworkers [59] which replace the evaporation step by a hydrothermal treatment step inducing as well sedimentation-aggregation. A superior 3D ordered macroporous structure has been found over the range of hundreds of micrometers. This method is simpler, faster but still suffers from a lack of ordering in some cases albeit order is not always necessary for catalytic applications.

7.2.2 *Hierarchically Ordered Porous Materials*

For almost two decades now, the development of synthetic hierarchical materials gained more and more interest in the materials community and beyond. Today, it is still a growing field and synthesis routes to hierarchical porous materials are currently updated in literature [60–66]. In these reviews, the combination of templating strategies, i.e., cooperative or evaporation induced self-assembly, hard templates, replica, biotemplates, phase separation, emulsion techniques provide a wide panel of possibilities for the synthesis of novel hierarchical structures. Regarding the tremendous efforts in this area, the dual-templating approach using structure directing agents with the additional presence of colloidal crystals is remaining, to date, the most adequate technique when looking to a high control of structural macroporosity and mesoporosity. Albeit the conception of hierarchical materials by other templating or either template-free routes is highly desirable, they still tend until now to provide less architectural control than the combination of colloidal crystals and structure directing agents. Thus, we decided to give some insight on the development of soft-hard templating chemistry combining surfactant templating chemistry and colloidal crystal templating with a focus on the textural control at each pore scale.

7.2.2.1 Surfactant/Colloidal Templating Approach

With the parallel development of the mesostructuring and macrostructuring routes during the end of the 1990s, it did not take long to assist to the first example of silica with bimodal porosity. Antonietti et al. [67] first demonstrated the possibility to physically include anionic polystyrene spheres (either with sulfate or sulfonate groups) in mesoporous silica. This was performed by adopting a nonionic surfactant route, using an amphiphilic triblock-copolymer template consisting of an apolar polystyrene block (S) and a polar polyethylene oxide block head (E) (SE10/10), in which various amounts of anionic polymer spheres aqueous suspensions were added to the solution synthesis. Based on polarization microscopy, authors claimed that the microspheres must be highly functionalized in order to just act as a filler and avoid any interactions with the micelles during the assembly process. Despite the dispersion of polymer spheres do not alter the soft templating process, randomly packed macropore and mesopore networks were obtained. The same year, hierarchically ordered macroporous-mesoporous silica films were fabricated by the team of Stucky et al. [68]. They smartly combined a top-down approach using poly (dimethyl siloxane) (PDMS) stamp as micromold (~1000 nm) with a bottom-up approach using polystyrene spheres (from bangs laboratory: ~100 nm) and F-127 macromolecular template (~5–10 nm mesopores and presumably ~1 nm micropores) in order to fabricate materials organized on three length scales. Specifically, a PS colloidal suspension prealably filled the micromold channels by capillary effect and was able to self-organize at ambient temperature upon evaporation of the solvent within the confinement of the micromold channel network. Then, the block copolymer-precursor solution imbibed the microsphere voids network via capillary action and silica condensation-precipitation yield an inorganic–organic solid. Finally, after removing the micromold, the hybrid material was calcined at 450 °C in air for 2 h. Albeit limited to thin films dedicated to optoelectronic applications, this pioneering work shed light on the possibility to obtain, with respect to pure mesostructured silica, a cooperative assembly process between the block copolymer micelles and the silicate species in confined spaces of the polystyrene sphere network.

In 2003, Sen et al. applied a bottom-up approach inspired from Stucky and co-workers to realize fully synthetic 3D hierarchically ordered silica materials [69] using pluronic block copolymers (F127 and P123), n-butanol as co-surfactant [70, 71] and non-cross-linked anionic polystyrene spheres. From this nice example, a parametric study of the synthesis focused on the nature and concentration of the surfactant (P123 or F127), co-surfactant (n-butanol and n-pentanol) and macrotemplate removal (either by calcination or toluene extraction/calcination) was undertaken by the same group [72]. First, ordered arrays of polymer spheres packed by centrifugation were investigated by SEM and Hg porosimetry. Using electron microscopy, the emulsifier-free polymerization technique has been confirmed as an efficient method to obtain nearly dispersed and uniform spherical polystyrene particles spanning size between 485 and 805 nm with $\pm$10 nm radial dispersion. Unlike electron microscopy, Hg porosimetry allowed to clearly identify two sets of pore openings that were assigned to octahedral (O) and tetrahedral (T) interstitial sites within the polystyrene monolith (structural units are shown Fig. 7.6).

As a consequence, diffusion of the micellar solution silicate species in the polymer void network is due to attractive interactions between the anionic surface charge of the beads and the cationic silicate species (pH < 2). Authors reported that in the presence of the organic–inorganic material (before calcination), "T" sites are completely filled by the composite whereas around half the void volume of "O" sites is still free. Solvent diffusion and stabilization into confined "O" sites could

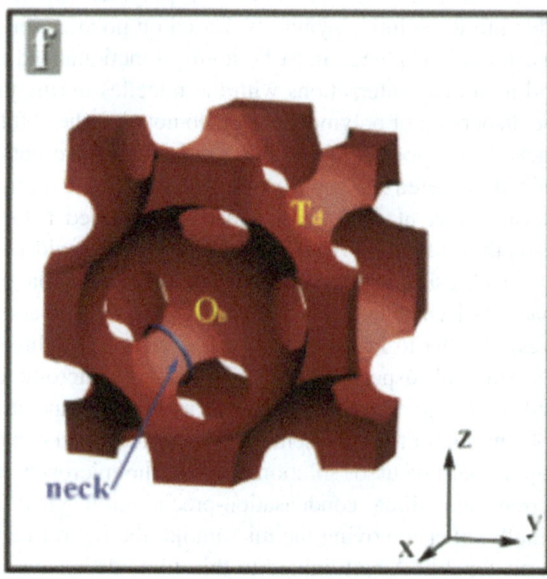

Fig. 7.6 Illustration of the inverse opal structure adopting a face centered cubic lattice. The unit cell is consisting of two different structural units: an *octahedral interstice* (*Oh*) originally surrounded by six polymer *spheres*, and a *tetrahedral interstice* (*Td*) surrounded by four polymer *spheres*. These units are all connected by *necks*. Adapted with permission from Ref. [78]. Copyright (2002) American Chemical Society

explain partial shrinkage and formation of isolated macropores. After polymerization and co-assembly process leading to the mesophase located in these confined spaces, removal of the beads by calcination or either toluene extraction/calcination led to honeycomb macrostructures with different extent of macropores interconnectivity. Besides, amorphous silica regions were observed around the windows, which originate from the condensation of silicate species without micelles (Fig. 7.7a). Toluene-treated samples are found more suitable for improving surface area and windows between adjacent macropores. All samples exhibit a combination of microporous (PEO chains in silica walls) and mesoporous (PPO chains) surface areas but lower values were found compared to their pure mesoporous SBA-15 and SBA-16 counterparts.

As illustrated by TEM, macroporous walls of the hierarchical materials were filled by worm-like or some hexagonal mesopores parallel to the empty sphere surface when using pluronic P123 with respectively n-pentanol and n-butanol (Fig. 7.7a, b). F127 templated materials led to parallel mesopore channels using

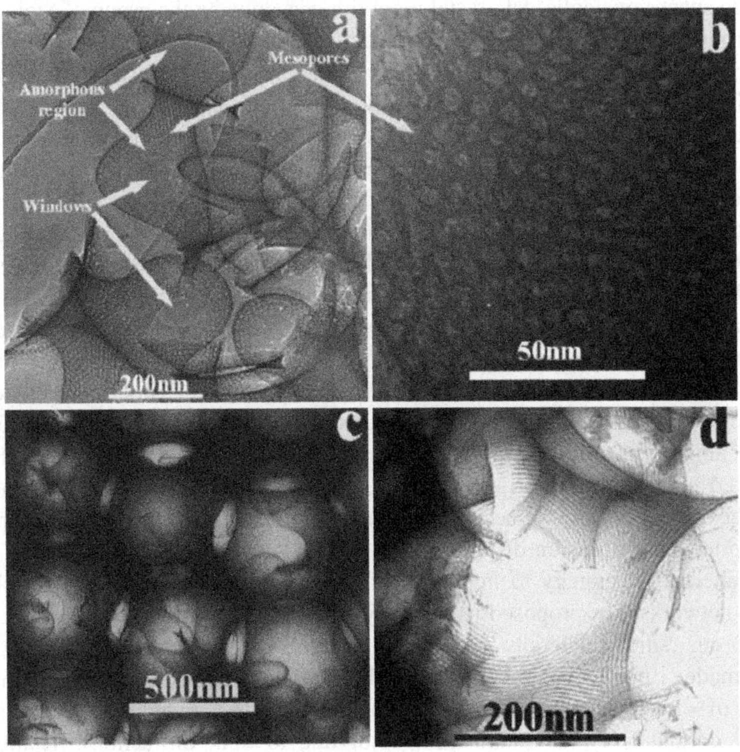

Fig. 7.7 Low and High magnification TEM micrographs of hierarchical macroporous-mesoporous silica using **a**, **b** Tetramethyl orthosilicate (TMOS)-F127-PS precursor solution, and **c**, **d** TMOS-P123-PS precursor solution. Adapted with permission from Ref. [72]. Copyright (2002) American Chemical Society

n-pentanol and hole structure with n-butanol (Fig. 7.7c, d). However, poor mesopore ordering was found for all materials as observed by the low XRD patterns exhibiting a broad 100 X-ray line. Such mesostructures with short range ordering and symmetry are often observed in the literature [44, 73–76]. Nevertheless, by a careful selection of the co-surfactant (i.e., ethanol), Anderson et al. [77] further improved the quality of F-127/PS dual-templated hierarchical porous silica. A cubic pore structure, exhibiting multi-oriented mesopores near the surface of the macropores, was found along with an extent of macropore interconnectivity. All the above studies show that the confinement effects within the voids of the polymer network obviously induce geometric constraints on the organization of mesopores. Second, the surface properties of polymer spheres could modify the phase behavior of the micellar assembly at the vicinity of the curved macropore surface. This is important to take into account when considering an eventual application, since the relative alignment of pores at each length scale is important. In membranes used for separation or in many reactions in heterogeneous catalysis, the relative orientation of mesopores determines their accessibility and plays a critical role on the diffusion of species. In 2007, a first step to understand the orientation of the mesopores in confined spaces of a colloidal crystal monolith was done by the group of Stein et al. [78]. Their first idea was based on the fact that block copolymer (Brij 56) with shorter chains than P123 or F127 would not be affected by the structural confinement inside the polymer sphere network. However, they could not obtain an ordered mesophase in the presence of PMMA while reproducing the same synthesis protocol used for the pure mesoporous Brij 56 templated silica [79]. Such discrepancies between the two solids might result from a slight interaction between Brij 56 and the PMMA surface that hampers the formation of the hexagonal mesophase. This indicates that preferred orientations of planes of the desired mesophase on the PMMA curved surface due to interfacial interactions could result eventually in an unwanted local phase transition. As for the team of Anderson et al., Stein et al. obtain cubic and 2D hexagonal pore arrangement perpendicular to the macropore surface with the help of a co-surfactant (i.e., dodecane) whereas P123-templated material still produces layered mesopores parallel to the macropore surface along with a hole structure in the octahedral voids. Authors claimed that templated materials from Brij-56 are more advantageous for catalytic applications since they could offer more accessible mesopores at the surface of the macropores and shorter inner diffusion distances (Fig. 7.8).

An alternative strategy to improve diffusion of species within the hierarchical porous networks was proposed recently by Deng et al. In this study, macroporous mesoporous silica materials can be prepared with tailorable window sizes. Home-made poly(Styrene-co-3-(TriMethoxySilyl)PropylMethacrylate) microspheres of ~1 μm diameter were first sedimented for a week in order to form a highly ordered colloidal crystal (pearlescent color due to light diffraction). Thanks to capillary forces, the polymer monolith was impregnated by the P123-TEOS-ethanol solution and subsequent evaporation induced self-assembly allowed to form the hybrid mesophase within the polymer network interstices. In a typical synthesis, all the internal voids of the microspheres were fully accessible by the solution since

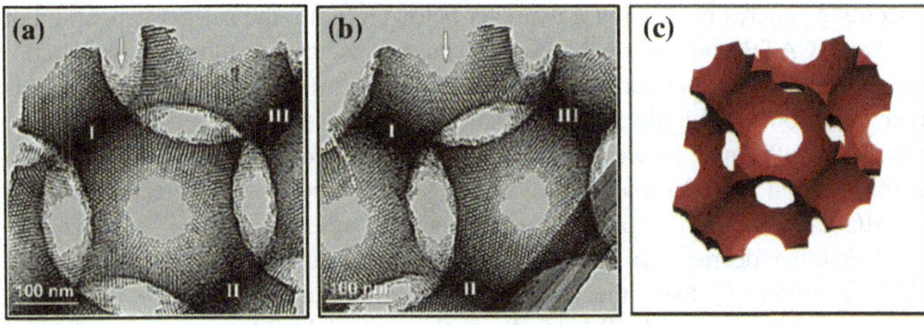

Fig. 7.8 TEM micrographs of 3D ordered macroporous-ordered mesoporous (3DOM/m) silica (2D hexagonal: P6 mm) obtained from Brij 56-$C_{12}H_{26}$-TMOS precursor solution. Adapted with permission from Ref. [78]. Copyright (2002) American Chemical Society

TEOS could swell the colloidal crystal that led to an in situ expansion of the monolith. Then, Deng and coworkers were able to control the extent of opening of the windows upon increasing the HCl amount, or either the prehydrolysis time and sintering temperature during the synthesis. Increase in HCl amount and pre-hydrolysis time parameters improved the hydrolysis ratio of TEOS into more hydrophilic cationic Si species which inhibit the swelling effect of TEOS. Therefore, the internal void volume is less accessible to the TEOS-P123-ethanol solution and windows can be produced from polymer spheres imprinting. Besides, sintering the polymer monolith at 90 °C is leading to the formation of windows that connect each macropore, particularly when a long sintering time is applied. The same phenomenon was earlier observed by Stein et al. [80]. They found that the diameter of the windows can be tuned by sintering the spheres in the colloidal crystal together forming necks that increase in diameter as the length of the thermal treatment is increased. However, a too high sintering temperature (i.e., 100 °C) contributed to the partial collapse of the macrostructure. This latter phenomenon arose from the low glass transition temperature of the poly(St-co-TMSPM) colloidal crystal, so that the sintering treatment should be kept at temperature slightly lower than 100 °C to prevent poly(St-co-TMSPM) microspheres from deformation. Other examples of controlled synthesis of hierarchical porous silica derived from the parent SBA-15 and MCM-41 have been given by several reports using sedimentation–aggregation technique. In 2005, such a route has been proposed by Ihm et al. to obtain a highly ordered macroporous skeleton with a mesoporous MCM-41 hexagonal phase located in the macropore walls [59]. The method is based on the addition of PS spheres (200–350 nm) and sodium silicate in an aqueous solution of CTAB. Sedimentation is occurring during the hydrothermal treatment at 100 °C during 12 h. SEM micrographs clearly show, after calcination of the hybrid solid, an ordered macroporous skeleton with adjacent macropores connected by windows. The same team subsequently studied the impact of the macrosphere size (from 110 to 800 nm) on the obtained morphology of the porous material [81]. They concluded that the bead size is playing a critical role on the formation of a 3D porous macrostructure.

Concerning spheres possessing a diameter below 200 nm, the sol-gel deposition of the silicate-CATB precursor is blocked because of the small volume of voids confined in the polymer network and led to the macrostructure collapse. With beads larger than 600 nm, highly ordered macrostructures were observed together with isolated "donut" rings. However, the direct application of this one-pot synthesis to macrostructured SBA-15 materials, which possesses a better hydrothermal stability than MCM-41, led to a disordered arrangement of macropores [82]. Authors suggested that the decrease in the SBA-15 macrostructure quality with respect to MCM-41 originated from the difference in textural morphology of the parent mesoporous MCM-41 and SBA-15. Recently, a reproducible preparation method allowing the production of tunable macrostructured mesoporous silica derived from SBA-15 has been put into evidence [44]. Authors brought clear insights on the identification of parameters governing structural and textural properties of macrostructured mesoporous silica derived from SBA-15 via the study of the influence of the polymer beads: TEOS ratios (P:TEOS) and aging conditions (time and temperature) that enabled densification/aggregation of the organic–inorganic network. The research of right conditions allow to succeed in guiding the SBA-15 rod-like morphology having randomly packed macropores to a derived SBA-15 having a well-organized honeycomb macrostructure with a low periodicity of the mesostructure (Fig. 7.9). In such dual-templated synthesis route [44, 59], the polymer sphere packing is less ordered with respect to the liquid infiltration technique where polymer spheres are prearranged by centrifugation before insertion of the precursors. Hence, because of the lower degree of confinement between spheres during the sedimentation-aggregation process, the P123 mesophase structure is less perturbed, even at the vicinity of the macropore surface. As illustrated in Fig. 7.9, such phenomenon could favor a perpendicular hole structure rather than the parallel alignment of mesopores and could be highly valuable regarding mass transport and accessibility properties.

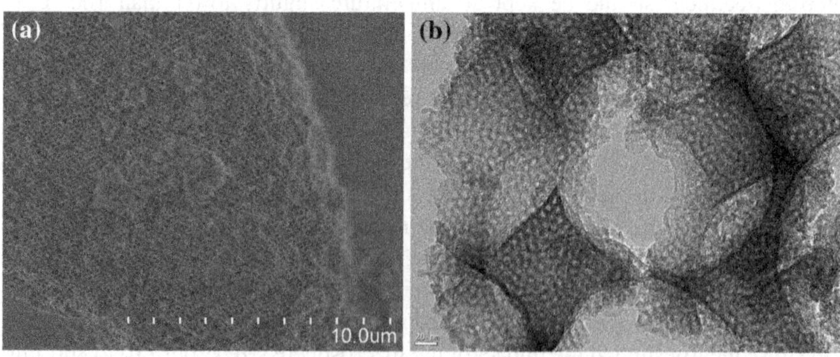

Fig. 7.9 **a** Low magnification and **b** high magnification TEM micrographs of hierarchical macroporous mesoporous silica materials (TEOS-P123-PS precursor solution) obtained by hydrothermal sedimentation-aggregation technique [44]

7.3 Emerging Catalytic Applications of 3D Hierarchically Ordered Porous Silica Materials

Hierarchical materials containing both interconnected macro-/mesoporous structures are expected to present enhanced properties compared to materials with unimodal macropore sizes due to increased mass transport through the material as well as maintaining specific surface areas of high values. Consequently, potential applications of functionalized 3D macroporous mesoporous silica have been the subject of recent investigations that we propose to sum-up with selected catalytic applications.

7.3.1 Heterogeneous Catalysis

A catalyst is generally formed by an active phase supported on a porous support. In this line, silica is one of the most used support (with alumina) in heterogeneous catalysis due to its resistance to corrosion, low cost, and easy availability. Its physical properties (i.e., specific surface area, pore shape/size/interconnectivity/volume) are key parameters to obtain a satisfying dispersion of the active phase. Then, the large development of the mesostructuration and macrostructuration routes to obtain solids with bimodal porosity has rapidly attracted the attention of the catalysis community in various fields of reactions including energy, environment, fine chemistry, and green chemistry. The main role of macropores is to provide fast transport of reactants and products through the material's framework. The smaller pores (mesopores and/or micropores) are giving a high specific surface area to achieve a high dispersion of the active phase and can be as well size selective to the target molecule. Thus, rational design of the multiporous structure can maximize the catalyst performance and further decrease the amount of catalytic phase. In this paragraph, we will discuss the impact of the hierarchical macro-mesostructure on the catalytic performance during oxidation-hydrogenation reactions, biofuel and Fischer–Tropsch syntheses (Fig. 7.10).

7.3.1.1 Oxidation Reactions Dedicated to Chemicals Synthesis and Air Depollution

In 2011, Kamegawa et al. prepared tetrahedrally coordinated Ti-containing 3D ordered macroporous silica with a mesoporous phase (Ti-MMS) and were the first to investigate their catalytic performance in model reactions such as the epoxidation of linear R-olefins of different lengths. For comparison, they used Ti-silica which presents only mesopores (Ti-MS) [83]. A series of Ti-MMS materials have been prepared using organic surfactant and PMMA colloidal crystals as template by a process similar to the team of Stein et al. [5]. The materials present well-defined

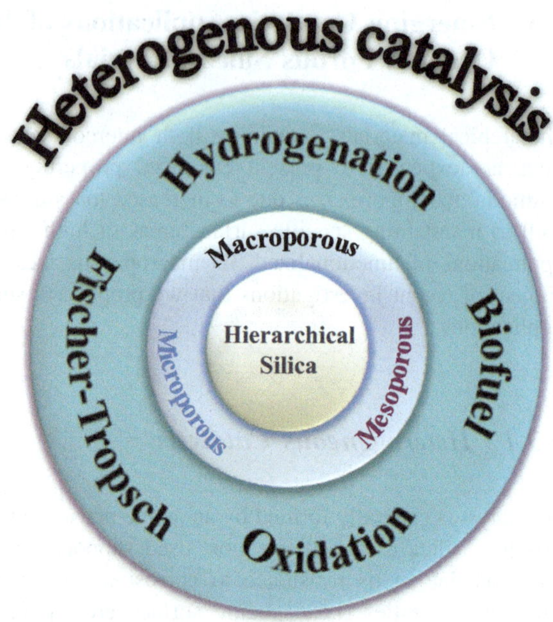

Fig. 7.10 Schematic view of the emerging applications of 3D hierarchical silica materials in heterogeneous catalysis

macroporous structure with high surface area (1053 m^2/g) with approximatively uniform macropores ~300 nm and mesopores smaller than 5 nm with thick silica walls (~70–80 nm). Authors claimed that the addition of the macroporous structure is reducing the mesoporous domains of silica which lead to the formation of many apertures and the shortening of the mesoporous channels. Due to the advantageous macrostructure which allow an efficient diffusion of reactants to the catalytically active sites, the reaction rate is significantly increased in the epoxidation of long chain R-olefins for Ti-hierarchical materials with respect to Ti-MS mesoporous counterparts.

In the recent work of Yanga et al. [84], Au and MnO_x particles are supported respectively on 3D ordered macroporous mesoporous silica (220 $m^2 \cdot g^{-1}$), surfactant-free porous silica (8 $m^2 \cdot g^{-1}$) and fumed (non-porous) silica (200 $m^2 \cdot g^{-1}$) in order to be tentatively compared in the catalytic elimination of a refractory cyclic organic molecules (toluene) known as an important volatile organic compound. First initiated by Haruta and Hutchings [85–88], gold catalysis has become an exciting area of research and gold catalysts represent now one of the best choice for many oxidation reactions. 3D hierarchical silica spanning macropore size of ~200 nm and mesopore size of ~5 nm were prepared following an impregnation/filtration technique using PMMA as macrotemplate and P123 as mesostructuring agent. Then, different amounts of MnO_x and Au nanoparticles were supported on the 3DOM material using respectively incipient wetness preparation and polyvinyl alcohol-protected reduction methods. Over the three series of catalysts prepared, TEM analysis revealed particle sizes deviation comprised between

22.8 and 29.2 nm for MnO_x and 3.6–5.3 nm for Au nanoparticles. After a full comparison of MnO_x and Au physicochemical properties, temperature programmed reactions were performed under toluene (1000 ppm) diluted in a O_2/He mixture. Albeit superior catalytic properties were clearly observed for the 0.93Au/11.2MnO_x/macroporous mesoporous SiO_2 compared to the nonporous and sol-gel counterparts, it has been proven difficult to isolate the hierarchical pore structure impact on the catalytic behavior improvement. Based on the initial rates calculations and characterization results, the best performance of the hierarchical catalyst was attributed to the stronger interaction between Au and MnO_x nanoparticles stabilized throughout the siliceous framework and to a higher oxygen O_{ads} concentration.

In 2015, a clear and smart demonstration of the hierarchical pore structure advantage over single pore structure was given by Parlett et al. who studied the two-step oxidative reaction of cinnamyl alcohol into the stable cinnamic acid product [89]. Efficient selective oxidations of cinnamyl alcohol and cinnamylaldehyde are usually obtained by the use of respectively supported Pd and Pt catalysts. Despite the two reactions worked well separately, it has been proven very difficult to sequentially transform cinnamyl alcohol into cinnamic acid since undesired interactions are occurring between reactants and undesired catalytic function or between the two catalytic sites. Thus, an optimal catalyst design would ensure that cinnamyl alcohol is oxidized over Pd before encountering Pt sites, while the reactively formed cinnamaldehyde would subsequently reach Pt sites for the selective production of cinnamic acid in the second oxidation step. Parlett et al. achieved this goal by a fastidious but smart preparation of 3D ordered macroporous mesoporous bi-functional catalysts, which contain two catalytic functions (i.e., Pt and Pd) individually located in separated pore networks. Hierarchical silica was synthesized using a modified true liquid crystal templating technique (i.e., P123 as surfactant) and employing nonpolar polystyrene microspheres as macrotemplate [90, 91]. Then, following a careful template extraction and functionalization steps of the initial P123-PS-silica composite, Pt and Pd noble metal species were able to be dispersed respectively into the mesostructure and the macrostructure of the siliceous framework. Successful achievement of the spatially separated Pd and Pt was revealed by HAADF-STEM and EDX analyses on hierarchical support presenting ~400 nm macropores filled by a muti-oriented mesoporous phase (~3.5 nm). First, conversion reaction profiles for cinnamyl alcohol clearly indicate that the porous architectures of the support fitted well with the localization of the particles in the macrostructure and/or mesostructure. Then, direct evidence of the superior catalytic properties of the spatially orthogonal Pd-Pt hierarchical catalyst was highlighted by increased turn-over frequencies for the formation of aldehyde and acid products. Not only that the spatial control of the two metal active sites (Pd in maropores and Pt in mesopores) improved the activity and selectivity but also diffusion properties of this architecture boosted the one-pot oxidation of the alcohol into the corresponding acid.

7.3.1.2 Biofuel Synthesis

Biodiesel Synthesis

In a post-petroleum period, catalysis will be a key science to overcome the engineering and scientific barriers to economically feasible routes to biofuels. Non-corrosive heterogeneous catalysts that are easily recovered and recyclable could greatly improve the environmental impact of commercial biofuel processing, eliminating the need for aqueous quench cycles usually required for homogeneous catalysts [92, 93]. The work of Dhainaut et al. [90] highlights some of the recent development in both transesterification and esterification reactions through utilizing tailored porous solids as high surface area supports to enhance triglycerides and free fatty acids diffusion to the active acid or basic groups [94–96]. In this line, the macroporosity effect inside SBA-15 materials was reported by Dhainaut et al. toward model esterification and transesterification reactions. A series of macroporous–mesoporous materials derived from SBA-15 (MM-SBA15-X where X is equal to the PS/TEOS ratio) were prepared by incorporation of different amounts of polystyrene beads into the P123-TEOS-HCl solution. The bare SBA-15 and derived macro-mesoporous SBA-15 materials maintain all high surface areas and uniform mesoporosity. Upon macropores incorporation, a disordered interconnected macroporous skeleton is being formed and a progressive decrease in the average mesopore domain size is observed by HRTEM. Subsequent functionalisation with propylsulfonic acid yields the first organized, macro-mesoporous solid acid catalyst. The resulted solid acid catalysts were tested in palmitic acid esterification and glyceryl trioctanoate transesterification reactions. These new materials show enhanced conversions and TOFs for triglyceride transesterification and fatty acid esterification over sulfonic acid parent SBA-15. This improved catalytic behavior could arise from two plausible mechanisms. The bulk diffusion of species is improved by the presence of larger pores which act as highways and/or by decreasing the mesopore domain size upon macropores incorporation, sulfonic acid sites are becoming more accessible at the entrance of the macropores.

Fischer-Tropsch Synthesis

The Fischer–Tropsch synthesis reaction has been investigated to convert CO_2 into sulfur-free clean fuels using Co-doped heterogeneous catalysts [97]. Very recently, Bae et al. studied the effect of spherical macro-mesoporous Co/SiO_2 structure on the hydrogenation of CO to hydrocarbons [98]. The macro-mesoporous spherical silica supports (MMS) were obtained following a dual-templating method using an evaporation-driven self-assembly of 200 nm PS beads, P123 surfactant, and silica precursor inside emulsion droplets [99]. A successful bimodal pore structure on the MMS was attributed to the simultaneous formation of mesopores (~4 nm) from the surfactants microemulsion and the macropores spanning size between 20 and 100 nm from the polymer spheres imprinting. Additional preparation of mesoporous SBA-15 and SiO_2 (silica gel) as references were made to be further compared during the polymerization reaction. 20 wt.% Cobalt and 0.1 wt.% Pt species were added on all supports by a stepwise incipient wetness impregnation method. Authors observed that the particle size of cobalt increases when macroporosity is present in the siliceous support. Materials catalytic activity and stability were then investigated during CO hydrogenation. Co/MMS (405 $m^2 \cdot g^{-1}$) and Co/SBA-15 (670 $m^2 \cdot g^{-1}$), possessed similar conversion of CO and outperformed the Co/SiO_2 analog. Complementary stability tests suggested that the deactivation of Co/SBA-15 was faster with respect to the hierarchical catalyst. Authors attributed the enhanced catalytic activity and stability of the Co/MMS catalysts to its large confining capacity for hydrocarbons which can be easily removed from macropores during regeneration by feeding of liquid octane during the Fischer–Tropsch synthesis reaction.

An alternative and greener route to hierarchical silica with meso-macropores was prepared using chitosan as biotemplate by the group of Witoon and applied to CO hydrogenation [100]. The support is characterized by the presence of mesopores with 9 nm diameter and macropores which are in the range of 100–150 nm. From these supports, cobalt catalysts were prepared using the classical incipient wetness impregnation method. As reported above [98], the cobalt particle size increases when supports with larger pores are used and the sizes of the particles vary from ~9 to 21 nm. During the Fischer–Tropsch process, the pore size of the support and the cobalt particle size strongly impact on the catalytic performance. Specifically, the selectivity to C_{5+} is affected by the Co crystallite size. Authors claimed that the large pore diameter affects the CO conversion for small and large cobalt particles and this is due to the better transport of reactants and products.

7.4 Conclusion and Future Challenges

The present manuscript highlights the important progress made towards the preparation of siliceous architectures following structure-directed methodologies based on sol-gel chemistry. More specifically, the dual-templating method associating amphiphilic molecules and polymer spheres is a powerful method to shape the textural/morphological properties of the siliceous materials at the meso- and macroscale. Further nano-engineering of the surfactant and polymer structure, the study of their interaction with the silica precursor (with the help of theoretical simulations) would be crucial for the rational development of highly controlled hierarchical porous architectures. For instance, the independent control of the mesopore/micropore size in hierarchical materials is of high importance in heterogeneous catalysis and in other domains such as separation/adsorption but still remains a big challenge. The future development of commercial catalytic processes will require a better understanding of how each reactant can diffuse in the material's particle and interact with the active phase, particularly when dealing with bulky and viscous molecules. Nevertheless, while the generation of smart 3D macrostructures with a mesoporous phase can be obtained following various synthetic procedures presented here, there is still restricted applications due to limitation in scaling-up, especially due to the use of monodispersed polymer beads. In this line, eco-efficient conception of hierarchical materials is one of the future directions to be taken. Because the extraction of natural biopolymers like cellulose or chitosan is already being conducted on an industrial scale, there is a real opportunity for exploring the use of such macromolecules in materials science, as these biopolymers can directly replace synthetic petrosourced polymers used for the synthesis of inorganic materials. Expected progress in the elaboration of bio-templated hierarchical materials can open new perspectives for siliceous but also non-siliceous material synthesis, with applications opened to other relevant catalytic applications. Indeed, such hierarchical solids will easily find application in soot oxidation, hydrodesulphurization for automotive depollution, catalytic reforming and in many different fields than heterogeneous catalysis (Piezoelectric/Ferroelectric materials, CO_2 capture, Drug delivery, SOFC).

References

1. Oliver S, Kuperman A, Coombs N, Lough A, Ozin GA (1995) Lamellar aluminophosphates with surface patterns that mimic diatom and radiolarian microskeletons. Nature 378:47–50
2. Yanagisawa T, Shimizu T, Kuroda K, Kato C (1990) The preparation of alkyltrimethylammonium-kanemite complexes and their conversion to microporous materials. Bull Chem Soc Jpn 63:988–992
3. Beck JS, Vartuli JC, Roth WJ, Leonowicz ME, Kresge CT, Schmitt KDC, Chu TW, Olson DH, Sheppard EW, McCullen SB, Higgins JB, Schlenke JL (1992) A new family of mesoporous molecular sieves prepared with liquid crystal templates. J Am Chem Soc 114–27:10834–10843
4. Velev OD, Jede TA, Lobo RF, Lenhoff AM (1997) Porous silica via colloidal crystallization. Nature 389:447–448
5. Holland BT, Blanford CF, Do T, Stein A (1999) Synthesis of highly ordered, three-dimensional, macroporous structures of amorphous or crystalline inorganic oxides, phosphates, and hybrid composites. Chem Mater 11–3:795–805
6. Dickey FH (1949) The preparation of specific adsorbents. Proc Natl Acad Sci 35–5:227–229
7. Monnier A, Schüth F, Huo Q, Kumar D, Margolese D, R. Maxwell S, Stucky GD, Krishnamurty M, Petroff P, Firouzi A, Janicke M, Chmelka BF (1993) Cooperative formation of inorganic-organic interfaces in the synthesis of silicate mesostructures. Sci New Ser 261–5126:1299–1303
8. Huo Q, Margolese DI, Ciesla U, Feng P, Gier TE, Seger P, Leon R, Petroff PM, Shüth F, Sucky GD (1994) Generalized synthesis of periodic surfactant/inorganic composite materials. Nature 386:317–321
9. Bagshaw SA, Prouzet E, Pinnavaia TJ (1995) Templating of mesoporous molecular sieves by nonionic polyethylene oxide surfactants. Science 269:1242–1244
10. Tanev PT, Pinnavaia TJ (1996) Mesoporous silica molecular sieves prepared by ionic and neutral surfactant templating: a comparison of physical properties. Chem Mater 8–8: 2068–2079
11. Zhao D, Feng J, Huo Q, Melosh N, Fredrickson GH, Chmelka BF, Stucky GD (1998) Triblock copolymer syntheses of mesoporous silica with periodic 50 to 300 angstrom pores. Science 279:548–552

12. Wan Y, Zhao D (2007) On the controllable soft-templating approach to mesoporous silicates. Chem Rev 107:2821–2860
13. Kleitz F, Choi SH, Ryoo R (2003) Cubic Ia3d large mesoporous silica: synthesis and replication to platinum nanowires, carbon nanorods and carbon nanotubes. Chem Commun 2136–2137
14. Atluri R, Hedin N, Garcia-Bennett AE (2009) Nonsurfactant supramolecular synthesis of ordered mesoporous silica. J Am Chem Soc 131–9:3189–3191
15. Benamor T, Vidal L, Lebeau B, Marichal C (2012) Influence of synthesis parameters on the physico-chemical characteristics of SBA-15 type ordered mesoporous silica. Microporous Mesoporous Mater 153:100–114
16. Galarneau A, Cambon H, Di Renzo F, Ryoo R, Choi M, Fajula F (2003) Microporosity and connections between pores in SBA-15 mesostructured silicas as a function of the temperature of synthesis. New J Chem 27:73–79
17. Choi M, Heo W, Kleitz F, Ryoo R (2003) Facile synthesis of high quality mesoporous SBA-15 with enhanced control of the porous network connectivity and wall thickness. Electronic Chem. Commun. 12:1340–1341
18. H Lee, Kim JH, Stucky GD, Shi Y, Pak C, Kim JM (2010) Morphology-selective synthesis of mesoporous SBA-15 particles over micrometer, submicrometer and nanometer scales. J Mater Chem 20:8483–8487
19. Zhao D, Huo Q, Feng J, Chmelka BF, Stucky GD (1998) Nonionic triblock and star diblock copolymer and oligomeric surfactant syntheses of highly ordered, hydrothermally stable, mesoporous silica structures. J Am Chem Soc 120:6024–6036
20. Ravikovitch PI, Neimark AV (2001) Characterization of micro- and mesoporosity in SBA-15 materials from adsorption data by the NLDFT method. J Phys Chem B 105: 6817–6823
21. Fan J, Yu C, Wang L, Tu B, Zhao D, Sakamoto Y, Terasaki O (2001) Mesotunnels on the silica wall of ordered SBA-15 to generate three-dimensional large-pore mesoporous networks. J Am Chem Soc 123:12113–12114
22. Liu J, Zhang X, Han Y, Xiao FS (2002) Direct observation of nanorange ordered microporosity within mesoporous molecular sieves. Chem Mater 14:2536–2540
23. Fulvio PF, Pikus S, Jaroniec M (2005) Tailoring properties of SBA-15 materials by controlling conditions of hydrothermal synthesis. J Mater Chem 15:5049–5053
24. Kim TW, Ryoo R, Kruk M, Gierszal KP, Jaroniec M, Kamiya S, Terasaki O (2004) Tailoring the pore structure of SBA-16 silica molecular sieve through the use of copolymer blends and control of synthesis temperature and time. J Phys Chem B 108:11480–11489
25. Kruk M, Jaronie M (1999) A unified interpretation of high-temperature pore size expansion processes in MCM-41 mesoporous silicas. J Phys Chem B 103:4590–4598
26. Kresge CT, Leonowicz ME, Roth WJ, Vartuli JC, Beck JS (1992) Ordered mesoporous molecular sieves synthesized by a liquid-crystal template mechanism. Nature 359:710–712
27. Schmidt-Winkel P, Lukens Jr WW, Zhao D, Yang P, Chmelka BF, Stucky GD (1999) Mesocellular siliceous foams with uniformly sized cells and windows. J Am Chem Soc 121:254–255
28. Johansson EM, Córdoba JM, Odénet M (2009) Synthesis and characterization of large mesoporous silica SBA-15 sheets with ordered accessible 18 nm pores. Mater Lett 63: 2129–2131
29. Schmidt-Winkel P, Glinka CJ, Stucky GD (2000) Microemulsion templates for mesoporous silica. Langmuir 16:356–361
30. Boissière C, Martines MAU, Tokumoto M, Larbot A, Prouzet E (2003) Mechanisms of pore size control in MSU-X mesoporous silica. Chem Mater 15:509–515
31. Sun J, Zhang H, Ma D, Chen Y, Bao X, Klein-Hoffmann A, Pfänderb N, Su DS (2005) Alkanes-assisted low temperature formation of highly ordered SBA-15 with large cylindrical mesopores. Chem Commun 2343–2345

32. Zhang H, Sun J, Ma D, Weinberg G, Su DS, Bao X (2006) Engineered complex emulsion system toward modulating the pore length and morphological architecture of mesoporous silicas. J Phys Chem B 110:25908–25915
33. Kruk M, Cao L (2007) Pore size tailoring in large-pore SBA-15 silica synthesized in the presence of hexane. Langmuir 23:7247–7254
34. Yi J, Kruk M (2015) Pluronic-P123-templated synthesis of silica with cubic Ia3d structure in the presence of micelle swelling agent. Langmuir 31:7623–7632
35. Van Grieken R, Martínez F, Morales G, Martín A (2013) Nafion-modified large-pore silicas for the catalytic acylation of anisole with acetic anhydride. Ind Eng Chem Res 52: 10145–10151
36. Zubrzycki R, Ressler T (2015) Influence of pore size of SBA-15 on activity and selectivity of H3 [PMo12O40] supported on tailored SBA-15. Microporous Mesoporous Mater 214: 8–14
37. Templin M, Franck A, Du Chesne A, Leist H, Zhang Y, Ulrich R, Schadler V, Wiesner U (1997) Organically modified aluminosilicate mesostructures from block copolymer phases. Science 278:1795–1798
38. Deng Y, Yu T, Wan Y, Shi Y, Meng Y, Gu D, Zhang L, Huang Y, Liu C, Wu X, Zhao D (2007) Ordered mesoporous silicas and carbons with large accessible pores templated from amphiphilic diblock copolymer poly(ethylene oxide)b–polystyrene. J Am Chem Soc 129:1690
39. Wei J, Wang H, Deng Y, Sun Z, Shi L, Tu B, Luqman M, Zhao D (2011) Solvent evaporation induced aggregating assembly approach to three-dimensional ordered mesoporous silica with ultralarge accessible mesopores. J Am Chem Soc 133:20369–20377
40. Bloch E, Llewellyn PL, Phan T, Bertin D, Hornebec V (2009) On defining a simple empirical relationship to predict the pore size of mesoporous silicas prepared from PEO-b-PS diblock copolymers. Chem Mater 21:48–55
41. Lokupitiya HN, Jones A, Reid B, Guldin S, Stefik M (2016) Ordered mesoporous to macroporous oxides with tunable isomorphic architectures: solution criteria for persistent micelle templates. Chem Mater 28:1653–1667
42. Studart AR, Gonzenbach UT, Tervoort E, Gauckler LJ (2006) Processing routes to macroporous ceramics: a review. J Am Ceram Soc 89:1771–1789
43. Ohji T, Fukushima M (2012) Macro-porous ceramics: processing and properties. Int Mater Rev 57:115–151
44. Zaki A, Xu J, Stoclet G, Casale S, Dacquin JP, Granger P (2015) Tunable hierarchical porous silica materials using hydrothermal sedimentation-aggregation technique 208: 140–151
45. Ozin GA, Yang SM (2001) The race for the photonic chip: colloidal crystal assembly in silicon wafers. Adv Funct Mater 11–2:95–104
46. Joannopoulos JD, Villeneuve PR, Fan S (1997) Photonic crystals: putting a new twist on light. Nature 386:143–149
47. Marlow F, Muldarisnur Sharifi P, Brinkmann R, Mendive C (2009) Opals: status and prospects. Angew Chem Int Ed 48:6212–6233
48. Cong H, Yu B, Tang J, Li Z, Liu X (2013) Current status and future developments in preparation and application of colloidal crystals. Chem Soc Rev 42:7774–7800
49. Stein A (2001) Sphere templating methods for periodic porous solids. Microporous Mesoporous Mater 44–45:227–239
50. Velev OD, Jede TA, Lobo RF, Lenhoff AM (1998) Microstructured porous silica obtained via colloidal crystal templates. Chem Mater 10:3597–3602
51. Abdullah M, Iskandar F, Shibamoto S, Ogi T, Okuyama K (2004) Preparation of oxide particles with ordered macropores by colloidal templating and spray pyrolysis. Acta Mater 52:5151–5156
52. Cho YS, Yi GR, Chung YS, Park SB, Yang SM (2007) Complex colloidal microclusters from aerosol droplets. Langmuir 23:12079–12085

53. Cao Y, Wang Y, Zhu Y, Chen H, Li Z, Ding J, Chi Y (2006) Fabrication of anatase titania inverse opal films using polystyrene templates. Superlattices Microstruct 40:155–160
54. Wijnhoven JEGJ, Vos WL (1998) Preparation of photonic crystals made of air spheres in titania. Science 80:802–804
55. Sadakane M, Ueda W (2016) Three dimensionally ordered macroporous (3DOM) perovskites mixed metal oxides. Perovskites and related mixed oxides: concepts and applications, vol 1. Wiley, pp 113–138
56. Stein A, Rchroden RC (2001) Colloidal crystal templating of three-dimensionally ordered macroporous solids: materials for photonics and beyond. Curr Opin Solid State Mater Sci 5:553–564
57. Vaudreuil S, Bousmina M, Kaliaguine S, Bonneviot L (2001) Synthesis of macrostructured silica by sedimentation-aggregation. Adv Mater 13:1310–1312
58. Vaudreuil S, Bousmina M, Kaliaguine S, Bonneviot L (2001) Preparation of macrostructured metal oxides by sedimentation-agregation. Microporous Mesoporous Mater 44–45:249–258
59. Oh CG, Baek YY, Ihm SK (2005) Synthesis of skeletal-structured biporous silicate powders through microcoloidal crystal templating. Adv Mater 17–3:270–273
60. Su BL, Leonard A, Yuan ZY (2005) Highly ordered mesoporous CMI-n materials and hierarchically structured meso–macroporous compositions. CR Chim 8:713–726
61. Yuan ZY, Su BL (2006) Insights into hierarchically meso–macroporous structured materials. J Mater Chem 16:663–677
62. Yang XY, Li Y, Lemaire A, Yu JG, Su BL (2009) Hierarchically structured functional materials: Synthesis strategies for multimodal porous networks. Pure Appl Chem 81: 2265–2307
63. Boissiere C, Grosso D, Chaumonnot A, Nicole L, Sanchez C (2010) Aerosol route to functional nanostructured inorganic and hybrid porous materials. Adv Mater 23:599–623
64. Yang XY, Leonard A, Lemaire A, Tian G, Su BL (2011) Self-formation phenomenon to hierarchically structured porous materials: design, synthesis, formation mechanism and applications. Chem Commun 47:2763–2786
65. Petkovitch ND, Stein A (2013) Controlling macro- and mesostructures with hierarchical porosity through combined hard and soft templating. Chem Soc Rev 42:3271–3739
66. Colombo P, Vakifahmetoglu C, Costacurta S (2010) Fabrication of ceramic components with hierarchical porosity. J Mater Sci 45:5425–5455
67. Göltner CG, Henke S, Weissenberger MC, Antonietti M (1998) Mesoporous silica from lyotropic liquid crystal polymer templates. Angew Chem Int Ed 37–5:613–614
68. Yang P, Deng T, Zhao D, Feng P, Pine D, Chmelka BF, Whitesides GM, Stucky GD (1998) Hierarchically ordered oxides. Science 282:2244–2246
69. Sen T, Tiddy GJT, Casci JL, Anderson MW (2003) One-pot synthesis of hierarchically ordered porous-silica materials with three orders of length scale. Angew Chem Int Ed 42:4649–4653
70. Subramanian G, Manoharan VN, Thorne JD, Pine DJ (1999) Ordered macroporous materials by colloidal assembly: a possible route to photonic bandgap materials. Adv Mater 11–15:1261–1265
71. Feng P, Bu X, Pine DJ (2000) Control of pore sizes in mesoporous silica templated by liquid crystals in block copolymer-cosurfactant-water systems. Langmuir 16:5304–5310
72. Sen T, Tiddy GJT, Casci JL, Anderson MW (2004) Synthesis and characterization of hierarchically ordered porous silica materials. Chem Mater 16:2044–2054
73. Lebeau B, Fowler CE, Mann S, Farcet C, Charleux B, Sanchez C (2000) Synthesis of hierarchically ordered dye-functionalised mesoporous silica with macroporous architecture by dual templating 10:2105–2108
74. Yang Z, Qi K, Rong J, Wang L, Liu Z, Yang Y (2001) Template synthesis of 3-D bi-modal ordered porous silica. Chin Sci Bull 46:1785–1789

75. Wang Z, Stein A (2008) Morphology control of carbon, silica, and carbon/silica nanocomposites: from 3D ordered macro-/mesoporous monoliths to shaped mesoporous particles. Chem Mater 20:1029–1040
76. WeiJ Wang H, Deng Y, Sun Z, Shi L, Tu B, Luqman M, Zhao D (2011) Solvent evaporation induced aggregating assembly approach to three-dimensional ordered mesoporous silica with ultralarge accessible mesopores. J Am Chem Soc 133:20369–20377
77. Loiola AR, Da Silva LRD, Cubillas P, Anderson MW (2008) Synthesis and characterization of hierarchical porous materials incorporating a cubic mesoporous phase. J Mater Chem 18:4985–4993
78. Li F, Wang Z, Ergang NS, Fyfe CA, Stein A (2007) Controlling the shape and alignment of mesopores by confinement in colloidal crystals: designer pathways to silica monoliths with hierarchical porosity. Langmuir 23:3996–4004
79. El-Safty SA, Hanaoka T (2003) Monolithic nanostructured silicate family templated by lyotropic liquid-crystalline nonionic surfactant mesophases. Chem Mater 15–15:2892–2902
80. Holland BT, Blanford CF, Do T, Stein A (1999) Synthesis of highly ordered, three-dimensional, macroporous structures of amorphous or crystalline inorganic oxides, phosphates, and hybrid composites. Chem Mater 11:795–805
81. Yun JS, Seong MY, Ihm SK (2008) The synthesis of MCM-41 with different macropore morphologies: Residual volume- and skeletal-structure. J Phys Chem Solids 69:1129–1132
82. Yun JS, Ihm S (2008) Synthesis of mesoporous SBA-15 having macropores by dual-templating method. J Phys Chem Solids 69:1133–1135
83. Kamegawa T, Suzuki N, Che M, Yamashi H (2011) Synthesis and unique catalytic performance of single-site Ti-containing hierarchical macroporous silica with mesoporous frameworks. Langmuir 27:2873–2879
84. Yang H, Denga J, Xie S, Jiang Y, Dai H, Au CT (2015) Au/MnOx/3DOM SiO2: highly active catalysts for toluene oxidation. Appl Catal A 507:139–148
85. Haruta M (2005) Catalysis: Gold rush. Nature 437:1098–1099
86. Hutchings GJ (2005) Catalysis by gold. Catal Today 100:55–61
87. Haruta M, Kobayashi T, Sano H, Yamada N (1987) Novel Gold Catalysts for the Oxidation of Carbon Monoxide at a Temperature far Below 0°C. Chem Lett 16:405–408
88. Hutchings GJ (1985) Vapour phase hydrochlorination of acetylene: correlation of catalytic activity of supported metal chloride catalysts. J Catal 96:292–295
89. Parlett CMA, Isaacs MA, Beaumont SK, Bingham LM, Hondow NS, Wilson K, Lee AF (2015) Spatially orthogonal chemical functionalization of a hierarchical pore network for catalytic cascade reactions. Nat Mater Lett 1–6
90. Dhainaut J, Dacquin JP, Lee AF, Wilson K (2010) Hierarchical macroporous-mesoporous SBA-15 sulfonic acid catalysts for biodiesel synthesis. Green Chem 12:296–303
91. Wainwright SG, Parlett CMA, Blackley RA, Zhouc W, Lee AF, Wilson K, Bruce DW (2013) True liquid crystal templating of SBA-15 with reduced microporosity. Microporous Mesoporous Mater 172:112–117
92. Narasimharao K, Lee AF, Wilson K (2007) Catalysts in production of biodiesel: a review. J Bio Mater Bioenergy 1:19
93. Sivasamy A, Cheah KY, Fornasiero P, Kemausuor F, Zinovievand S, Miertus S (2009) Catalytic applications in the production of biodiesel from vegetable oils. ChemSusChem 2:278–300
94. Shah P, Ramaswamy AV, Lazar K, Ramaswamy V (2004) Synthesis and characterization of tin oxide-modified mesoporous SBA-15 molecular sieves and catalytic activity in transesterification reaction. Appl Catal A 1–2:239–248
95. Dacquin JP, Lee AF, Pirez C, Wilson K (2012) Pore-expanded SBA-15 sulfonic acid silicas for biodiesel synthesis. Chem Commun 48:212–214
96. Pirez C, Caderon JM, Dacquin JP, Lee AF, Wilson K (2012) Tunable KIT-6 mesoporous sulfonic acid catalysts for fatty acid esterification. ACS Catal 2–8:1607–1614

97. Khodakov AY, Chu W, Fongarland P (2007) Advances in the development of novel cobalt Fischer–Tropsch catalysts for synthesis of long-chain hydrocarbons and clean fuels. Chem Rev 107:1692–1744
98. Koo HM, Phu TT, Yi GR, Shin CH, Chung CH Bae JW (2016) Effect of the ordered meso–macroporous structure of Co/SiO2 on the enhanced activity of hydrogenation of CO to hydrocarbons. Catal Sci Technol 6:4221–4231
99. Cho YS, Kim YK, Kim SH, Lim DC, Lee JG, Baek YK, Yi GR (2014) Spherical meso macroporous silica particles by emulsion-assisted dual-templating. Mater Express 4(2): 91–104
100. Witoon T, Chareonpanich M (2012) Synthesis of hierarchical meso-macroporous silica monolith using chitosan as biotemplate and its application as polyethyleneimine support for CO_2 capture. Mater Lett 81:181–184
101. Soler-Illiaa GJ de AA, Sanchez C, Lebeau B, Patarin J (2002) Chemical strategies to design textured materials: from microporous and mesoporous oxides to nanonetworks and hierarchical structures. 102:4093–4138
102. Sakamoto Y, Diaz I, Terasaki O, Zhao D, Perez-Pariente J, Kim JM, Stucky GD (2002) Three-dimensional cubic mesoporous structures of SBA-12 and related materials by electron crystallography. J Phys Chem B 106:3118–3123
103. Kleitz F, Bérubé F, Guillet-Nicolas R, Yang CM, Thommes M (2010) Probing adsorption, pore condensation, and hysteresis behavior of pure fluids in three-dimensional cubic mesoporous KIT-6 silica. J Phys Chem C 114:9344–9355
104. Xin C, Zhao N, Zhan H, Xiao F, Wei W, Sun Y (2014) Phase transition of silica in the TMB-P123-H_2O-TEOS quadrucomponent system: a feasible route to different mesostructured materials. J Colloid Interface Sci 433:176–182

Chapter 8
Porous Silicon: From Optical Sensor to Drug Delivery System

Chiara Piotto and Paolo Bettotti

Nowadays sensors are everywhere around us. We are monitoring nearly all aspects of our daily life: from health to security, from domotic to environmental aspects. The availability of smaller, more efficient and intelligent systems is highly valued to increase the quality and the accuracy of such analysis and of our well-being.

Applications of particular interest are those related with healthcare: both sensing devices for analytic and diagnostic purposes, as well as targeted drug delivery are active research fields with possible important advancements in our dailylife.

Among all the possible types of sensor, optoelectronic ones are a broad class of devices that exploit the variation (or sometime the generation) of an electromagnetic field upon interaction with a functional layer that "*captures*" the analytes.

Several different technologies have been proposed. They differ in both the read-out scheme and the type of interaction between the molecule to be detected and the functional interfaces: from fiber optics [1] to integrated photonics [2–5]; from Surface Plasmon Resonance [6–8] to fluorescence [9–11] and electrochemistry [12, 13]. Despite the large differences between the basic mechanisms of all these technologies, a key and common element that largely determines the final properties of the sensor is the effectiveness and the specificity of the recognition done by the interfacial layer between the transducer and the analyte. In fact, the Limit of Detection (LoD), that is the minimum amount of analyte detectable, is determined not only by how much sensitive the transduction mechanism is (i.e. the generation of a measurable signal stimulated by the analyte binding event) but also by the specificity of the recognition mechanism. Signals generated by non-specific absorption (i.e. by molecules different from the target ones) create a background that often hinders the specific one and promotes the risk of false positive measurements.

C. Piotto · P. Bettotti (✉)
Nanoscience Laboratory, Department of Physics, University of Trento, Trento, Italy
e-mail: paolo.bettotti@unitn.it

P. Bettotti (ed.), *Submicron Porous Materials*,
DOI 10.1007/978-3-319-53035-2_8

Many of the aforementioned technologies have demonstrated single molecule detection limit [14–20] in laboratory environment but, still, they show limited applications with real samples. Most often, this is due to a limited specificity of the binding layer and to an important contribution from a non-specific signal generated from adsorption/binding of unwanted species present in the sample.

Porous silicon (PSi) is a versatile material that, despite being known since several decades, is still matter of lively research and new technological applications are constantly demonstrated. The richness of the research it generates is due to the great flexibility in the porous nanostructures that can be realized exploiting a simple electrochemical corrosion process.

Since 90s PSi gained much attention after the demonstration of its efficient light emission once it is reduced down to nanosize. Later, PSi has been studied to fabricate optical filters and, more recently, the optical response of such photonic structures has been exploited to create highly sensitive sensors. Even more recently, its porous structure and the fact that PSi is considered a biocompatible and biodegradable material, has motivated its use for drug delivery (DD) applications.

PSi enables the development of smart DD systems that exploit both its optical and structural properties. In particular, complex behaviour and release mechanisms can be envisioned by matching the opto-electronic properties of PSi with its rich surface chemistry and large specific area obtainable. Thus, the control over the response of the PSi structure to external stimuli is strictly matched with the possibility to obtain complex and on-demand drug release mechanisms.

The first part of the chapter reports about the design and the fabrication of PSi optical sensors, while the second part describes the fundamental key points to develop a nanoporous DD system and describe some of the recent results obtained.

8.1 Porous Silicon

8.1.1 Fabrication of Porous Silicon

PSi is fabricated starting from crystalline silicon through an electrochemical corrosion process in hydrofluoric acid solutions. PSi has been discovered in '57 at Bell Labs [25] and, for decades, it was considered a non-stochiometric oxide obtained as a side-results of corrosion processes in electronic industry. PSi was then poorly investigated and considered only as sacrificial material up to '90 when Canham discovered that bright visible photoluminescence is obtained from it under UV illumination [26]. Since then, and upon the development of applied photonics, PSi was heavily investigated: in the early days it was considered for its use as active material to fabricate silicon-based light emitting devices. Later it raised a strong interest thanks to its peculiar porosification mechanism that enables the realization of complex porosity profiles. These have been used to fabricate multilayered structure with tailored optical properties [27]. More recently, the use of PSi in bio-related applications has become more and more common as scientists realized that it owns most of the attractive properties required by a porous sensor host. In fact PSi permits to tune its pores size to a large extent (from nm to μm), it can be rendered luminescent and its optical properties are greatly modified by tuning the porosity profiles. Finally, it is a common and biocompatible material.

Some of the most striking advantages of PSi compared to many other porous materials are related with the great control over porous structure enabled by its fabrication method.

The following key parameters are of particular importance:

- pores growth preferentially along selected crystalline directions;
- silicon doping level is used to select the order of magnitude of the pore size and a finer control is achieved by tuning the electrochemical parameters;
- given certain etching conditions, pore size is proportional to the applied current;
- the already porosified layer is (mostly) passivated against corrosion.

Pores growth occurs preferentially along high atomic density directions, thus in the common case of (100) Si wafer orientation, they grow perpendicular to wafer surface (because of the driving force of the electric field applied) and run parallel one to each others. Generally, neighbor pores are not interconnected to each other, unless the case of very high porosity or when truly nanoporous structures are formed: in these cases, pores do not follow precise crystalline directions and create a 3D random network. (100) oriented substrates are by far the most investigated because of the high quality of the pore's surface and limited pores branching. (110) and (111) oriented substrates were investigated too: in these cases the pores form a complex and three-dimensional porous framework with quite well-organized pores, nevertheless they have found limited use so far [28].

A schematic chemical dissolution reaction path has been proposed at the beginning [29, 30] and it requires holes to be available at the interface to weaken the bonds of superficial silicon atoms. This model describes only qualitatively the formation of pores under average etching conditions and it does not account for all the possible structures obtainable in PSi. For a more detailed description of the silicon electrochemistry see the dedicated monographs [30, 31].

The main difference between p-type and n-type Si relies on the fact that the semiconductor-liquid junction is polarized in the opposite way when acceptor or donor doping is considered. Thus, while p-type Si etching proceeds straightforward in forward bias condition, etching of n-type Si requires a slightly more complex setup. In fact, in n-type substrates holes are minority carriers and they have to be artificially generated to sustain the electrochemical corrosion (at least in lightly doped subtrates with resistivity $>10^{-1}\,\Omega\,\text{cm}$). This is commonly obtained by back illuminating the sample and by driving the holes towards the interface under an applied electric field.

Figure 8.1 show a typical iV curves for p-type Si (the same shape is achieved for illuminated n-type Si). The curve shows a peculiar peak above a certain value of the current density. This is the so-called electropolishing peak and it indicates the current density required to homogeneously and completely etch (dissolve) the

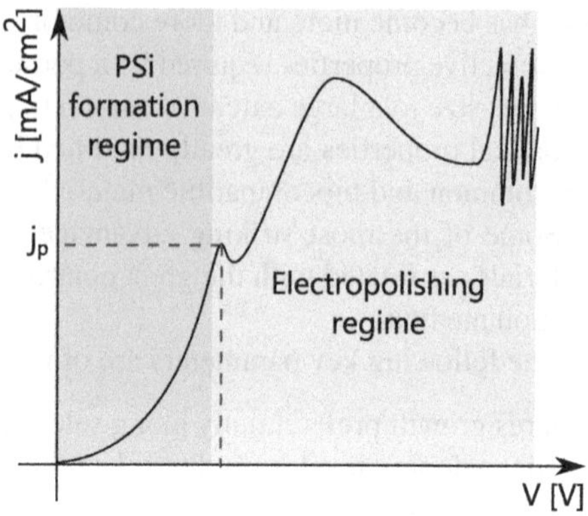

Fig. 8.1 Typical iV curve for p-type Silicon under forward bias. The region where pores grow in a stable manner is limited in the range of current density below the polishing limiting value j_p

reactive interface exposed to the acid. PSi is obtained only for current densities below such threshold. Further peaks (oscillations) appear at larger currents. Etching of PSi is a really complex process and extensive literature has been published to account for its details [32–39].

Recently, it has been demonstrated that high-quality PSi samples can be obtained also in moderately doped n-type Si without illumination [40, 41].

As general rules, doping level determines the average size of the pores, with macropores (with diameter in the μm range) formed in lightly doped substrates ($\rho \sim 10\,\Omega\,\mathrm{cm}$), mesopores (100's nm) in moderately doped substrates ($\rho \sim 10^{-2}$ $\Omega\,\mathrm{cm}$) and nanopores in heavily doped wafers ($\rho \sim 10^{-3}\,\Omega\,\mathrm{cm}$).[1]

Once Si resistivity and composition of HF solution are fixed, pore size is proportional to the applied current. A peculiar aspect of PSi dissolution is that the already porosified layer is passivated against further electrochemical dissolution. This fact can be qualitatively described as a cooperative effect when the pore density is high enough to overlap the space charge regions of the neighbor pores and passivate the wall in between them from further electrochemical dissolution (as no carriers are available within these regions). The important result is that a time-dependent current profile is then transformed into a corresponding porosity variation. This is shown in Fig. 8.2a that reports the cross section of a PSi sample: darker layers correspond to high porosity regions, while low porosity layers are imaged as lighter bands. The porous layer passivation is demonstrated by the periodicity of the structure that does not present significant enlargement of the first etched layers, compared to the last ones. Figure 8.2b reports a top view of PSi surface.

Recently, the metal assisted etching method has been proposed to fabricate PSi. In this case, islands of noble metals are used to catalitically induce pore growth in spatially localized sites [42]. This method demonstrated to be a powerful method and it widened the type of porous structures achievable on a given silicon substrate.

[1] In PSi technology pore size is often named differently from the usual IUPAC nomenclature: macropores are pores of micron size, mesopores account for 100s nm pores and nanopores indicates pores up to 10s nm.

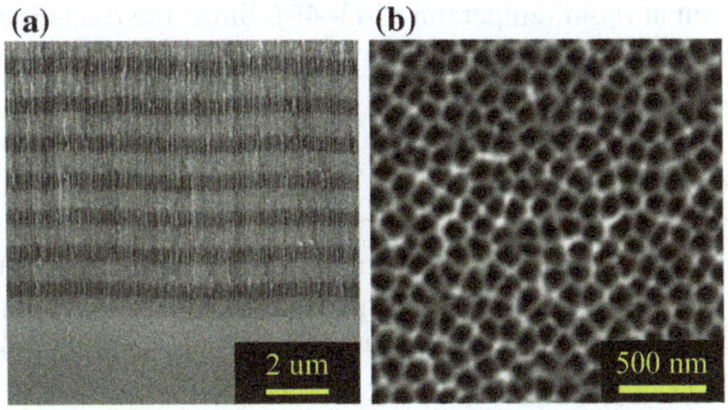

Fig. 8.2 **a** Scanning electron image cross section of a PSi multilayer sample. A periodic current profile has been applied to fabricate the sample and it is reflected into a periodic variation of the pore size. **b** *Top view* of a PSi surface

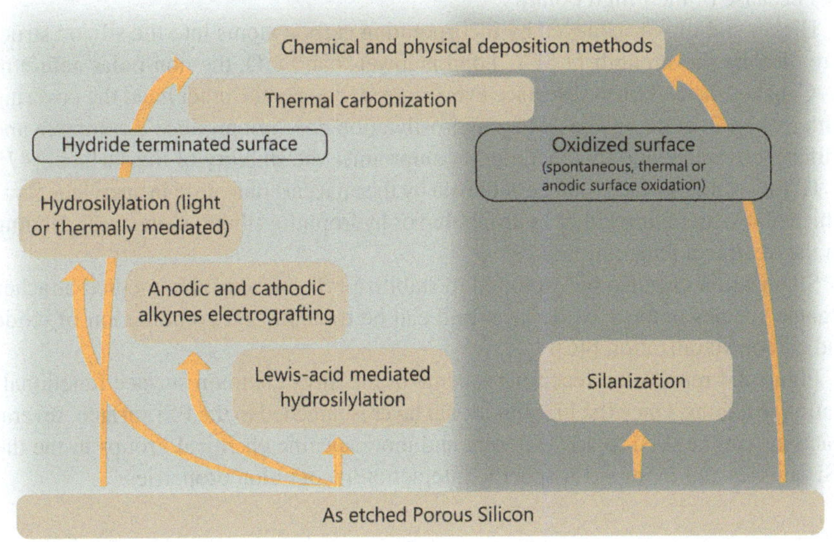

Fig. 8.3 Main strategies to functionalize PSi. Depending on the state of the PSi surface, the methods act on either hydrophobic (hydride terminated, *left part*) or hydrophilic (oxidized, *right part*) surfaces. While silanization, hydrosilylation and electrografting create a monolayer coverage, oxidation, thermal carbonization and deposition methods permit to create thicker layer

8.1.2 Functionalization of Porous Silicon

As etched PSi surface is covered with hydrides species (Si–H) that are readily oxidized in ambient environment. Thus, both for sensing as well as for DD applications, a proper stabilization of the surface and, eventually, controlled surface functionalization strategies are needed to tune and control the behaviour of the PSi samples.

The hydrophobic as-anodized PSi cannot meet this requirement as the hydride-terminated surface undergoes oxidation in contact with atmospheric oxygen or water even at room temperature [43–46]. Since the oxidation changes the chemical, optical and structural properties of PSi, surface stabilization is a crucial step in the PSi fabrication process.

Silicon chemistry is rather well known and different functionalization strategies have been proposed so far. The most common methods are shown in Fig. 8.3 and can be classified in six main categories namely: oxidation, hydrosilylation, thermal carbonization, electrografting, silanization and deposition methods.

Thermal, anodic, liquid phase and ozone oxidation are widespread methods to produce a hydrophilic and stable surface, since the bond strength of the produced Si–O bonds is considerably greater than both Si–Si and Si–H bonds. Although oxide layer provides enhanced stability, the Si–O bonds are susceptible to hydrolysis because of their high polarity.

Higher stability is achieved by incorporating carbon atoms into the silicon structure: despite the strength of Si–C bond is lower than Si–O, the non-polar nature of Si–C makes the carbonized surface even more stable. On the other hand the covering with carbon-containing species by hydrosilylation does not produce a complete and uniform coverage (and this fact might compromise the stability of the surface) [47].

Highly stable surfaces can be achieved by thermal carbonization in acetylene flow, which allows to obtain either hydrophobic or hydrophilic silicon surfaces depending on the treatment temperature.

Silanization can also be exploited to stabilize PSi surfaces: this method attaches organic species onto an oxide layer and can be thought as a combination of oxide and carbon stabilization [46].

Figure 8.4 reports the reactions scheme of the most common surface functionalization reactions. Once the first functional layer is attached to the PSi surface, several reactions can be used to add different and more specific chemical groups to the the PSi samples and to impart it specific interaction/recognition properties.

8.2 Introduction to Porous Sensors

Porous materials own great possibilities and are ideal candidates in sensing technology. In fact, their large surface area can be exploited to tune the interactions with analytes as well as to modify the surface interactions between the porous matrix and the analytes.

During the last decades, the improved control on material nanostructuring and on their characterization have boosted the interest on porous biosensors and several new concepts and improvements have been demonstrated.

A fine control over material structure (from nano to macro scale) is of great importance to improve the capability of the host to act as an optimum substrate for sensing applications. The extreme miniaturization of the sensors and the development of lab-on-a-chip devices are based on a detailed control over surface interactions as well as on the capability to pre-separate (and, eventually, pre-concentrate) the molecules

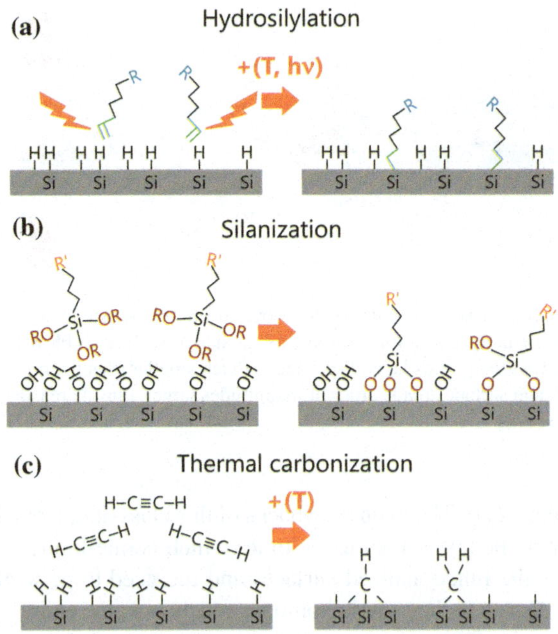

Fig. 8.4 Most common chemical reactions to functionalize PSi surface. **a** Hydrosilylation requires the breaking of an unsaturated carbon bond, stimulated by either light or thermal treatment. Coverage is always smaller than a complete monolayer. **b** Silanization proceeds through hydrolisys and condensation reactions of functional silane groups. A different extent of coverage can be obtained by playing with the silane chemistry. A drawback of this method is that multisubstituted silane tends to polymerize, thus the thickness of the layer is not controlled at molecular scale. **c** Thermal carbonization has been recently proposed and exploits the reactivity of alchenes to stabilize the PSi surface through Si–C bonds. So far this is the method that creates the most stable interfaces but the chemistry of the siliconoxycarbides species formed is poorly controlled

to be detected within the microsystem itself. The size reduction of these devices has two main advantages:

- a reduction of the volume of analyte required for the analysis,
- a shorter analysis time and the possibility to develop point of care devices with nearly real-time response.

A broad classification of porous sensors is done depending if sample flows over (FO) or through (FT) the sensing element (see Fig. 8.5).

In FO the sample interacts with the sensor surface by flowing over its active area. This means that flat surfaces can be used, assuming they bring functional groups able to recognize the analytes. FO has some advantages compared to FT. In principle, any type of material can be used as transducer, upon a proper surface functionalization; there are no tight constraints on the geometry of the material; active surfaces are easily integrated with standard microfluidic systems; highly sensitive optical elements are demonstrated with single molecule sensitivity (for a recent review of opti-

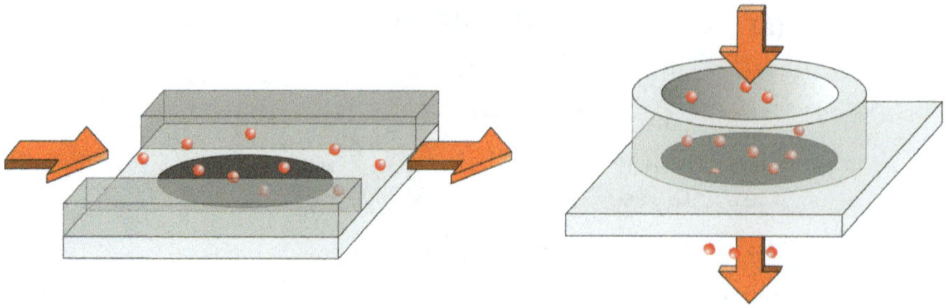

Fig. 8.5 **a** Flow over approach: the sample flows over the active sensor area. The binding events are rate-limited by the molecular diffusion towards the active surface. **b** Flow through approach: the sample flows across the porous layer. Pore size is much smaller than the microfluidic channels of the FO approach and surface area is order of magnitudes larger, thus increasing the binding rate and the overall sensitivity

cal sensors see, e.g. [21]). The main drawbacks of this class of sensors deals with: the limited active area, the rather long assay time (which is rate-limited by the analyte diffusion towards the functionalized surface) and the need to perfectly control the microfluidics to distribute the sample homogeneously over the entire sensors area to maximize its response [22].

On the other hand, in FT sensors the sample flows across the transducer layer. Thus, by carefully tuning the size of the pores, the interaction of the analyte with the sensor surface can be greatly increased and higher sensitivity can be obtained [23, 24]. Generally, pore size is much smaller than the thickness of the microfluidic channels used in FO and the specific area of the active surface is order of magnitudes greater. Both these facts increase the probability of capturing analyte molecules. Moreover, as described in detail later on, porous sensors do not rely on evanescent field interaction but on a much stronger and direct interaction of the electromagnetic field with the sensor surface, thus boosting the sensitivity to very large values.

8.3 PSi Optical Sensors

Several detection schemes have been proposed for PSi based biosensors. Most of them exploit either optical or electrical detections. Electrical detection is based on variation of the conductivity of the porous sponge and it demonstrated very high sensitivity against different gaseous [48, 49] and liquid species [50]. In this chapter, we focus on two optical detection schemes (interferometry and polarimetry) because of their wide applicability and their broad use in biological-related characterization techniques.

8.3.1 Interferometric Sensors

The basic mechanism optical sensors are based on, is that their optical response changes after the binding of the analyte. In fact, the addiction of dielectric material (the analyte) modifies the average refractive index of the PSi and the corresponding interference pattern. The basic idea is shown in Fig. 8.6: the interferometric spectra of porous thin film (black curve, left panel) is modified by the functionalization of the inner pores surface (red curve, central panel) and is further shifted towards the red side of the spectra after the binding of the analyte (green curve, right panel).

The period of the fringes also shrinks because of the increased average refractive index of the layer. Thus, a simple thin film acts as the basic interferometer, albeit with a limited sensitivity (*S*). *S* is defined as the magnitude of the wavelength shift versus the refractive index change induced by the binding of the analyte (measured in nm/RIU).

Optical microcavities (MC) are the most widespread investigated approach to increase *S*. In these cases, the shift of their sharp resonant peak is measured as a consequence of the molecular binding. MC are formed by etching two Bragg reflectors (which are dielectric mirrors made by bi-layers those thickness is designed to produce a controlled interference and to reflect specific wavelengths) separated by a layer that brokes the periodicity. The width of thee reflectivity stop band of each Bragg mirror is maximized when the $\lambda/4$ condition is satisfied:

$$\lambda/4 = nd \tag{8.1}$$

where λ is the wavelength, n the layer refractive index and d the physical layer thickness. The so-called cavity layer is placed in the middle of such structure (such optical device corresponds to an integrated Fabry–Perot interferometer). The addition or removal of dielectric material from the porous sponge modifies the interference condition and shifts the optical response of the MC towards the red or the blue part of the spectra, respectively. The role of the defect layer is to introduce a sharp resonance that eases the detection of small spectral shift and, thus, increases the *S*.

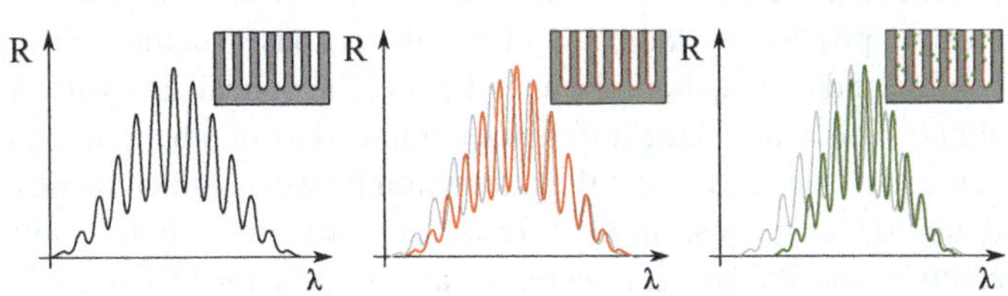

Fig. 8.6 Basic mechanism of optical sensors: the interferometric pattern of a thin film is modified by the addition of dielectric material. Both the functionalization process as well as the assay increase the average refractive index of the layer. This modification is detected with both a redshift and an increased period of the interferometer

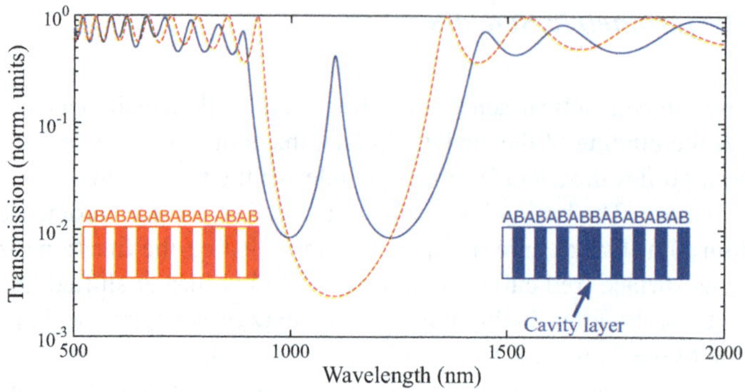

Fig. 8.7 Reflectance spectra of a DBR (*red, dotted line*) and of a MC (*blue, continuous line*). MC is obtained by adding a defect layer in the *middle* of the periodic sequence (the BB element). The sequence of the layers is sketched by the strings. A and B correspond to alternating layers with different porosity (and thus refractive index). For this simulation $n_A = 1.7$ and $n_B = 2.7$

Figure 8.7 shows the reflectance spectra of a Distributed Bragg Reflector (DBR) and of the same structure with an added "*defect*" layer (made by insertion of an extra B layer) that introduces the cavity state within the reflectance stop band (bandgap region).

A key factor that is often disregarded when working with MC formed in porous materials is the role of the overlap between the optical field and the analyte that increases the device *S* compared to other photonic sensors based on evanescent field.

The Q-factor of a resonance is a measure of its sharpness and is defined as:

$$Q = \lambda/\Delta\lambda = 1/\tau \tag{8.2}$$

where λ is the resonance wavelength, $\Delta\lambda$ is the width of the peak and τ is the photon lifetime within the cavity. Large Q-factors are useful because they increase the resolution of the measurement and allow a precise determination of the peak shift (thus *S* increases, too). The highest Q-factor are achieved in 2D and 3D structures that confine photons along multiple directions and minimize losses of photons. Despite a 1D MC does not provide the largest Q-factor, the porous structure of the material permits to confine and overlap both the analytes and the optical cavity modes within the same volume. This is a striking difference compared to all other photonic sensors that exploit guided modes and where the interaction between guided modes and analytes is mediated by an evanescent field. In fact, an evanescent field is intrinsically a weak interaction and its strength decreases as the Q-factor increases. Moreover, the evanescent field-mediated interaction using guided mode structures acts at short distances and requires the analyte to be bound to the device surface. On the other hand a porous material (with properly sized pores) forces the molecule to be within the interaction range of the optical modes and thus, it greatly increases the sensor response.

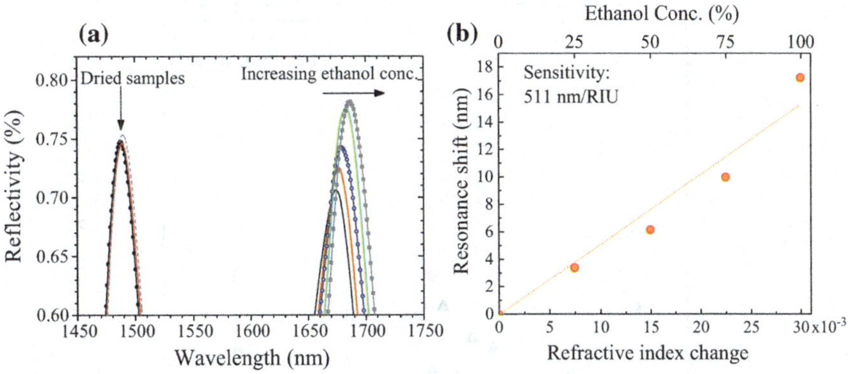

Fig. 8.8 **a** Close view of the peak shift of PSi MC immersed in different alcoholic solutions. The same sample has been used to test all the solutions after being dried. **b** Plot of the sensitivity $[\Delta\lambda/RIU]$

For the majority of the sensors based on refractive index measurements, the following holds:

$$\Delta\lambda = \eta\Delta n_s \frac{\lambda}{n_{eff}} \tag{8.3}$$

where η is the fraction of the field (evanescent) interacting with the sensor surface, Δn_s is the refractive index shift, n_{eff} is the effective refractive index experienced by the resonant mode.

On the other hand, *S* alone is not enough to quantify the minimum response detectable, because this is determined by the resolution of the measuring system as well as by the noise of the experimental setup. The limit of detection (LoD) is a better figure of merit to quantify the minimum detectable response:

$$LoD = \frac{S}{R} \tag{8.4}$$

where R is the system resolution.

The best LoD for an evanescent field sensor is found as a trade-off among two counterposed elements: large Q-factors increase R but decrease *S*, while strong interacting fields increase *S* but decrease R. For a more detailed description see, for example [51, 52].

Figure 8.8 shows the *S* obtained using PSi MC to detect refractive index changes of aqueous solutions of ethanol. Figure 8.8a reports a close view of the resonances peaks: after each measurement the MC has been left to dry and then it was tested on a different solution. The peaks in the dried conditions give an estimation of the reproducibility of the measurements (a set of 5 measurements on the same MC gives a standard deviation of 1 nm, corresponding to about 0.002 RIU). Cavities that differ for the number of porous layers have been fabricated. Ideally, the Q-factor scales

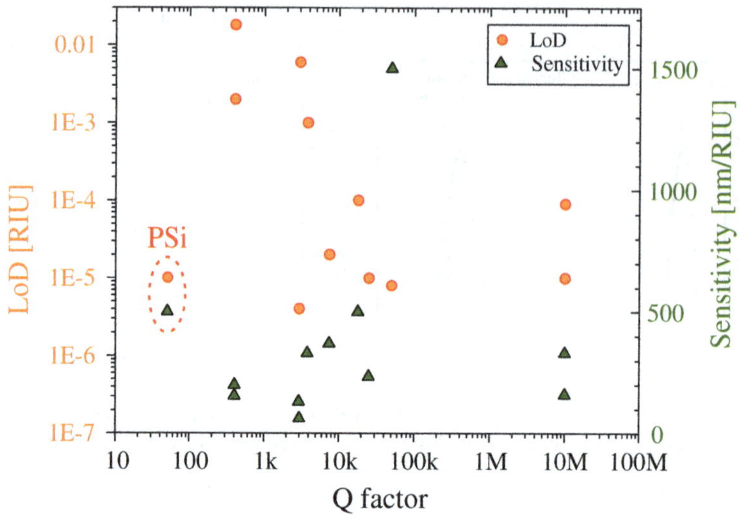

Fig. 8.9 Comparison among several photonic crystal MC-based optical sensors and the experimental result of PSi MC. Despite having the lowest Q-factor, the PSi MC shows one of the lowest LoD and highest S (adapted from [53])

proportionally to the number of layers that compose each Bragg reflector but, in a real sample, the effects of even tiny fabrication imperfections (such as interface roughness and layer thickness mismatch) pile up and reduce the interference. Thus, there is always an optimal number of layers that determine the maximum Q-factor achievable. For PSi MC with pore size of the order of 100 nm the highest sensitivity is achieved on samples having mirrors composed by about 10 periods each one. Figure 8.8b shows the resonance shift versus refractive index variation and is used to calculate S (511 nm/RIU).

In general the overall S of an optical sensor is hard to predict because it depends in a complex way on several elements. Figure 8.9 reports a comparison between several optical sensors based on photonic crystals (adapted from [53]) and a PSi MC. Two factors are evident from the graph:

1. there is no clear trend between the Q-factor and the LoD of the sensor;
2. even if the PSi MC has the smallest Q-factor, it reports among the higher S and smaller LoD.

An alternative to MC is the use of Fano resonances [54]. Fano resonance arises from the interference between two different scattering paths: one is with a discrete resonant state and the other is with a continuous background [54]. The Fano lineshape can be highly asymmetric and the slope of the asymmetric side of the resonance is used to increase the sensitivity (analogous to a high Q-factor resonance) [55].

8.4 Polarimetric Sensors

A different optical detection method is the one based on the polarimetric measurement. The highly anisotropic structure of PSi samples introduces the so-called *form birefringence*: photons propagating at oblique incidence (with respect to the pores axis) and polarized along different directions interact with different "*local*" dielectric structures. In fact photons "*feel*" different refractive indexes and travels at different speeds across the porous layer. Some of the components will be slowed down and the overall polarization of the light exiting from a PSi layer will be rotated compared to the impinging one. The dephasing acquired during the propagation along the PSi layer is:

$$\Delta\phi = \frac{2\pi}{\lambda} \cdot d \cdot \Delta n(\lambda) \tag{8.5}$$

where $\Delta\phi$ is the dephasing, λ is the wavelength, d is the physical thickness of the PSi layer and $\Delta n(\lambda)$ is the birefringence (which, in turn, is a function of the wavelength).

The simultaneous measurement of perpendicularly polarized light beams produces an interferometric spectrum that contains information about the material birefringence. Also in this case the binding of molecules within the pores is revealed by modification of the interferogram.

An example of sensing with polarimetric-based PSi sensor is reported in Fig. 8.10. These measurements report the first demonstration of flow through sensing using

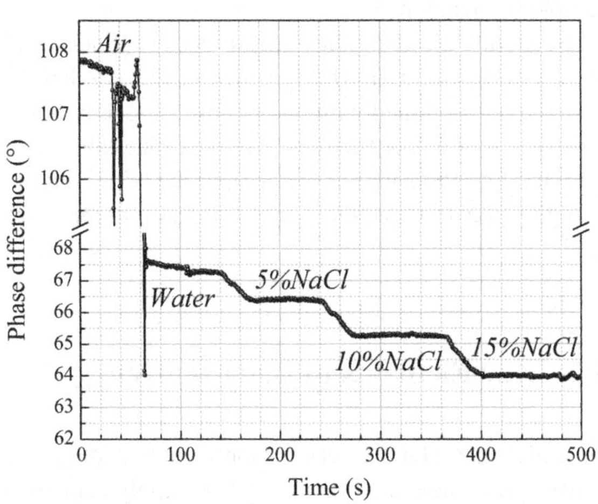

Fig. 8.10 Example of flow through sensing using free standing PSi membranes. The phase difference decreases by flowing solutions with larger refractive indexes

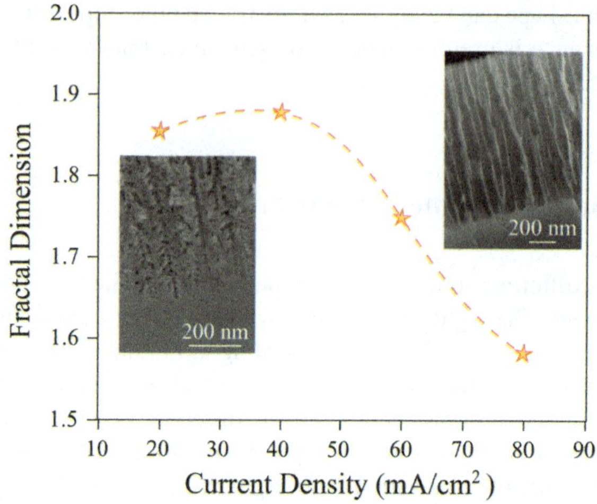

Fig. 8.11 Fractal dimension of pores inner surface at different etching currents

free standing PSi membranes. The LoD of such type of sensor is of the order of 10^{-4} RIU. Even if this value is smaller than the one obtained using MC, the alignment of a polarimetric sensor is by far easier and the overall setup is simpler (unfocused optical beams and no precise alignment needed).

Polarimetric approach had been initially proposed for PSi samples etched in (110) oriented substrates, because of the oblique alignment of the pores with respect to the sample surface maximizes the birefringence [56]. Unfortunately such oblique etched nanochannels have a rather branched structure that induces a strong depolarization of the light. In turn, this fact acts as a noise in the measurement and lowers S. The effect of the pore branching can be evaluated considering the fractal dimension of the pore surface. Figure 8.11 shows how the fractal dimension is reduced at larger etching currents: smoother pores reduce the light scattering and increase the overall sample birefringence.

By greatly reducing the pore surface roughness, and thus the depolarization, very large S in the order of 10^{-6} [nm/RIU] has been demonstrated [57]. This value rivals with the one obtained with more complex techniques (such as Surface Plasmon Resonance) and does not require complex nanofabrication technologies. Furthermore, free standing PSi samples are compatible with flow through assays which are known to reduce the non-specific signal compared to the more common flow over approach [23].

8.5 Key Elements in Drug Delivery

Drug delivery (DD) refers to approaches and systems designed to transport and deliver pharmaceutical compounds to safely achieve the desired therapeutic effect. This concept is heavily integrated with bioavailability, pharmacokinetics and route

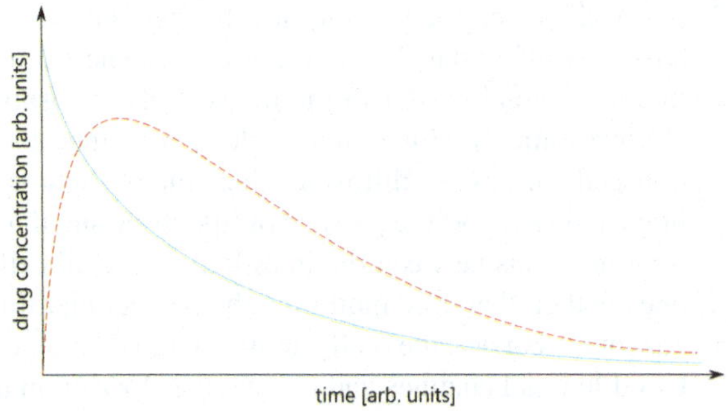

Fig. 8.12 Plasma concentration-time curves after a single administration of the same drug, by intravenous (*continuous green line*) and oral (*dotted red line*) routes

of administration. Bioavailability refers to the extent and rate at which the active moiety enters the systemic circulation, thereby accessing the site of action. Pharmacokinetics, instead, describes how the body affects a specific drug after administration through the mechanisms of absorption and distribution, the chemical changes of the substance in the body (e.g. by enzymes), and the routes of excretion of the drug metabolites.

Oral delivery remains the preferred route of administration because it is non-invasive and among the less expensive method, it can be self-administered and it improves patient compliance. The high surface area and plethora of absorptive mechanisms available in the intestine make it the primary target for drugs absorption [58–60]. Furthermore, it can provide a long-term, continuous release of the drugs using lower, and thus safer concentrations resulting in much better efficacy and fewer side effects compared to the current intermittent injection therapy (See Fig. 8.12) [61].

However, numerous drugs (more than 90% of the new candidates [43]) suffer from poor bioavailability when orally administered and tend to be eliminated from the gastrointestinal tract (GIT) before reaching the systemic blood circulation. This limit is due to a number of reasons [43, 46, 59, 61–63]. The most important are:

1. low drug stability in the gastrointestinal (GI) environment, resulting in compounds degradation prior to their absorption;
2. low drug solubility which results in low dissolution rate and elimination of a fraction of the drug prior to absorption;
3. poor permeability across the GI wall;
4. high intestinal and first hepatic metabolism.

These problems are strongly related to the physiology of the GIT, which poses formidable barriers [58, 59, 64]:

1. chemical barrier: the pH value changes from highly acidic (pH 1.2–3.0) in the stomach, to neutral or slightly alkaline (pH 6.5–8.0) in the intestine. Several active ingredients are susceptible to the pH change which may cause degradation of the drugs.

2. physical barrier: the first physical barrier is represented by the mucus[2] that may adversely affect the absorption of drugs. The epithelial cells themselves constitute the second barrier to the transcellular transport of the drugs. Hydrophilic pharmaceuticals with a large molecular weight do not easily enter the cells via transcellular passive diffusion, since the phospholipid bilayers of the cell membrane only support the passage of lipophilic small molecule drugs. Moreover the tight junctions between the intestinal epithelial cells make them tightly bind to one another, therefore inhibiting the paracellular transport of large drugs.
3. enzymatic barrier: the orally administered drugs face the proteolytic degradation by GI luminal enzimes and by enzymes located in the cell membrane during its transport through the intestinal cells. Additionally, the liver enzyme system can further reduce the amount of the absorbed drug across the intestinal epithelium that enters systemic circulation [66].

Current approaches to improve the bioavailability of orally administered drugs include the use of enzyme inhibitors and the exploration of special formulation ingredients, such as absorption enhancers (excipients to enhance drug solubility) and mucoadhesive polymers [59, 67, 68]. Once in the bloodstream, the orally administered drugs have the same fate of those administered intravenously.

The intravenous injection is the preferred route for chemotherapy and to treat neurodegenerative diseases. Its main advantage is the avoidance of the absorption phase, therefore the drug bioavailability is practically equal to the dose (see Fig. 8.12). The main hindrance is that several drugs are hydrophobic, thus, only slightly soluble in blood. Other side effects which have to be taken into account when dealing with drugs in the bloodstream, are enzymatic degradation, phagocytosis in the reticulo-endothelial system (RES) (also referred to as MPS, mononuclear phagocyte system) and the obstacles offered by the body-barriers such as the vascular-endothelium (VEB) and the blood–brain barrier (BBB) [69].

8.6 General Considerations on Nanostructured DDS

An exciting way to improve both oral and intravenous administration is the use of Drug Delivery Systems (DDS); that is micro- and nano-vehicles made of biocompatible and biodegradable materials. DDS are designed to carry the drug, while preserving its integrity and to deliver it to the desired target place [70–73]. Nanostructured drug carriers allow the delivery of both small-molecule drugs as well as of large nucleic acids and proteins. Generally, nanomaterials are defined as systems with one or more external dimensions in the size range 1–100 nm [74], but despite this size restriction, nanotechnology commonly refers to structures that are up to several hundred nanometers in size [75]. In DD, however, useful carrier dimensions range from

[2]The mucus is a complex aqueous mixture of glycoprotein, lipid, salts and cellular debris produced by epithelial cells, it affords protection for the underlying tissues from various environmental insults and the effects of enzymes or other chemical agents [65].

truly nano- to micro-systems (in the range of 100 μm). For practical reasons, in this perspective, nanotechnology often includes microtechnology [73].

Nano- and micro-DD show many advantages compared to traditional therapeutic methods, such as: speeding up of the treatment, protecting the drug from degradation, minimizing side effects to other parts of the body not exposed to the medicine. Finally they ease the passage of the drugs across VEB and BBB. Several efforts have been made to reach these goals and to design targeted DDS that enable the delivery of the proper amount of active ingredients at the desirable sites (such as tumors, diseased tissues,etc.) while minimizing unwanted side effects of the drugs on other tissues [76].

Controlled DDS are designed to deliver drugs at predefined periods of time and at predetermined rates. This fact permit to prolong the therapeutic effect, to the minimize number of treatments and to reduce the fluctuations of the drug concentration during the treatment and to achieve a steady state level [77].

When dealing with nano-DDs that are injected in the bloodstream, the targeting can be either active or passive [70]. Passive targeting is often used to treat tumors because it exploits the unique pathophysiological characteristics of tumor vessels, which are highly disorganized and dilated with a high number of pores. They result in enlarged gap junctions between endothelial cells (tumor microvasculature typically contains pores ranging from 100 to 1000 nm in diameter, whereas in healthy tissues the tight intercellular junctions less than 10 nm) and in a compromised lymphatic drainage [72]. The "*leaky*" vascularization, referred to as the enhanced permeability and retention (EPR) effect, allows to selectively target the tumor by creating DD nanostructures greater than the intercellular gap of the healthy tissue but smaller than the pores found within the tumor vasculature [70].

Active targeting requires the conjugation of the carrier system with a specific ligand [78]. For example, it is well known that tumors cells express specific types of antigens, so that if the nanocarriers are equipped with the proper antibodies, the drug will be released to the tumor site exploiting the antigen-antibody reaction. Active targeting can be used in oral DD too. Various types of epithelial cells can be targeted relying on ligand-receptor binding [63]; moreover, the fact that the nature of mucous in the normal and diseased states is different, can be exploited to deliver the drugs only in the diseased area [79].

Alternatively to passive and active targeting, DD can be triggered by specific microenvironments in the body (e.g. changes in pH, temperature, and enzymatic activities) or manipulated by external events (e.g. light, electric field, magnetic field and ultrasound). This goal can be achieved using the so-called *smart* (also called intelligent or responsive) DDS. These carriers are made of materials that respond to chemical (pH changes), physical (light, temperature) or biological (enzymes) stimuli so that the active ingredients are released only after the carriers have received the proper stimulus [80, 81].

Nowadays, there are more than two-dozen therapeutic nanoparticle products approved for clinical use, although these first-generation nanoproducts are relatively simple and only provide clinical benefits across a narrow range of clinically validated drugs. Toward the development of next-generation nanoparticles, the introduction of controlled release properties and targeting ligands is expected to create safer and more effective therapeutic nanoparticles [75, 82].

To develop such drug carriers, it is important to understand how they interact with the human body. The orally administered carriers, if not directly destroyed due to the variable environments along the first part of the GIT, arrive in the intestine; if not eliminated by the intestinal enzymes, two different strategies can be adopted:

1. the vehicles adhere either to the epithelial tissue or to the mucus layer and release their content. Then the active ingredients penetrate the GI walls and, if not eliminated by the hepatic first-pass metabolism, they reach the systemic circulation.
2. after being in contact with the mucosal surface, the carriers cross the GI barrier to deliver their content inside epithelial cells, blood or even target organ. Several studies have been carried out to understand the possible mechanisms for the particles to pass through the GI (and other physiological) barriers. The possible mechanisms are (1) paracellular passage: particles pass between intestinal epithelial cells due to their extremely small size (smaller than <50 nm); (2) endocytotic uptake: particles are absorbed by intestinal enterocytes through endocytosis (for particles smaller than <500 nm); and (3) lymphatic uptake: particles are adsorbed by M cells of the Peyer's patches (particle size <5 mm) [61, 63, 83].

Either after the intestinal uptake or directly by injection, the nanocarriers come in contact with all blood's elements (proteins, lipids, salts, immunitary system cells, ...); the challenge is to make the carriers able to reach the target site, without being eliminated ahead of time.

Endocytosis and phagocytosis can be exploited by the targeted cells to uptake the nanocarriers, but nanovehicles are often identified as foreign by the immune system, so that they are engulfed via endocytosis or phagocyitosis by RES. In general, most DDs want to avoid phagocytosis of the immune system since, once inside immune cells, they are destroyed and cannot deliver their content. Some DDs actually exploit the phagocytosis to release their content—after being destroyed-inside the immune cells. These, in turn, transport the active ingredients to their final destination.

The understanding of how endocytosis and phagocytosis work is a key factor to develop effective nano-DDS that either suppress or exploit these mechanisms.

8.7 Carrier Dimensions and External Functionalization

The factors that could affect the clearance and biodistribution of nanoparticles, such as particle properties and external surface functionalization, should be carefully considered for the optimal design of therapeutic nanoparticles.

8.7.1 Particle Dimension

The carrier size determines the in vivo distribution, biological fate, toxicity and the targeting ability of nanoparticle systems [84]. It is reported that carrier size plays a key role in the adhesion to and interaction with the biological cells [61, 68]; moreover, it affects the vehicles movement in and out the vasculature [75].

The optimal particle size depends on the route of administration and the targeted cells: if released into circulation, the carriers dimensions can be up to 10 μm, but usually they are much smaller because the smallest capillaries in the body are 5–6 μm in diameter and the size of the particles should be small enough to avoid aggregation and embolism. In oral DD the maximum size for particles is in the order of few millimeters, but usually carriers dimensions are in the order of microns [58].

Generally, nanoparticles have relatively higher intracellular uptake compared to microparticles and they are available to a wider range of biological targets due to their small size and relative mobility. For example, it has been demonstrated that 100 nm size nanoparticles showed 2.5 fold greater uptake, compared to 1 μm and 6 fold higher uptake, compared to 10 μm microparticles in Caco-2 cell line. Similar results were obtained when these formulations of nano- and microparticles were tested in a rat in situ intestinal loop model. The efficiency of uptake of 100 nm size particles was 15–250 fold greater than larger size (1 and 10 μm) microparticle [85]. As noted above, the small size of nanoparticles is an advantage due to EPR in tumours [86].

Once in the bloodstream, nanoparticles smaller than 10 nm can be rapidly cleared by the kidneys or through extravasation, while larger nanoparticles may have higher tendency to be cleared by cells of the MPS [82].

The particle dimension can influence the stability of nanoparticles: smaller particles have greater tendency to aggregate during storage and a proper surface stabilization has to be used to stabilize them.

Particle dimensions affect also the drug release profile: smaller particles have a larger surface area-to-volume ratio; therefore, most of the drug associated with small particles would be at or near the particle surface, leading to a faster drug release. In contrast, larger particles have large cores which allow more drug to be encapsulated per particle and give slower release. Thus, the control of the particle size provides a means of tuning drug release rates [72, 87].

8.7.2 External Surface Functionalization

The carriers outer surface should be designed taking into account that it interacts with the physiological environment; for this reason, several parameters (such as: the carriers solubility and stability against aggregation, the crossing of physiological barriers an mitigation of non-discriminant interactions with proteins and biomolecules) have to be considered to maximize the efficacy and minimise toxicity of the treatment [86, 88]. As an example, once particles enter the blood circulation, they are in contact with all kinds of blood cells and plasma proteins; therefore their surface may experience non-specific protein adsorption (opsonization), thereby making them more visible to phagocytic cells [82].

It has been found that the polyethylene glycol (PEG) coating creates a hydrophilic surface that can prevent agglomeration, reduce protein adsorption, prevent phagocytosis by immunodefensive cells, such as macrophages and increase circulation time [43, 82].

The charge of the outer layer can be used to promote specific and desirable interactions. For example, cationic nanoparticles have a higher internalisation rate, due to their electrostatic attraction with the negative charge of cell membrane lipids [86].

Concerning oral DD, an important challenge is to maximize the retention time in the intestine to increase the absorption probability. Bioadhesive materials (such as polyvinyl alcohol (PVA) or PEG) are used to improve their adhesion (and to increase the transit time of the particulate drug carrier in the GIT, resulting in prolonged time window for the release of the payload) and absorption into the intestinal cells. Moreover, enhanced absorption of the drug could occurs because of the increased effective surface area in contact with the intestinal mucosa upon material swelling. Swollen material fills the crevices of the mucous membrane and yields to a high local concentration of the drug [89–91].

8.8 Type of Nanocarriers

8.8.1 Organic DDS

When dealing with organic materials for DDs, a useful classification can be done relaying on the geometry of the resulting nanostructure:

- **Nanospheres** are matrix systems in which the drug is uniformly dispersed (see Fig. 8.13a). Solid-Lipid nanoparticles (SLNs) possess a solid lipid core matrix,

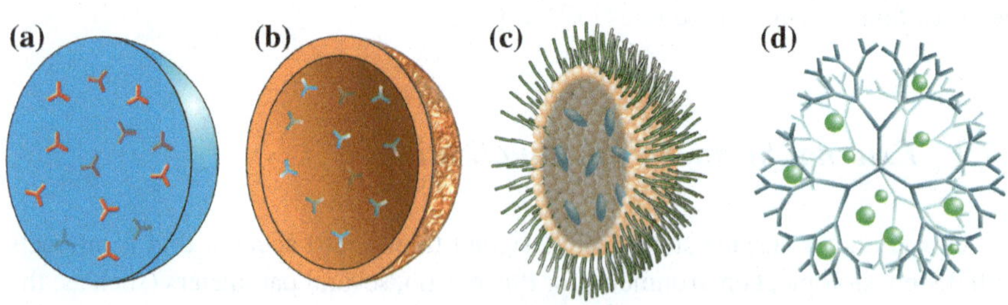

Fig. 8.13 Most of the nano-DD structures are based on the realization of containers exposing the proper surface to the environment they have to interact with. This figure reports the basic realization of the most used structures, which are: **a** nanospheres, **b** nanocapsules, **c** -phylic and/or -phobic (also called direct/inverse) micelles and **d** dendrimers

while the core of the nanostructured lipid carriers (NLC) is composed of a mix of solid and of liquid lipids. This structure improves the drug loading and reduces the drug expulsion. Their inner core is lipidic, enabling the loading of hydrophobic drugs [87, 92–96].

Polymeric nanospheres posses a core structure which can be varied by changing the composition of the hydrophobic and the hydrophilic blocks on the polymer chains. Hydrogels are obtained if the cross-linked networks of hydrophilic polymers absorb several water molecules (up to 20% of their weight in water) while keeping the 3D structure of the polymer network [70–72, 77, 82, 87, 95].

- **Nanocapsules** are nano-vesicular systems in which drugs are enclosed to a cavity, surrounded by a membrane (see Fig. 8.13b). The lipid nanocapsules (LNCs) have a liquid lipid core in which the lipophilic drugs are embedded, covered with an essentially lipid solid film [96]. In polymeric nanocapsules, the drug is confined into an oily or aqueous core (to enable the encapsulation of both hydrophilic and lipophilic active ingredients) surrounded by thin polymer membranes [70, 71, 82, 87, 95, 96].
- **Micelles** are systems in which amphiphilic molecules self-aggregated to minimize their surface energy in a given environment (see Fig. 8.13c): the centre of the micelles is "phobic" and the outer surface is "philic," so to match the solvent it is immersed in. In the case of lipids the micelles are mainly made of phosphatidylcholine and bile salts, whereas polymeric micelles can be obtained using amphiphilic copolymers [70–72, 77, 87, 95–97].
- **Liposomes and Polymerosomes** are hollow shell nanoparticles constituted by a double layer membrane. Liposomes are spherical lipid vesicles constituted by one or more bilayers consisting of amphiphilic lipid molecules concentrically oriented around an aqueous compartment [70, 72, 87, 95–97]. Polymersomes differ from liposomes by the fact that the membrane is made of amphiphilic block copolymers. In both cases this vesicle membrane encloses a hydrophilic environment, enabling the encapsulation of water-soluble molecules. Hydrophobic drugs can be segregate in the vesicle membrane [72, 77, 78, 87, 95, 96].
- **Dendrimers** are monodisperse symmetric polymeric macromolecules with highly branched structures around an inner core (see Fig. 8.13d). Their structures are comprised of three components: (a) a focal core, (b) several layered building blocks formed by repeated units and (c) multiple peripheral functional groups. Usually, the interior of a dendrimer shows hydrophobic characteristics while the exterior surface is hydrophilic [70–72, 87, 95, 97].

8.8.2 Inorganic DDS

Metal nanoparticles. Gold, silver, platinum, palladium gadolinium and nickel are the most used metals to obtain both nanospheres (in the range 1–200 nm) and nanocapsule (hollow metal nanoshells). Metallic nanoparticles a have large surface area, thereby, incorporating a high drug dose (a 2 nm Au nanoparticle could, in principle, be conjugated with hundred drug molecules [72]). Their surface can be modified with different functional groups that enable both drug loading as well as targeted DD. However, the toxicity of metallic nanoparticles is of concern [70, 72, 77, 87, 95, 98].

Carbon-based Nanocarriers. Two kinds of nanostructures that have received much attention in recent years are hollow, carbon-based, cage-like architectures: nanotubes and fullerenes. Carbon Nanotubes are self-assembling sheets of atoms rolled up in tubular shape. They can be classified into two categories based on their structure: single-walled carbon nanotubes with a single cylindrical graphite sheet (diameters in the range 0.7–1.5 nm) and multi-wall carbon nanotubes designed by 2–30 concentric tubes with diameters in the 2–10 nm for the inner tubular layer and extra thickness of about 0.7 nm for each extra layer, resulting in diameters ranging from several nanometers to tens of nanometers. Fullerenes are carbon allotrope made up of tens of carbon atoms with a polygonal structure, resulting in structures whose diameters are on the order of 1 nm [70–72, 77, 87, 95, 97, 98]. The size, geometry and surface characteristics of these structures make them appealing for drug carrier usage, upon a proper surface functionalization to render them biocompatible. Some studies suggest a possible toxicity of these materials [72, 95].

Ceramic nanocarriers. These porous particles are fabricated from silica, alumina and titania. Their high surface area and large pore volume make these carriers able to load great amounts of drug molecules. Unlike organic materials, ceramic nanoparticles in aqueous situations usually do not change size or porosity and are very stable against chemical (pH) and physical (heat) stress. Among the ceramic nanocarriers, those made of silica have attracted much research attention. They are typically synthesized through polymerization of silica species on surfactant template in basic or acidic conditions. After removal of the surfactant template by high temperature calcination or organic solvent extraction, the mesoporous silica-based framework is produced. Since the preparation of mesoporous silica is based on a sol-gel process, the particle size of mesoporous silica can be controlled very precisely during the synthesis process. MCM-41 (Mobil Composition of Matter-41) has been first employed as a drug delivery matrix; SBA (Santa Barbara Amorphous), HMS (hollow mesoporous sphere) and MSU (Michigan State University) can be used for drug delivery applications [46, 72, 77, 78, 86, 87, 95, 98–103].

8.9 Porous Silicon

8.9.1 Why Porous Silicon?

The research of PSi for drug delivery applications has mainly started after the year 2000, although some reports about silicon-based nanopore technology were published earlier. Currently, injectable DDS are the most studied and reported applications, but since the first report in 2003, interest in oral delivery also arose [104].

PSi nano- and micro-materials are broadly investigated as drug delivery vehicles, first of all because of the high surface-to-volume ratio that allows a large payload of drug molecules. Porosity can be more than 80% and the pSi internal surface per unit volume can be on the order of 500 $\frac{m^2}{cm^3}$ [105].

As ceramic vehicles, PSi carriers are fabricated to resist the heat stress and the harsh conditions of the GIT, maintaining their physicochemical properties [106].

Pores morphology depends on several fabrication parameters, thus feasible nanostructure can be tuned over a wide range: pore diameters from few nanometers up to several micrometers can be obtained separately and the pore walls of macropores (pore diameter >50 nm) can be decorated with micropores (pore diameter < 2 nm). The PSi that is mostly investigated for DD applications is the mesoporous one, with pore diameter ranging from 2 to 50 nm [43]. Moreover not only the pore size can be tuned, but also the particle size and shape [106].

In oral DD applications microparticles are normally used, whereas nanoparticles are used in the case of intravenous administration or when their mucoadhesive and mucopenetrating properties need to be exploited.

As highlighted previously, particle size has a crucial role in carrier biodistribution: it has been demonstrated that for intravenous administration, approximately 80% dose of PSi nanoparticles was accumulated in the liver and spleen 24 h after the injection; whereas using microparticles, the silicon content in the spleen was almost threefold compared to the one in the liver, demonstrating that PSi microparticles mainly accumulated in the liver other than in the spleen [105].

Along with size, both the the efficiency and the specificity of DD carriers are also shape-dependent [75, 86]. Hemispherical-shaped particles are more beneficial to surmount some biological barriers such as the cellular membranes, to escape from the MPS uptake and to migrate towards the vasculature wall, compared to the spherical ones [43]. According to another study, non-spherical particles exhibit more complex motions with tumbling and rolling in flowing blood, decreasing the cellular adhesion probability [107]. The adhesion of orally delivered carriers on the mucus layer, is promoted using flat shapes because: (1) they increase the contact area available for interaction with the epithelial lining of the GI tract and, (2) they decreases the force exerted on the devices from the fluid flow in the GI tract [107].

8.9.2 Drug Loading

In general, drug loading into a carrier material can be performed in different ways, but the fabrication method of the carrier may limit the alternatives. In many cases, the drug needs to be added during the synthesis of the carrier material ("incorporation method") and this fact may restrict the usable drug concentrations and leads to relative low drug loads. Furthermore, the chemical environment in the synthesis can be too harsh for sensitive compounds, such as peptides and proteins. In the case of mesoporous carrier materials, it is advantageous to perform the drug loading after the fabrication of the carrier ("adsorption–absorption method") [78, 84, 87]. In these cases, the loading methods are classified into two groups: immersion and impregnation [43].

A commonly used procedure is simple immersion of the carrier into the loading solution. A certain amount of the drug is dissolved in a desired solvent and the volume of the solution is substantially higher than the pore volume of the particles. After the adsorption equilibrium is reached, the particles are separated from the solution by filtration or centrifugation.

In the "impregnation" or "incipient wetness method," a controlled amount of drug solution is added to the particles and allowed to infuse through capillary forces into the pores. The excess solvent is then removed by evaporation, while the drug remains in the pores.

Common advantages of these methods are: the loading can be performed at room temperature and the drug to be loaded is not exposed to a harsh chemical environment. The "immersion" method gives a better control over the drug loading, but the "impregnation" one is more applicable in the case of expensive drug molecules or small amounts of sample, because the drug is completely loaded. In addition, the amount of the drug in the loaded material is easily determined in advance. However, the uniformity of the loading and the possible crystalline fraction on the external surface of the particles are more difficult to control than in the immersion method.

The control of the drug fraction on the external surface of the particles is complicated and some additional washings of the surface might be needed. This is important, because even quite small fractions of drugs on the particle surface can significantly change the dissolution behaviour and lead to erroneous results, if the dissolution rate of the crystalline drug on the particle surface is significantly different from the one in the pores [104].

Several factors are involved in the loading process, such as pH, temperature, time, drug concentration. Moreover, the appropriate choice of the solvent and the chemical properties of the carrier internal surface are essential to avoid drugs degradation [44, 46, 104, 108].

Despite its apparent simplicity, drug loading by immersion has a complex set of interactions, which all affect the final product.

The most important interactions are carrier–drug, drug–solvent and carrier–solvent, but also drug–drug and solvent–solvent interactions may play an important role in the loading processes. The drug-loading process begins with a monolayer adsorption on the surface of the carrier, which is determined mainly by drug–carrier, drug–solvent and carrier–solvent interactions. For efficient monolayer adsorption, the affinity of the drug molecule to the carrier surface should be much stronger than drug–solvent interactions. In the subsequent multilayer adsorption, the drug–drug interaction becomes predominant.

Three different adsorption mechanisms can affect drug loading into porous materials:

1. physical adsorption (physisorption),
2. chemical adsorption (chemisorption),
3. physical trapping of the payload.

The primary mechanism by which adsorption takes place has a significant effect on the loading process and especially on the drug release. Physisorption refers to an adsorption process in which the molecule is adsorbed on the surface with weak physical interactions and it is based on three different interactions:

1. attractive electrostatic forces between permanent charges like dipoles, quadrupoles or higher order multipoles (Keesom forces),
2. attractive polarization forces in which a permanent multipole (charge) induces multipole on another (nonpolar) compound (Debye forces),
3. attractive force between two instantaneously induced dipoles arising from a transient fluctuation of electron density in molecules (London forces).

Physisorption is reversible and allows the quick release of the loaded compound with minimal activation, so that it is often used when the aim is to enhance the solubility of the payload compound [45]. Silicon oxide surfaces tend to present a negative surface charge to an aqueous solution that can be exploited to hold positively charged molecules; another strategy is to functionalize the pores with functional groups that electrostatically trap the drugs [109].

Chemisorption provides a stronger (covalent) chemical interaction between the drug molecule and the surface of the carrier. This is favourable when the aim is to achieve a sustained release and makes it possible to control the release rate of the drug by tailoring the dissolution rate of the carrier material itself. However, there are also issues in the covalent bonding approach: first of all, only one monolayer of the adsorbate can be chemisorbed onto the carrier, limiting the maximum payload. Thus, the surface area of the carrier determines the maximal loading capacity. Secondly, the release mechanism of the attached drug molecule should be known to ensure that no chemical degradation of the payload molecules takes place.

Typically, the internal surface is modified with molecules containing carboxylic acid or amine groups and the biomolecules are conjugated to these groups either via an amide bond or cross linked to them. For example, thermally oxidized PSi has been silanized with aminopropyltriethoxysilane (APTES) and the load has been subsequently linked to the amine group by cross-linking with glutaraldehyde. In another study, hydrogen-terminated PSi was hydrosilylated by undecylenic acid, that has been conjugated to a peptide via activation by N-Hydroxysuccinimide (NHS) and 1-ethyl-3-(3-dimethylaminopropyl) carbodiimide (EDC) [105].

Physical trapping of payloads is a rarely used procedure, but it can be applied basically using two different methods: either the drugs can be trapped in the PSi by loading the molecules into the pores and subsequently reducing the pore size by thermal oxidation (as silicon oxide has a larger molar volume and in oxidized PSi pores shrink), or some large compounds (proteins or polymers) are adsorbed on the external surface to block the pore opening [45].

8.9.3 Solid-State Properties of Loaded Compounds

The incorporation of drugs into PSi has been shown to affect the solid-state properties of loaded compounds [110].

There are four possible physical states where the drug compound loaded in porous materials can be found [43]: (a) a thin liquid-like layer next to the pore wall, which does not undergo any phase transitions due to the strong interaction with the pore wall (i.e. Stern or Delta layer); (b) the amorphous drug in the middle of the pores; (c) the nanocrystalline drug in the middle of the pores and (d) the bulk crystalline fraction on the external surface of the particles. At least part of the drug always exists in the liquid-like state and the thickness of this layer depends on the strength of the surface–drug interactions. The existence of the other states depends on: the diameter of the pores, the size of the drug molecules, the arrangement of the drug molecules in the pores and the filling factor of the pores.

It has been demonstrated that in pores few times larger than the drug molecule, the formation of crystalline material is hampered by the tightly confined volume, thus, the drug retains its amorphous form; the amorphous state can be dispersed on the pore walls of the material leaving the centre of the pore free or it can at least partially fill it [44, 46].

Besides the fractions of the drug inside the pores and on the external surface, it is important to investigate other physicochemical characteristics of the drug loaded into mesoporous materials, such as the amount and integrity of the loaded drug and the type of drug-carrier interactions [43, 46].

In order to measure the loading degree of the carrier (the maximum loading degree is defined as the ratio of the drug mass to the total mass of the drug loaded carrier, when the pores are totally filled [46]), extraction and thermogravimetry (TG) are the most commonly applied methods.

With extraction, the drug is released from the carrier and the drug concentration of the release medium is measured. High performance liquid chromatography (HPLC) is frequently employed for drug concentration assay since it is also able to gather information about the possible degradation of the payload molecules. UV/vis spectroscopy is an alternative technology for the measurement of the drug concentration for drug molecules with strong optical absorption. However, in these cases, care should be taken that all of the payload material has been completely extracted from the carriers.

In TG, the weight of the sample is measured as a function of temperature. The mass of the loaded drug is determined because the thermal decomposition of the organic drug takes place at lower temperatures than the decomposition of the carrier.

Drug-carrier interactions can be determined with Fourier Transform Infrared Spectroscopy (FTIR) or by defining the adsorption isotherms (either TG or Differential Scanning Calorimetry (DSC)) of the loaded drug, which can provide important information about the possible chemisorption or physisorption on the carrier surface.

Knowledge on the state of the drug and the fractions of the drug inside the pores and on the external surface in the mesoporous carrier are of utmost importance. If a poorly soluble drug is located on the external surface of the carrier particle it can block the release of the drug from the pores and have a negative impact on the release rate of the drug. Drug material located outside the pores can be distinguished from the drug inside the pores by the size of the crystallites. In the pores, the crystalline size is restricted by the pore walls whereas outside the pores such restraints do not exist and crystallites tend to grow significantly larger. To detect the crystalline phases and the size of the crystallites in the loaded materials, X-ray diffraction (XRD) and DSC are used.

In XRD, the peak width increases with decreasing crystallite size. In DSC, the drug material inside the pores can be distinguished from the one on the external particles surface, as the drug inside the pores does not contribute to the normal (bulk) melting endotherm. If the drug is crystallized inside the pores, this causes a depressed melting endotherm due to the small crystallite size. In contrast, if the drug is in an amorphous form, no melting endotherm is observed.

In addition, other complementary techniques, such as nitrogen adsorption/ desorption (BET and BJH), helium picnometry and Nuclear Magnetic Resonance can be used to obtain information about the loaded drugs [43, 44, 46, 104].

Concerning the loading capacity, values as high as 60 %(w/w) can be obtained in carriers with very high porosity. Thermal carbonized PSi (TCPSi) carriers have been shown to be able to adsorb compounds (ibuprofene) up to 41.5 %(w/w) [108]. Another study demonstrates that the loading capacity of the PSi microparticles for dexorubicin is 45 mg/g [111]. In addition to the surface properties, the chemical nature of the drug and of the loading solution are critical and heavily affect the loading efficiency [44].

8.9.4 Drying

After loading, the particles are separated from the solution by filtration or centrifugation and dried. A suitable drying method is very important for the loaded compounds and different drying techniques from freeze or vacuum drying to thermal drying have been used [104].

8.9.5 Release

In general, the drug release rate depends on: (1) drug solubility; (2) desorption of the surface-bound or adsorbed drug; (3) drug diffusion through the nanoparticle matrix; (4) nanoparticle matrix erosion or degradation and (5) the combination of erosion and diffusion processes. Hence, solubility, diffusion and biodegradation of the particle matrix govern the release process [78].

If the drug is uniformly distributed in the matrix, drug release occurs by diffusion or erosion of the matrix. If the diffusion of the drug is faster than matrix erosion,

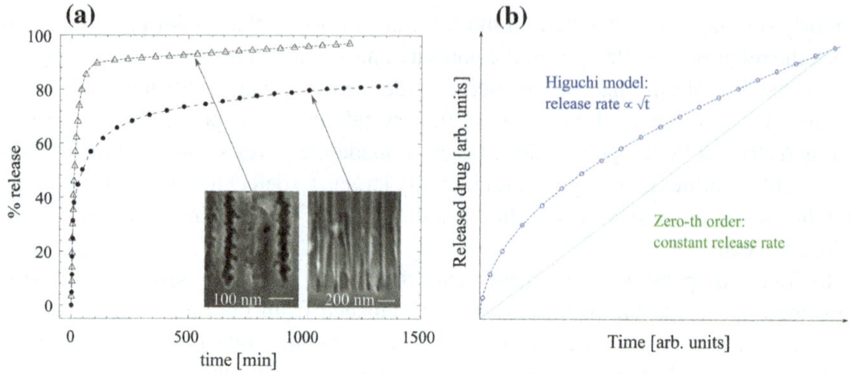

Fig. 8.14 Drug release dynamics can be modified by playing with pore structure and surface interactions. **a** Experimental data showing how the release of a model molecule is affected by different PSi structures. **b** Release dynamics can assume different profiles: 0th-order and Higuchi models are shown as example

then the mechanism of release is largely controlled by a diffusion process. The rapid, initial release, or 'burst,' is mainly attributed to weakly bound (or adsorbed) drug. It is evident that the loading method has an effect on the release profile. If the drug is loaded by the incorporation method, then the system has a relatively small burst effect and a sustained release characteristic. Moreover, when porous nanocarriers are used both drug loading and release can be tuned by changing structural parameters such as: pore density, size, distribution.

Through precise control of the drug carrier architecture, the release of the drug can be tuned to achieve the desired kinetic profile, as reported in Fig. 8.14. Figure 8.14a reports two different release dynamics from PSi microcarrier that differ for their pore size.

The ideal release profile for most drugs would follow a steady release rate, so that the drug levels in the body remain constant while it is being administered. Nevertheless, such zeroth-order profile is extremely difficult to achieve in porous DD systems and it has been demonstrated only in few peculiar cases [70]. Generally, nonlinear dynamics are observed: a saturated-like behavior is often expected (as shown in Fig. 8.14a), because the drug release rate is proportional to the amount of drug that remains in the carrier. Several models have been so far proposed to fit the experimental data. Among the most common we mention:

1. Higuchi model. Proposed in '61 to describe drug release from matrix system. It is based on few assumptions: the drug concentration in the carrier is much higher than its solubility, perfect sink conditions, negligible edge effects, no matrix dissolution, constant diffusivity:

$$\frac{M_t}{M_\infty} = k\sqrt{t} \tag{8.6}$$

where left part of the equation is the fraction of the drug released at time t and k is a constant related with drug concentration and diffusivity. This equation can be derived from second Fick's law in the limit of short time release (for $M_t < 60\% M_\infty$). A cartoon description of this release model is depicted in Fig. 8.14b [112].

2. Power law. This model is slightly more generalized than the previous as:

$$\frac{M_t}{M_\infty} = kt^n \tag{8.7}$$

the exponent n characterizes the type of release mechanism (fickian, case-II, etc.)

3. Weibull model is another empirical model often used:

$$\frac{M_t}{M_\infty} = 1 - e^{-\alpha t^\beta} \tag{8.8}$$

both α and β are fitting parameters of this stretched exponential function (which can be understood as a continuous sum of exponential decays). Even if it does not have a physical meaning, Monte Carlo simulations demonstrated that diffusion of molecules from porous material follows this law [113].

Regarding mesoporous materials, drugs are released in a controlled manner through several combinations of mechanisms including desorption, diffusion and degradation at different stages [105]. In particular, concerning desorption and diffusion (and neglecting the possible carriers degradation), the following four steps take place consecutively after immersing the drug carriers in biological media: (1) absorption of aqueous medium into the mesoporous system driven by capillary forces; (2) dissolution of drug molecules into the release medium inside the pores; (3) diffusion of drug molecules out of the mesopores owing to concentration gradients; (4) diffusion and convectional transport of drug molecules in the release medium. Among the several factors influencing the drug release rate from mesoporous carriers, pore size and drug-internal surface interactions are the most important to control release kinetics [46, 105]. The functionalization of the outer layer, drug loading methods and release environment are other important factors determining the drug release rate.

If a fast release of the payload molecules is desired, there should not be any strong interactions between the payload molecules. The pore size should also be optimized to facilitate the diffusion of the molecules out from the pores, which favours larger pores, but in contrast, smaller pores may be preferred as they prevent the formation of ordered structures of the payload compound [46]. The critical factor to enhance the dissolution behaviour is to reduce the lattice energy (high Gibbs free energy) of well ordered crystalline structures by producing amorphous or disordered structures [44, 62]. If the pores of PSi are large enough for the fast diffusion of the molecules out of the pores, a fast release is obtained and it typically obeys a first order kinetic. However, depending on the pore diameter and morphology, the diffusion of the drug molecules might be the limiting step of the drug release [43]. Generally, the release of poorly soluble drugs can be fitted with Higuchi model, suggesting that drug release is a diffusion limited process [46].

8.9.6 Smart PSi

For sustained drug delivery applications, the drug release rate should be prolonged and this can be effectively achieved by covalently bind the molecules on the pore walls of the inorganic mesoporous carrier [43].

Moreover, the prominent advantage of PSi nanoparticles as drug delivery carriers is the near zero premature controlled release by entrapping drugs inside their pores. This is the prerequisite for achieving long blood circulation time and targeting cells or tissues for stimuli responsive drug release via intravenous administration. Hence, the drug release takes place only when PSi nanocarriers are triggered by external or internal stimuli that are manipulated at a desired location and time. For example, the drug release can be triggered and fine-tuned by the incorporation of responsive polymers creating a so-called 'capping' or 'gatekeeper' concept. This kind of approach can be attained by employing a pH-responsive, chitosan-based hydrogel film to cap the pores of the PSi layer after being loaded with insulin [105].

Some smart PSi carriers demonstrated controlled release of multiple drugs for sequential combination therapy: two anticancer drugs were separately loaded into the pores and into the outer shell [114]. In another study, an anticancer drug has been covalently attached to the inner pore walls of mesoporous silicon and its oxidation-triggered release has been demonstrated [111].

8.10 Conclusions

This chapter reports about the fabrication and use of PSi material in optical sensing and drug delivery applications. Despite being known since decades, PSi still shows intriguing possibilities to develop smart systems for bio-related applications. The large tunability of its porous structure, together with its luminescent properties and biocompatibility, make it an ideal material to realize complex nanodevices. These will be able to sense the different biological environments they pass through and can be programmed to react to specific stimuli.

The first part of the chapter deals with the PSi fabrication using the classical eletrochemical method. This is the most simple yet powerful technique to produce PSi sample and free standing membranes. Then, the two most common optical sensing methods are described: MCs and polarimetric approaches are presented and their pros & cons are highlighted.

In the second part, a review of the issues associated with conventional drug delivery is highlighted and it is shown that DDS can be used for improving both oral and intravenous drug administration. In particular these systems allow to achieve a controlled and targeted DD; another attractive aspect is that it is possible to engineer smart systems.

After a brief introduction on the various DDS, the properties of the PSi that make it a promising vehicle for DD, have been discussed: good biocompatibility, low toxicity and high pore volume necessary for hosting large amounts of drugs. Moreover, the size and shape of both the particles and the pores can be easily controlled.

Particles stabilization is an essential, but not sufficient step. Surface functionalization is crucial for DD applications and an attractive PSi feature is the possibility to tailor its surface properties (wettability, surface charge and presence of functional groups), that are strongly related to drug loading, release and pharmacokinetics.

The methodologies concerning drug loading and quantification of the loaded drug have been discussed, highlighting that to obtain reliable results the use of many different quantification methods is recommended.

Another important aspect that has been discussed deals with the physical state in which the drug compound is loaded in porous materials: it depends on several factors and affects the drug release profile.

The last topic that has been reviewed is the release kinetics. The advantage of using DDS is that, through precise control of the drug carrier architecture, the release of the drug can be tuned to achieve a desired kinetic profile.

Acknowledgements The authors thanks prof. M. Scarpa for the continuous support and collaboration, Dr. N. Kumar for his collaboration in the microfluidic measurements. This work was supported by the Italian Ministry of University and Research through the "Futuro in Ricerca" project RBFR12OO1G NEMATIC.

References

1. Pissadakis S, Selleri S (2015) Optofluidics, sensors and actuators in microstructured optical fibers. Woodhead Publishing
2. Washburn AL, Bailey RC (2011) Photonics-on-a-chip: recent advances in integrated waveguides as enabling detection elements for real-world, lab-on-a-chip biosensing applications. Analyst 136(2):227–236
3. Bates KE, Lu H (2016) Optics-integrated microfluidic platforms for biomolecular analyses. Biophys J 110:1684–1697
4. Mukundan H, Anderson AS, Grace KW, Grace KM, Hartman N, Martinez JS, Swanson BI (2009) Waveguide-based biosensors for pathogen detection. Sensors 9(7):5783–5809
5. Taitt CR, Anderson GP, Ligler FS (2005) Evanescent wave fluorescence biosensors. Biosens Bioactuators 20(12):2470–2487
6. Zeng S, Baillargeat D, Ho H-P, Yong K-T (2014) Nanomaterials enhanced surface plasmon resonance for biological and chemical sensing applications. Chem Soc Rev 43(10):3426–3452
7. Cao J, Sun T, Grattan KTV (2014) Gold nanorod-based localized surface plasmon resonance biosensors: a review. Sens Actuators B Chem 195:332–351
8. Estevez CM, Otte MA, Sepulveda B, Lechuga LM (2014) Trends and challenges of refractometric nanoplasmonic biosensors: a review. Anal Chim Acta 806:55–73
9. Seidel M, Niessner R (2008) Automated analytical microarrays: a critical review. Anal Bioanal Chem 391(5):1521–1544

10. Bally M, Halter M, Voeroes J, Grandin MH (2006) Optical microarray biosensing techniques. Surf Interf Anal 38(11):1442–1458
11. Cretich M, Chiari M (eds) (2016) Peptide microarrays. Methods and protocols. Springer
12. Kimmel DW, LeBlanc G, Meschievitz ME, Cliffel DE (2012) Electrochemical sensors and biosensors. Anal Chem 84(2):685–707
13. Pan L, Yu G, Zhai D, Lee HR, Zhao W, Liu N, Wang HL, Tee BC-K, Shi Y, Cui Y, Bao ZN (2012) Hierarchical nanostructured conducting polymer hydrogel with high electrochemical activity. Proc Natl Acad Sci Am 109(24):9287–9292
14. Wu C, Chiu DT (2013) Angew Chem Int Ed 52(11):3086–3109
15. Zijlstra P, Paulo PMR, Orrit M (2012) Optical detection of single non-absorbing molecules using the surface plasmon resonance of a gold nanorod. Nat Nanotech 7(6):379–382
16. Ament I, Prasad J, Henkel A, Schmachtel S (2012) Single unlabeled protein detection on individual plasmonic nanoparticles. Nano Lett 12(2):1092–1095
17. Baaske MD, Foreman MR, Vollmer F (2014) Single-molecule nucleic acid interactions monitored on a label-free microcavity biosensor platform. Nat Nanotech 9(11):933–939
18. Kang S, Nieuwenhuis AF, Mathwig K, Mampallil D, Lemay SG (2013) Electrochemical single-molecule detection in aqueous solution using self-aligned nanogap transducers. ACS Nano 7(12):10931–10937
19. Lemay SG, Kang S, Mathwig K, Singh PS (2013) Single-molecule electrochemistry: present status and outlook. Acc Chem Res 46(2):369–377
20. Guo X (2013) Single-molecule electrical biosensors based on single-walled carbon nanotubes. Adv Mater 25(25):3397–3408
21. Wade JH, Bailey RC (2016) Applications of optical microcavity resonators in analytical chemistry. Annu Rev Anal Chem 9(3):1–25
22. Lynn NS, Sipova H, Adama P, Homola J (2013) Enhancement of affinity-based biosensors: effect of sensing chamber geometry on sensitivity. Lab Chip 13:1413–1421
23. Kumar N, Froner E, Guider R, Scarpa M, Bettotti P (2014) Investigation of non-specific binding in nanoporous flow-through and flow-over based sensors. Analyst 139:1345
24. Zhao Y, Gaur G, Mernaugh RL, Laibinis PE, Weiss SM (2016) Comparative kinetic analysis of closed-ended and open-ended porous sensors. Nanoscale Res Lett 11:395
25. Uhlir Jr A (1956) Electrolytic shaping of germanium and silicon. Bell Syst Tech J 35:333–347
26. Canham LT (1990) Silicon quantum wire array fabrication by electrochemical and chemical dissolution of wafers. Appl Phys Lett 57(10):1046–1048
27. Ghulinyan M, Oton CJ, Gaburro Z, Bettotti P, Pavesi L (2003) Porous silicon freestanding coupled microcavities. Appl Phys Lett 82:1550
28. Lehmann V, Rönnebeck S (1999) The physics of macropore formation in low-doped p-type silicon. J Electrochem Soc 146(8):2968–2975
29. Smith RL, Collins SD (1992) Porous silicon formation mechanisms. J Appl Phys 71(8):R1–R22
30. Sailor Michael J (2012) Porous Silicon in practice. Wiley-VCH, Weinheim, Germany
31. Lehmann Volker (2002) Electrochemistry of silicon. Wiley-VCH, Weinheim, Germany
32. Lehmann V, Föll H (1990) Formation mechanism and properties of electrochemically etched trenches in n-type silicon. J Electrochem Soc 137(2):653–659
33. Lehmann V (1993) The physics of macropore formation in low doped n-type silicon. J Electrochem Soc 140(10):2836–2843
34. Lehmann V, Stengl R, Luigart A (2000) On the morphology and the electrochemical formation mechanism of mesoporous silicon. Mater Sci Eng B 69:11–22
35. Lehmann V, Rönnebeck S (1999) Physics of macropore formation in low-doped p-type silicon. J Electrochem Soc 146(8):2968–2975
36. Föll H, Christophersen M, Carstensen J, Hasse G (2002) Formation and application of porous silicon. Mater Sci Eng R 39(4):93–142
37. Bettotti P, Dal Negro L, Gaburro Z, Pavesi L, Lui A, Galli M, Patrini M, Marabelli F (2002) P-type macroporous silicon for two-dimensional photonic crystals. J Appl Phys 92:6966–6972

38. Carstensen J, Christophersen M, Föll H (2000) Pore formation mechanisms for the Si-HF system. Mater Sci Eng B 69:23–28
39. Christophersen M, Carstensen J, Feuerhake A, Föll H (2000) Crystal orientation and electrolyte dependence for macropore nucleation and stable growth on p-type Si. Mater Sci Eng B 69:194–198
40. Ouyang H, Christophersen M, Viard R, Miller BL, Fauchet PM (2005) Macroporous silicon microcavities for macromolecule detection. Adv Funct Mater 15:1851–1859
41. Kumar N, Gennaro S, Sasikumar PVW, Soraru GD, Bettotti P (2014) Self detachment of free-standing porous silicon membranes in moderately doped n-type silicon. Appl Phys A 116:251–257
42. Huang Z, Geyer N, Werner P, de Boor J, Gösele U (2010) Metal-assisted chemical etching of silicon: a review. Adv Mater 23(2):285–308
43. Lehto V-P, Salonen J, Santos HA, Riikonen J (2013) Nanostructured silicon-based materials as a drug delivery system for water-insoluble drugs. In: Drug delivery strategies for poorly water-soluble drugs. Wiley, pp 477–508
44. Salonen J, Laitinen L, Kaukonen AM, Tuura J, Björkqvist M, Heikkilä T, Vähä-Heikkilä K, Hirvonen J, Lehto JV-P (2005) Mesoporous silicon microparticles for oral drug delivery: loading and release of five model drugs. J Control Release 108(2):362–374
45. Anglin EJ, Cheng L, Freeman WR, Sailor MJ (2008) Porous silicon in drug delivery devices and materials. Adv Drug Deliv Rev 60(11):1266–1277
46. Xu W, Riikonen J, Letho V-P (2013) Mesoporous systems for poorly soluble drugs. Int J Pharm 453(1):181–197
47. Cattaruzza F, Cricenti A, Flamini A, Girasole M, Longo G, Mezzi A, Prosperi T (2004) Carboxylic acid terminated monolayer formation on crystalline silicon and silicon nitride surfaces. A surface coverage determination with a fluorescent probe in solution. J Mater Chem 14:1461–1468
48. Gaburro Z, Bettotti P, Saiani M, Pavesi L, Pancheri L, Oton CJ, Capuj N (2004) Role of microstructure in porous silicon gas sensors for NO_2. Appl Phys Lett 85:555
49. Liao J, Li Z, Wang G, Chen C, Lva S, Lia M (2016) ZnO nanorod/porous silicon nanowire hybrid structures as highly-sensitive NO_2 gas sensors at room temperature
50. Harraz Farid A, Ismail Adel A, Bouzid Houcine, Al-Sayari Saleh A, Al-Hajry Ali, Al-Assiri Mohammad S (2015) Phys Stat Sol A 212(8):1851–1857
51. White IM, Fan X (2008) On the performance quantification of resonant refractive index sensors. Opt Express 16(2):1020–1028
52. Zhu H, White IM, Suter JD, Dale PS, Fan X (2007) Analysis of biomolecule detection with optofluidic ring resonator sensors
53. Zhang YN, Zhao Y, Lv R-Q (2015) A review for optical sensors based on photonic crystal cavities. Sensors Actuators A 233:374–389
54. Avrutsky I, Gibson R, Sears J, Khitrova G, Gibbs HM, Hendrickson J (2013) Linear systems approach to describing and classifying Fano resonances. Phys Rev B 87:125118
55. Lee K-L, Huang J-B, Chang J-W, Wu S-H, Wei P-K (2015) Ultrasensitive biosensors using enhanced Fano resonances in capped gold nanoslit arrays. Sci Rep 5:8457
56. Kovalev D, Polisski G, Diener J, Heckler H, Knzner N, Timoshenko YuV, Koch F (2001) Strong in-plane birefringence of spatially nanostructured silicon. Appl Phys Lett 78(7):916–918
57. Bettotti P, Kumar N, Pavesi L, Alvarez J, Hill D (2012) A polarimetric sensor based on nanoporous free standing membranes. In: Proceedings of IEEE sensors, Taipei
58. Sant S, Tao SL, Fisher OZ, Xu Q, Peppas NA, Khademhosseini A (2012) Microfabrication technologies for oral drug delivery. Adv Drug Deliv Rev 64(6):496–507
59. Park K, Kwon IC, Park K (2011) Oral protein delivery: Current status and future prospect. React Funct Polym 71(3):280–287
60. Ahmed A, Bonner C, Desai TA (2002) Bioadhesive microdevices with multiple reservoirs: a new platform for oral drug delivery. J Control Release 81(3):291–306

61. Win KY, Feng S-S (2005) Effects of particle size and surface coating on cellular uptake of polymeric nanoparticles for oral delivery of anticancer drugs. Biomaterials 26815:2713–2722
62. Santos HA, Salonen J, Bimbo LM, Letho V-P, Peltonen L, Hirvonen J (2011) Mesoporous materials as controlled drug delivery formulations. J Drug Deliv Sci Technol 21(2):139–155
63. Ponchel G, Irache J-M (1998) Specific and non-specific bioadhesive particulate systems for oral delivery to the gastrointestinal tract. Adv Drug Deliv Rev 34(2):191–219
64. Ariga K, Lvov YM, Kawakami K, Ji Q, Hill JP (2011) Layer-by-layer self-assembled shells for drug delivery. Adv Drug Deliv Rev 63(9):762–771
65. Khanvilkar K, Donovan MD, Flanagan DR (2001) Drug transfer through mucus. Adv Drug Deliv Rev 48(2):173–193
66. Bimbo LM, Mäkilä E, Laaksonen T, Lehto V-P, Salonen J, Hirvonen J, Santos HA (2011) Drug permeation across intestinal epithelial cells using porous silicon nanoparticles. Biomaterials 32(10):2625–2633
67. Carvalho FC, Bruschi ML, Evangelista RC, Gremião MPD (2010) Mucoadhesive drug delivery systems. Braz J Pharm Sci 46(1):1–17
68. das Neves J, Bahia MF, Amiji MM, Sarmento B (2011) Mucoadhesive nanomedicines: characterization and modulation of mucoadhesion at the nanoscale. Expert Opin Drug Deliv 8(8):1085–1104
69. Tasciotti E, Liu X, Bhavane R, Plant K, Leonard AD, Price BK, Cheng MM-C, Decuzzi P, Tour JM, Robertson F (2008) Mesoporous silicon particles as a multistage delivery system for imaging and therapeutic applications. Nat Nanotech 3(3):151–157
70. Hughes GA (2005) Nanostructure-mediated drug delivery. Nanomedicine 1(1):22–30
71. Sagadevan S, Periasamy M (2014) A review on role of nanostructures in drug delivery system. Rev Adv Mater Sci 36(2):112–117
72. Bamrungsap S, Zhao Z, Chen T, Wang L, Li C, Fu T, Tan W (2012) Nanotechnology in therapeutics: a focus on nanoparticles as a drug delivery system. Nanomedicine 7(8):1253–1271
73. Park K (2007) Nanotechnology: what it can do for drug delivery. J Control Release 120(1–2):1
74. EU adopted a definition of a nanomaterial in 2011: recommendation on the definition of a nanomaterial (2011/696/EU). http://eur-lex.europa.eu/legal-content/EN/TXT/?uri=CELEX:32011H0696
75. Farokhzad OC, Langer R (2009) Impact of nanotechnology on drug delivery. ACS Nano 3(1):16–20
76. De Jong WH, Borm PJA (2008) Drug delivery and nanoparticles: applications and hazards. Int J Nanomed 3(2):133
77. Hoffman AS (2008) The origins and evolution of 'Controlled' drug delivery systems. J Control Release 132(3):153–163
78. Singh R, Lillard JW (2009) Nanoparticle-based targeted drug delivery. Exp Mol Pathol 86(3):215–223
79. Florence AT, Hillery AM, Hussain N, Jani PU (1995) Nanoparticles as carriers for oral peptide absorption: studies on particle uptake and fate. J Control Release 36(1):39–46
80. Mura S, Nicolas J, Couvreur P (2013) Stimuli-responsive nanocarriers for drug delivery. Nat Mater 12(11):991–1003
81. He J, Qi X, Miao Y, Wu H-L, He N, Zhu J-J (2010) Application of smart nanostructures in medicine. Nanomedicine 5(7):1129–1138
82. Swami A, Shi J, Gadde S, Votruba AR, Kolishetti N, Farokhzad OC (2012) Nanoparticles for targeted and temporally controlled drug delivery, pp 9–29
83. Lavelle EC, Sharif S, Thomas NW, Holland J, Davis SS (1995) The importance of gastrointestinal uptake of particles in the design of oral delivery systems. Adv Drug Deliv Rev 18(1):5–22
84. Mahanraj VJ, Chen Y (2006) Nanoparticles—a review. Trop J Pharm REs 5(1):561–573
85. Panyam J, Labhasetwar V (2003) Biodegradable nanoparticles for drug and gene delivery to cells and tissue. Adv Drug Deliv Rev 55(3):329–347

86. Siefker J, Karande P, Coppens M-O (2014) Packaging biological cargoes in mesoporous materials: opportunities for drug delivery. Expert Opin Drug Deliv 11(11):1781–1793
87. Caban S, Aytekin E, Sashin A, Capan Y (2014) Nanosystems for drug delivery. OA Drug Des Deliv 18(2):1–7
88. Nakki S, Rytkonen J, Nissinen T, Florea C, Riikonen J, Ek P, Zhang H, Santos HA, Narvanen A, Xu W (2015) Improved stability and biocompatibility of nanostructured silicon drug carrier for intravenous administration. Acta Biomater 13:207–215
89. Köllner S, Dünnhaupt S, Waldner C, Hauptstein S, de Sousa IP, Bernkop-Schnürch A (2015) Mucus permeating thiomer nanoparticles. Eur J Pharm Biopharm 97:265–272
90. Sarparanta MP, Bimbo LM, Mäkilä EM, Salonen JJ, Laaksonen PH, Kerttuli Helariutta AM, Linder MB, Hirvonen JT, Laaksonen TJ, Santos HA (2012) The mucoadhesive and gastroretentive properties of hydrophobin-coated porous silicon nanoparticle oral drug delivery systems. Biomaterials 33(11):3353–3362
91. Foraker AB, Walczak RJ, Cohen MH, Boiarski TA, Grove CF, Swaan PW (2003) Microfabricated porous silicon particles enhance paracellular delivery of insulin across intestinal Caco-2 cell monolayers. Pharm Res 20(1):110–116
92. Conniot J, Silva JM, Fernandes JG, Silva LC, Gaspar R, Brocchini S, Florindo HF, Barata TS (2014) Cancer immunotherapy: nanodelivery approaches for immune cell targeting and tracking. Front Chem 2:105
93. Ekambaram P, Abdul Hasan Sathali A, Priyanka K (2012) Solid lipid nanoparticles: a review. Sci Rev Chem Commun 2(1):80–102
94. Uchechi O, Ogbonna JDN, Attama AA (2014) Nanoparticles for dermal and transdermal drug delivery. In: Sezer AD (ed) Applications of nanotechnology in drug delivery. InTech, pp 193–235
95. Ochekpe NA, Olorunfemi PO, Ngwuluka NC (2009) Nanotechnology and drug delivery part 2: nanostructures for drug delivery. Trop J Pharm Res 8(3):275–287
96. Fonseca-Santos B, Gremiao MPD, Chirlli M (2015) Nanotechnology-based drug delivery systems for the treatment of Alzheimers disease. Int J Nanomed 10:4981
97. Sanvincens N, Marco MP (2008) Multifunctional nanoparticles—properties and prospects for their use in human medicine. Trends Biotechnol 26(8):425–433
98. Mirjalili F, Soltani M, Chen P (2012) Nanotechnology in drug delivery a systems. Int J Drug Del 4:275–288
99. Tzankov B, Yoncheva K, Lambov N (2014) Mesoporous silica nanoparticles as drug carriers. Pharmacia 61(1):27–37
100. Bharti C, Nagaich U, Pal AK, Gulati N (2015) Mesoporous silica nanoparticles in target drug delivery system: a review. Int J Pharma Investig 5(3):124–133
101. Teng Z, Han Y, Li J, Yan F, Yang W (2010) Preparation of hollow mesoporous silica spheres by a sol–gel/emulsion approach. Micropor Mesopor Mater 127(1):67–72
102. DeMuth P, Hurley M, Wu C, Galanie S, Zazhariah MR, DeShong P (2011) Mesoscale porous silica as drug delivery vehicles: synthesis, characterization, and pH-sensitive release profiles. Micropor Mesopor Mater 141(1):128–134
103. Wu X, Wang Z, Zhu D, Zong Z, Yang L, Zhong Y, Cui Y (2013) pH and thermo dual-stimuli-responsive drug carrier based on mesoporous silica nanoparticles encapsulated in a copolymer–lipid bilayer. ACS Appl Mater Interf 5(21):10895–10903
104. Salonen J, Kaukonen AM, Hirvonen J, Letho V-P (2008) Mesoporous silicon in drug delivery applications. J Pharm Sci 97(2):632–653
105. Mu C, Shen H (2016) Silicon nanoparticles and microparticles. In: Nanomaterials in pharmacology. Springer, pp 153–183
106. Santos HA, Riikonen J, Salonen J, Mäkilä E, Heikkilä T, Laaksonen T, Peltonen L, Lehto V-P, Hirvonen J (2010) In vitro cytotoxicity of porous silicon microparticles: effect of the particle concentration, surface chemistry and size. Acta Biomater 6(7):2721–2731
107. Fox CB, Kim J, Le LV, Nemeth CL, Chirra HD, Desai TA (2015) Micro/nanofabricated platforms for oral drug delivery. J Control Release 219:431–444

108. Heikkilä T, Salonen J, Tuura J, Kumar N, Salmi T, Murzin DYu, Hamdy MS, Mul G, Laitinen L, Kaukonen AM (2007) Evaluation of mesoporous TCPSi, MCM-41, SBA-15, and TUD-1 materials as API carriers for oral drug delivery. Drug Deliv 14(6):337–347
109. Kaasalainen M, Mkil E, Riikonen J, Kovalainen M, Jrvinen K, Herzig KH, Lehto VP, Salonen J (2012) Effect of isotonic solutions and peptide adsorption on zeta potential of porous silicon nanoparticle drug delivery formulations. Int J Pharm 431(1–2):230–236
110. Prestidge CA, Barnes TJ, Lau CH, Barnett C, Loni A, Canham L (2007) Mesoporous silicon: a platform for delivery of therapeutics. Expert Opin Drug Deliv 4(2):101–110
111. Wu EC, Park J-H, Park J, Segal E, Cunin F, Sailor MJ (2008) Oxidation-triggered release of fluorescent molecules or drugs from mesoporous Si microparticles. ACS Nano 2(11):2401–2409
112. Siepmann J, Peppas NA (2001) Modeling of drug release from delivery systems based on hydroxypropyl methylcellulose (HPMC). Adv Drug Deliv Rev 48:139–157
113. Kosmidis K, Argyrakis P, Macheras P (2003) Fractal kinetics in drug release from finite fractal matrices. J Chem Phys 119:6373–6377
114. Xu W, Thapa R, Liu D, Nissinen T, Granroth S, Närvänen A, Suvanto M, Santos HA, Lehto HV-P (2015) Smart porous silicon nanoparticles with polymeric coatings for sequential combination therapy. Mol Pharm 12(11):4038–4047

Chapter 9
Modeling Thermal Transport in Nano-Porous Semiconductors

M. Verdier, K. Termentzidis and D. Lacroix

9.1 Introduction

Nano-porous semiconductors can be used for several applications in biology, chemistry, optics, energy, etc. Starting with the pioneering work by Canham [1] on photo-luminescence of mesoporous silicon at room temperature, nano-porous semiconductors have drawn the attention of the scientific community. Their specific optical properties have been extensively discussed in review papers about photonics and optoelectronics [2, 3]. Chemical and biological applications of porous silicon are also numerous. Their ratio of surface exchange area to the control volume of the porous material is very large, up to 500 m^2/cm^3 according to Buriak [4], enabling a wide range of applications related to the surface chemistry. Furthermore, through its tailoring, these materials can be functionalized [5, 6] and also used for filtering, sensing or medical applications, etc. During the last decade, a new field of applications has arisen which is the thermal management of nanostructures [7]. In this framework, knowing how heat is transported within the material is a crucial issue since thermal properties are affected by the nanostructuration of materials. In this chapter, models that can be used to describe heat transport in nano-porous materials, mostly semiconductors, will be reviewed.

In the late 90s, photo-acoustic [8] and 3-ω [9] measurements of porous silicon films conclude on a strong reduction of their thermal conductivity (TC). For silicon

M. Verdier · K. Termentzidis · D. Lacroix (✉)
LEMTA, Université de Lorraine, UMR 7563, 54500 Vandœuvre-lès-Nancy, France
e-mail: david.lacroix@univ-lorraine.fr

M. Verdier
e-mail: maxime.verdier@univ-lorraine.fr

K. Termentzidis
e-mail: konstantinos.termentzidis@univ-lorraine.fr

M. Verdier · K. Termentzidis · D. Lacroix
LEMTA, CNRS, UMR 7563, 54500 Vandœuvre-lès-Nancy, France

P. Bettotti (ed.), *Submicron Porous Materials*,
DOI 10.1007/978-3-319-53035-2_9

porous layers with very high porosities, TC was observed to decrease by 2 to 3 orders of magnitude [9]. Physical reason that explains such a behavior in semiconductor materials is phonon boundary scattering. Phonons (normal modes of vibration) are energy carriers [10] in solids. Phonon transport and scattering can explain heat transfer at microscopic scales, like for photons all their scattering processes tend to limit energy transport. This is particularly the case in nano-porous materials which are heterogeneous and complex structures (dendritic, sponge like, etc.) with atomic size crystallites that are in the same length order as phonon mean free path (MFP).

TC lowering can be a serious drawback but also an interesting feature depending on the application. For optoelectronic devices like solar cells, an increased efficiency can be reached using mesoporous semiconductors such as silicon [11] or germanium. But it might deteriorate if the device temperature increases due to a poor global TC. On the other hand, nano-porous layers can be useful for insulating parts of microelectronic devices. Decreasing and controlling TC of nanostructured semiconductors are presently big challenges for thermoelectric (TE) materials and phononic crystals (PnCs) [12, 13]. For the latters, rapid progress in nanoengineering allows the manufacturing of structures made of "vertical pores" organized in a periodic array. Their TC can be accurately tailored through the choice of pore size and pitch between pores. For such devices, thermal properties are comparable or better to those of nanowire arrays; but they are more easier to elaborate [14]. Besides, that kind of porous material can exhibit particular features such as coherence effects, through the coherent reflection of phonons at internal surfaces [15]. In these "thermocrystals," heat transport can be controlled and application such as heat waveguides, or thermal diodes are expected opening path to several advanced applications. In connection with nano-porous media, the nanocomposite material's design and characterization are also attractive research topic where several recent studies focused on TC lowering through the nanoscale tailoring. The enhancement of the figure of merit *ZT* is often the purpose of these works and several models used to describe these materials can be easily transposed to nano-porous media [16–18].

Literature dealing with models and simulations of porous semiconductors thermal properties is recent. After the pioneering experimental works on mesoporous silicon [1], several papers were devoted to the modeling of p or n types porous silicon. For these porous media with dendritic structures, calculations of the TC were held using numerical techniques already developed in the field of radiative heat transfer like the discrete ordinate method (DOM) [19] or through steady-state Monte Carlo (MC) simulations [20]. In the case of periodic microporous films and phononic crystals numerical simulations have been also developed. Pioneering work by R. Prasher gave a description of these materials with an analytical approach [21] and a ballistic-diffusive approximation [22]. After that, more complex techniques were developed. Among them, they are the MC method proposed by Hao et al. [23] or the DOM approaches by Romano et al. [24] and Hsieh et al. [25]. But also hybrid methods lying on lattice dynamics and continuum mechanics were proposed by Reinke [26] or Dechaumphai and Chen [27]. Other studies were also done using modified kinetic theory approximation according to the Callaway-Holland model [28, 29] and using a MFP related to the porosity occurrence. In the case of very small pores, few molecular dynamic (MD) studies were carried out, for example, in the group of Galli [30, 31] and by Fang and Pilon [32] or by Şopu et al. [33] who used MD to compute TC of nano-porous germanium made with spherical pores. Such approaches using MC and MD modeling were also developed in our group [34–36].

As it is discussed above, nano-porous materials are interesting candidates in versatile and diverse applications [37]. A very recent molecular dynamics study of the thermal conductivity of silicon of Dettori et al. [38] showed a break down of the Eucken model (reliable at macroscale). In the latter, they have studied different porosities and pore diameters and also random pore distribution. The trend is that the porosity and the randomness of the pore distribution decrease rapidly the thermal transport, but in their study they did not add an amorphous shell around the pores, which is the most often in the case of porous materials. Several studies of the thermal conductivity of nano-porous materials via molecular dynamics exist in the literature, which most of them study cylindrical pores [39–44], except the work of He et al. [31], in which they have added an amorphous shell around their cylindrical pores. In our work, we modeled spherical pores with and without amorphous shell around them, and our modeling of the amorphous/crystalline interfaces is reproducible and respect the experimental measurements for the interfacial energy between the two phases.

Among the existing models dedicated to the assessment of nano-porous materials TC, one has to decide which one is the most suitable. Some of them can handle complicated geometries, other describe accurately amorphous phase on the edges of pores and other can provide reasonable TC appraisal with simple and fast numerical simulation. Whatever is the sought application the purpose of this chapter is to give an overview and examples of the models and tools that can be used to calculate TC of nano-porous semiconductors. The chapter is organized as follows. First, the basis of nano-porous material thermal design is recalled as well as the analytical models used to predict their properties. Then, in the second section the MC modeling applied to phonon transport in porous media is addressed. In this section, modeling of the Boltzmann transport Equation for phonon is explained and the simulation results are discussed and compared for several kinds of nano-porous thin films TC assessment. Last, in the third section thermal properties of nano-porous structures is considered in the framework of molecular dynamic. Theory used to determine TC from atom motion is briefly recalled and some examples are provided.

9.2 Analytic Models for Nano-porous Material Thermal Properties

Hereafter are detailed models for thermal transport in porous and composite materials, namely media with pores or inclusions. Those models are generic, they mostly rely on the assumption of the effective medium. In such an approach micro and nano-transport of energy carriers in the structure is taken into account in bulk materials through the so-called mixture rules. The latter ones are more or less complex and usually depends on the combination of the considered physical property at small scale combined to the volume fraction the different phases [45].

9.2.1 Effective Medium Theory for Heat Conduction as a Function of Porosity

The very first attempts to describe heat conduction in composite materials were undertaken by Maxwell [46] and Lord Rayleigh [47], for a system of spherical particles embedded in a host medium [48]. These approaches were extended by Hasselmann and Johnson [49] to include thermal boundary resistance (TBR) also known as "Kapitza resistance". More recently, Nan [50] introduced a simplified multiple-scattering based model to assess TC of arbitrary "particulate/porous" composites, allowing to account for values of the TC that are beyond the usual bounds, defined by the thermal conductivities of the constituent phases. To do so, an overall effective thermal conductivity (ETC) was determined with a Green function approach, assuming ergodic hypothesis, i.e., the considered heterogenous material is statically homogeneous. According to Nan [50] for a porous media, calculations lead to an ETC that reads:

$$k_e = k_h \frac{k_p(1+2\alpha) + 2k_h + 2\phi\left[k_p(1-\alpha) - k_h\right]}{k_p(1+2\alpha) + 2k_h - \phi\left[k_p(1-\alpha) - k_h\right]} \tag{9.1}$$

where k_h is the bulk TC of the host medium, k_p is the TC of pore phase (for air k_p = 0.024 $\mathrm{Wm^{-1}K^{-1}}$), ϕ is the volume fraction of the pores, namely the porosity. $\alpha = 2\,Rk_h/d$ is a parameter that characterizes the energy losses at the pore boundaries, R is the TBR (Kapitza resistance) and d is the characteristic length of the pore (diameter for spherical inclusions). In the case of porous media, the pore phase is often a gas or vacuum. Thus, the effective TC reduces to the Eucken formula:

$$k_e^{\star} = \frac{k_e}{k_h} = \frac{2-2\phi}{2+\phi} \tag{9.2}$$

9.2.2 Effective Medium Theory for Heat Conduction as a Function Phonon MFP

The above-discussed models are appropriate for materials with sufficiently large pore sizes. However, they present important flaws when the pore size is of the same order or smaller than the phonon mean free path (MFP) [51]. In the latter case, Yang [16] showed that ETC varies with the radius of the nano-pores and must account on the ballistic nature of phonon transport. As a consequence, TC reduction is mainly due to the MFP changes due to the presence of the nano-pores. This arises from the fact that phonon-interface scattering rate increases with interface density Φ (surface area of the pores per unit volume of composite). For spherical pores of diameter d, $\Phi = 6\phi/d$ and the collisional MFP, Λ_{pores}, becomes [17]:

$$\Lambda_{pores} = \frac{4}{\Phi} = \frac{2d}{3\phi} \tag{9.3}$$

The smaller the pores are, for a given porosity, the smaller the MFP becomes, and thus greater scattering rate is obtained.

A first attempt to include the MFP in ETC is the modified effective medium (MEM) approach proposed by Minnich [17]. The starting point is the kinetic theory, which states that the bulk TC depends on the phonon MFP according to $k_{KT} \approx \frac{1}{3}Cv_g\Lambda$. Where $C(p,\omega)$ is the volumetric specific heat at a given frequency, $v_g(p,\omega)$ the group velocity and $\Lambda(p,\omega)$ the global phonon mean free path. In a nanocomposite material made of two distinct phase (host matrix and inclusion), two different MFP are involved: the one related to the bulk host phase, $\Lambda_{b,h}$ and the one related to the bulk inclusion medium, $\Lambda_{b,p}$. Considering the Matthiessen approximation, the effective MFP of the host medium depends on polarization p and frequency ω through the group velocity of the considered phonon and writes as

$$\frac{1}{\Lambda_{e,h}(p,\omega)} = \frac{1}{\Lambda_{b,h}(p,\omega)} + \frac{1}{\Lambda_{pores}} = \frac{1}{\Lambda_{b,h}(p,\omega)} + \frac{3\phi}{2d} \tag{9.4}$$

Equation (9.4) gives a new estimate of the effective TC related to the host medium $k_{e,h}$ that encompasses various effects related to the porosity. Similarly, the effective MFP of the pore inclusion medium is given by Eq. (9.5) [17, 52]:

$$\frac{1}{\Lambda_{e,p}(p,\omega)} = \frac{1}{\Lambda_{b,p}(p,\omega)} + \frac{1}{d} \tag{9.5}$$

Including these two effective thermal conductivities in Nan's formula, Eq. (9.1), leads to the overall ETC that now includes both the classical size effects and the porosity effects on the lattice TC. For nonspherical pores, overall ETC can be conveniently expressed in terms of the interface density Φ and the characteristic length of the pores d. Neglecting the TC of inclusions, to depict nano-porous media, the ETC reduces to

$$k_e^\star = \frac{k_e}{k_h} = \left[1/\left(1 + \frac{3\phi\Lambda_{b,h}}{2d}\right)\right]\left(\frac{2-2\phi}{2+\phi}\right) \tag{9.6}$$

Recently, Tarkanyan and Niarchos [53] also proposed an ETC value using phonon MFP in porous materials that can be generalized to a distribution of spherical porosities. Such an attempt is interesting to model real materials where pores are often poly dispersed in terms of size. In the case of homogeneous nano-porous media ETC becomes

$$k_e^{\star} = \frac{k_e}{k_h} = \left[(1-\phi) \Big/ \left(1 + \frac{3\phi \Lambda_{b,h}}{2d}\right)\right] \tag{9.7}$$

The ETC expressed by Eqs. (9.2, 9.6, and 9.7) will be used hereafter to compare numerical results to analytic models.

9.3 BTE for Nano-Porous Material Thermal Properties

9.3.1 Phonon Transport Modeling

Numerous techniques can be used to model heat transport at nanoscale. A growing number of models and simulation tools devoted to the thermal characterization of nanostructured material have emerged during the past years. In the review by Luo and Chen [54] most of the latters are recalls, the main ones are the kinetic theory (KT), the first principle calculations (DFT), the molecular dynamics (MD), the atomistic Green's function, and the multiscales simulations (MS). Except the kinetic theory, the first methods are usually used in the case of very small simulation domains (supercell made of hundred of atoms for DFT, nanostructures made of hundred thousands of atoms for MD). The others, labeled MS, can be used to model mesoscales structures and can be applied to real engineering problems in order to provide temperature fields or heat fluxes within the nanostructures. However, both are useful since atomistic modeling provides reference values for thermal properties appraisal at nanoscales or inputs for MS models. Among multiscales simulations, several techniques exist. They usually rely on the solution of the BTE in the relaxation time approximation (RTA). Some techniques are derived from those developed in the field of the radiative transfer simulations like the DOM that has been successfully applied for several nanostructures see work by Volz [55]. More recently, computing resources fast improvement allows other methods such as the MC techniques [20, 23, 56–59]. Each of them is an improvement of the previously developed MC methods. In the following subsection, the statistical description of BTE applied to a porous medium will be discussed.

9.3.2 Monte Carlo Modeling of the BTE

The Boltzmann transport equation for phonons describes the variations of the distribution function $f(t, \mathbf{r}, p, \mathbf{K})$ which depends on time t, location $\mathbf{r}$, polarization p, and wave vector $\mathbf{K}$. $f(t, \mathbf{r}, p, \mathbf{K})$ can also be defined as the mean particle number at time t in the $d^3\mathbf{r}$ volume around $\mathbf{r}$ with $\mathbf{K}$ wave vector and $d^3\mathbf{K}$ accuracy for a given polarization p. In the absence of external force, the BTE expression is [60]

$$\frac{\partial f}{\partial t} + \nabla_{\mathbf{K}}\omega \cdot \nabla_{\mathbf{r}} f = \left.\frac{\partial f}{\partial t}\right|_{collision,} \tag{9.8}$$

with the phonon group velocity $\mathbf{v}_g = \nabla_{\mathbf{K}}\omega$ and $f^0(T, p, \mathbf{K})$ the equilibrium thermodynamic phonons population for polarization p and wave vector $\mathbf{K}$, also known as the Bose-Einstein distribution function:

$$f^0(p, \mathbf{K}, T) = \frac{1}{\exp\left[\frac{\hbar\omega(p,\mathbf{K})}{k_B T}\right] - 1.} \tag{9.9}$$

The general formulation of the BTE is often simplified considering the RTA. In this case, the right-hand side term of Eq. (9.8) is written:

$$\left.\frac{\partial f}{\partial t}\right|_{collision} = \frac{f^0(p, \mathbf{K}, T) - f(t, p, \mathbf{r}, \mathbf{K})}{\tau(p, \mathbf{K}, T, \ell),} \tag{9.10}$$

where τ is the phonon scattering relaxation time that depends on the phonon dispersion properties, on the temperature and possibly on a characteristic length ℓ. Scattering relaxation times can be expressed directly from atomistic simulations (DFT) or modeled with semi-empirical analytic relations.

The previously discussed equations are solved in two steps when using MC modeling. First, the advective part of the BTE is addressed (left-hand side term of Eq. (9.8)) assuming that there is no scattering. In this frame, a statistically representative population of phonons moves within the simulation domain with respect to (i) its group velocity within the nanostructure, (ii) the internal reflections at pore surfaces and borders. Second, the phonon scattering step is achieved considering Eq. (9.10). The characteristic relaxation time is given according to the usually considered "Matthiessen" summation rule. It takes into account: normal (N), umklapp (U), and impurity (I) scattering rates according to Holland [61] formalism. The second stage of the simulation can be also achieved using total phonon lifetimes derived from DFT calculation. In the latter case, the anisotropy of the dispersion relation is accounted for. Such calculations are longer but can be, in principle, used for any kind of semiconductors.

Knowing that the simulation is frequency dependent, the dispersion properties of materials shall be taken into account. In present chapter, we used data extensively detailed in previous works for silicon [61–63] and germanium [61, 64]. These parameters are given in Appendix. The optical modes contribution to the TC is neglected here; this is a weak assumption near room temperature and below which is commonly done.

9.3.3 Numerical Handling of the BTE

Using the "two steps" method discussed above in Sect. 9.3.2, the phonon transport within nanostructures is modeled through the sampling of a particle population which is representative of the theoretical total number of phonons N_{tot}. The first point which is common to all the existing MC techniques is that the whole theoretical phonon population N_{tot} cannot be sampled due to the large number of quasi-particles to follow. As a consequence, representative ensembles of phonons, "bundles," with the same frequency and polarization are defined to reduce the number of sampled particles. Two methods can be considered to select these bundles. The first one, developed by Mazumder [56], lies on a sampling based on a numeral basis with a constant weighting factor $W = N_{tot}/N_{\mathrm{MC}}$ where N_{MC} is the number of sampled phonons in the MC simulation. Sampling is achieved when theoretical energy E_{tot} of the structure is equal to the sampled one $E_{\mathrm{MC}} = \sum_{N_{\mathrm{MC}}} W \times \hbar\omega$. Drawbacks related to the energy conservation during the scattering stage can occur with this method. Improvements have been achieved in order to restore energy taking into account creation/destruction schemes [20, 56], scattering rates [57, 58] or using genetic algorithms [65]. This approach remains efficient as long as the factor W is not too large.

A second method for phonon sampling, based on energy consideration, was proposed by Péraud et al. [59]. In the latter one, the energy of a sampled phonon bundle is prescribed and remains constant $e_{i,\mathrm{MC}} = e_{\mathrm{Bundle}}$ whatever the value of the ith sampled frequency ω_i is, whereas in the previous case it was equal to $e_{i,\mathrm{MC}} = W\hbar\omega_i$. The phonon sampling is done on the basis of a normalized cumulative distribution function based on phonon energy F_E until the total sampled energy $E_{\mathrm{MC}} = N_{\mathrm{MC}} \times e_{\mathrm{Bundle}}$ is equal to E_{tot}. Distribution functions are as follows:

$$F_N(j, T) = \frac{\sum_p \sum_{i=1}^{j} g_p D(p, \omega_i) f^0(p, \omega_i, T),}{\sum_p \sum_{i=1}^{N_b} g_p D(p, \omega_i) f^0(p, \omega_i, T)} \tag{9.11}$$

$$F_E(j, T) = \frac{\sum_p \sum_{i=1}^{j} g_p D(p, \omega_i) f^0(p, \omega_i, T) \hbar\omega_i}{\sum_p \sum_{i=1}^{N_b} g_p D(p, \omega_i) f^0(p, \omega_i, T) \hbar\omega_i,} \tag{9.12}$$

where g_p is the polarization degeneracy and N_b is the number of spectral bins used for the phonon frequency discretization. The second formulation, based on energy scheme for sampling, provides accurate results for the same number of drawn phonons.

When dealing with a MC technique the accuracy of the achieved results is a major issue. In the case of BTE modeling, uncertainties decrease as the number of sampled bundles increases as well as simulation time. This dilemma often works against the use of MC methods because consequent numerical resources are required to provide reliable results. Recent studies [58, 59] have suggested to sample only "useful phonons," i.e., those which are above a reference temperature T_{ref} chosen at the beginning of the simulation. This reference temperature is usually set just below the lowest temperature prescribed by boundary conditions. Thus, the total number of phonons beyond T_{ref} is $\widetilde{N}_{tot}(T) = N_{tot}(T) - N_{tot}(T_{ref})$; it reads as follows:

$$\widetilde{N}_{tot}(T) = \int_V \int_\omega \sum_p g_p D(p,\omega) \left[f^0(p,\omega,T) - f^0(p,\omega,T_{ref})\right] d\omega dV, \quad (9.13)$$

This technique is implemented in both normalized cumulative distribution functions relying on Eq. (9.12) to achieve phonon sampling. It is also implemented in the procedures that calculate the polarization, the energy and the derived pseudo-temperature of the phonon bundle ensembles. The latter points are not detailed here, but can be found in [57–59].

A spatial and a spectral discretizations are achieved. The spectral one is done using equally spaced N_b spectral bins taken over the whole phonon spectral range (depending on the material). The spatial discretization is achieved using cubic cells (N_c) that fit the domain geometry. For nanowires and other non-Cartesian geometries the shape of the cell can be adapted. The initial sampling of the phonon population is usually done for an uniform temperature in the sample.

The last point of the modeling is about boundary conditions (BC) and discretizations. Different kinds of BC can be set, such as the isothermal and the adiabatic ones. Isothermal BC are used to set the temperature at edges of the nanostructure. Adiabatic BC, which assumes specular or diffuse reflection of phonons, allow the modeling of thin films and nanowires. Temperature at boundaries can be set either assuming that a constant heat flux is emitted from the corresponding surface or using a fictitious cell, outside of the domain, which is kept to the prescribed temperature [20, 57]. It should be also noted that the "reference temperature" concept discussed in the previous section must be also accounted when BC are implemented.

The MC simulation tool can be easily parallelized and usually 10 to 20 nodes are enough to ensure sufficiently low standard deviations on temperatures and heat fluxes. Every calculation needs to be carried out until the steady state is reached. Typical time step for micron length structures is equal or lower to $\Delta t = 1$ ps. TC appraisal is done in a post-processing stage, averaging the stabilized phonon heat flux on a sufficiently long integration time (several nanoseconds).

9.3.4 Kinetic Theory Integral Formulation for Porous Media

The TC assessment can be also addressed with the kinetic theory (KT) for phonons. With this approach, k is expressed through an integral formulation as a function of the total relaxation time (according to the Matthiessen's rule) that includes all kinds of scattering events, it reads as follows:

$$k_{KT}(T) = \frac{1}{3}\sum_{p}\int_{0}^{\omega_{p,max}} C(p,\omega,T)v_g^2(p,\omega)\tau(p,\omega,T)d\omega, \tag{9.14}$$

This formulation can be modified to take into account the occurrence of nanoinclusions as suggested by Minnich and Chen [17] or of empty pores as discussed by Hopkins [28] and more recently by Tarkhanyan [53]. In the latter cases, an added scattering process takes into account internal boundaries. It can be derived from geometric considerations as

$$\frac{1}{\tau_{pores}} = \frac{v_g(p,\omega)}{\Lambda_{pores,}} \tag{9.15}$$

where Λ_{pores} is the characteristic length related to the collision between phonons and pores. As it will be discussed further, the MFP Λ_{pores} can be appraised on the basis of the medium pore organization through analytical expression or using a ray tracing technique derived from the MC calculations.

9.3.5 Effective Monte Carlo Modeling

Considering the two previously discussed methods (MC and KT), we propose an effective Monte Carlo model (EMC) which gives a reliable description of nanoporous structures thermal properties using an equivalent medium that does not include real pores like those plotted in Fig. 9.1. The main advantage of this new technique is to lower computation time, especially when high porosities are considered ($\phi > 30\%$). Using this method, the MC procedure detailed in Sect. 9.3.3 remains almost the same. The only difference lies on the fact that there is no more calculations of phonon path intersection with pore cavities. Instead, a new relaxation time τ_{pores}, given by Eq. (9.15), is used in a new collision procedure added to the "drift step" of the MC simulation. A scattering probability $P_{scat}(\tau_{pores})$ is calculated and a random number is drawn. If the latter one is smaller than $P_{scat}(\tau_{pores})$, the phonon bundle is back-scattered in the half-space behind its incoming direction. Frequency and polarization remain the same and the phonon final location is calculated using the remaining duration of the current time step. In the case of high pore density, within a time step, several scattering events that modify the phonon path might occur.

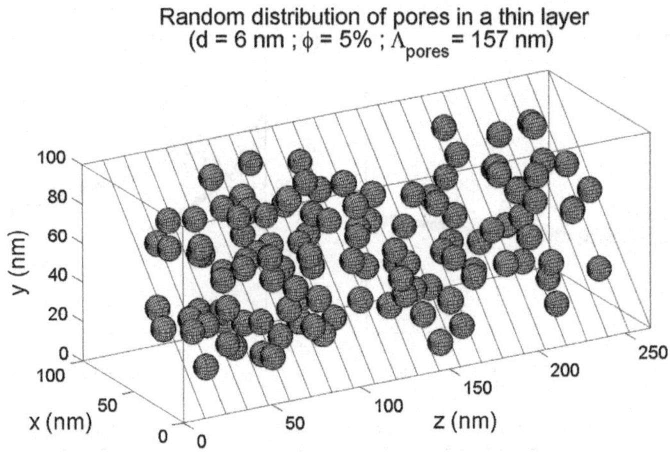

Fig. 9.1 Modeled porous materials, outlook of thin film (240 nm) with spherical pores of the same diameter (d = 12.4 nm) and porosity $\phi = 5\%$, $\Lambda_{pores} = 157$ nm; there is no interconnection between pores

In order to be reliable, this calculation needs an accurate value of Λ_{pores} which depends on pore nature, organization, and density. As discussed in Sect. 9.2, analytic approximation based on the effective medium theory can be used. Following, the approach suggested by Minnich [17], in the case of isotropically distributed spheres of diameter d, MFP is given by Eq. (9.3). For other simple pore geometries the MFP can be appraised on the basis of scattering cross-section considerations. In the case of cylindrical ones (present in phononic crystals PCs), various pore shape effects on TC have been studied by Liu and Huang [66].

In our group, we have developed an alternate technique to obtain the phonon MFP due to porosity scattering. This method is fast and simple, it lies on phonon MC ray tracing (MCRT) in a real porous media where pores are randomly (or not) set. The methodology has been already done to appraise $\tau_{boundary}$ in case of thin silicon nanowires [67]. The MC simulation tool is used to count the scattering events with pores $N_{scat\ pores}$ in each frequency bin for both TA and LA polarizations. A large number of phonons are drawn ($N_{ph} = 10^6$ in cubic cell with length equal to 100 nm) to ensure accurate statistical averaging. A single time step is required to achieve this calculation. Then using the probability law given in [57], one can obtain Λ_{pores}:

$$\Lambda_{pores}(p,\omega,d,\phi) = \frac{v_g \Delta t}{\ln\left(\dfrac{N_{ph}(p,\omega)}{N_{ph}(p,\omega) - N_{scat\ pores}(p,\omega)}\right)}. \tag{9.16}$$

This MFP does not change with polarization or frequency, which was expected, but it still depends on d and ϕ. Next, MCRT simulation results for a homogeneous nano-porous Si thin film with the geometrical characteristics $\phi = 5\%$, $d = 12.4$ nm,

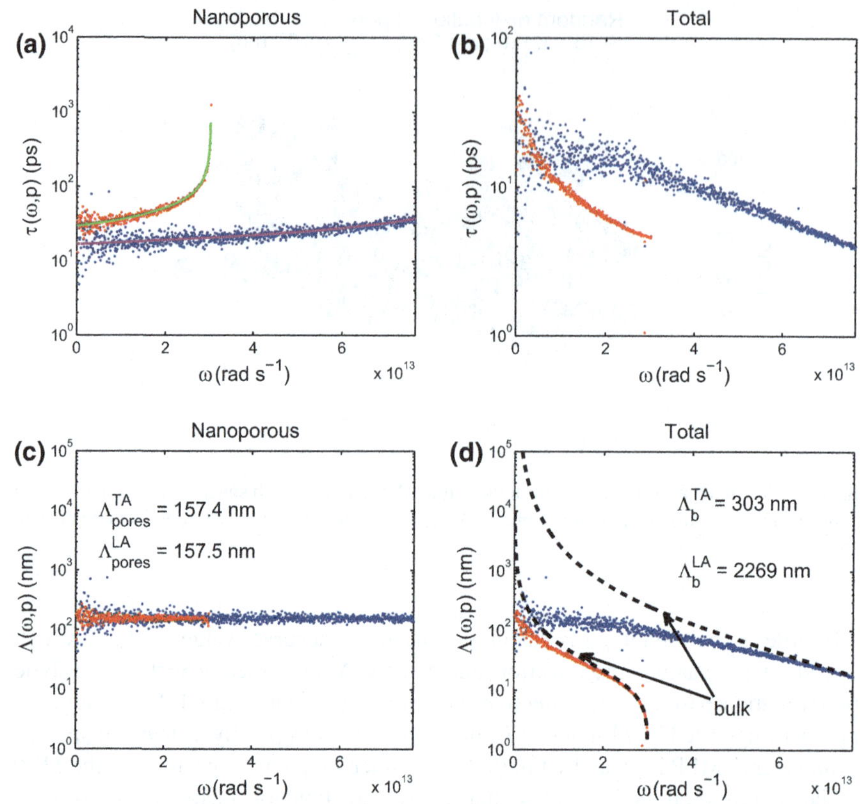

Fig. 9.2 MC simulations of phonon relaxation times and mean free path, for phonon/pore scattering and total scattering, in nano-porous materials versus frequency and polarization; $\phi = 5\%$, $d_{pores} = 12.4$ nm

and $T = 300$ K are provided in Fig. 9.2. In the sub-figures are plotted the relaxation time and the related MFP for collisions between phonons and pores (Λ_{pores}), and also the total ones which include normal, umklapp and impurity/defect scattering mechanisms (Λ_{np}). Red dots are related to the TA branches and blue dots to the LA one. All the sub-figures are plotted as a function of the phonon frequency ω. As expected, in Fig. 9.2c it can be noted that the phonon MFP due to pore collisions remains constant whatever the frequency or the polarization. The average value being equal to $\Lambda_{pores} \simeq 157$ nm. In Fig. 9.2d, the total MFP for both LA and TA branches is plotted the, black-dashed lines correspond to the bulk MFP variations with frequency (using Holland relaxation times [61] and Pop dispersion properties for Si [62]). It can be noted that porosity strongly affects LA phonons in the first half of the Brillouin zone. On the other hand, TA phonons are mostly affected for very low frequencies where the bulk mean free path is larger than the porous one. In this figure there are also given the average bulk MFP for LA and TA branches with respect to the Bose-

Einstein distribution times the DOS. A global MFP, for bulk, can be calculated as a Debye average:

$$\frac{1}{\overline{\Lambda}} = \frac{1}{3}\left(\frac{2}{\Lambda_{TA}} + \frac{1}{\Lambda_{LA}}\right), \tag{9.17}$$

For silicon a commonly admitted value is about $\overline{\Lambda}_b = 300$ nm, here at 300 K the calculated value derived from the analytic expression of relaxation times is equal to $\overline{\Lambda}_b = 452$ nm. This bulk value is coherent with one given in [17] where $\overline{\Lambda}_b =$ 268 nm or the ones proposed by Chen [68] which were between 409 and 2604 nm according the chosen dispersion model. Assuming a simple Matthiesen's rule $\Lambda_{np}^{-1} = \Lambda_b^{-1} + \Lambda_{pores}^{-1}$, the average MFP of the porous layer $\Lambda_{np} = 117$ mn is reduced here by a factor 4 as compared to bulk for a porosity of 5%. Obviously, it impacts the TC of nano-porous sample ($k \propto \Lambda$) as we will see in the result section.

Phonon MFP calculated by both MCRT and Eq. (9.3) are given for several porosities keeping d constant and several pore diameters using a constant value of ϕ. These results are reported in Table 9.1. It can be noted that both methods give similar results as long as porosity is lower than $\phi = 10\%$, the relative error between the models is less than 10%. When porosity gets larger, the discrepancies between both methods increases. When looking at Eq. (9.3), one can see that for high porosities (i.e., $\phi \to 1$) the MFP is constant and equal to $2d/3$. This has no meaning since the MFP is supposed to decrease as porosity increases with a limit value equal to 0. As a consequence, we suggest to compare MCRT calculation to the matrix MFP proposed by Liu [66] which makes an average MFP outside $\Lambda_{pores\ out}$ and inside $\Lambda_{pores\ in}$ in the case of a material made of inclusions embedded in a matrix. In the specific case of a porous material $\Lambda_{pores\ in} = 0$ and the MFP for collisions between phonons and pores Λ_{pores} reads as follows:

$$\Lambda_{pores} = (1-\phi)\Lambda_{pores\ out} + \phi\Lambda_{pores\ in} = \frac{2d(1-\phi)}{3\phi.} \tag{9.18}$$

The latter expression fulfills high and low porosity criteria and provides a nice agreement with the MCRT results.

9.3.6 *MC Modeling Results and Discussions*

In this result section, TC assessment of nano-porous thin films is discussed varying several parameters as temperature, size of pores, porosity, and porous MFP. Both direct and effective MC modeling approaches are used and compared to analytic models. Some insights on key parameters that rules the thermal transport in nano-porous semiconductors are also discussed.

Table 9.1 Phonon bundle MFP Λ_{pores} calculated by MCRT with uncertainty $\Delta\Lambda_{MCRT}$ or using Eq. (9.3) [17] or Eq. (9.18) [66]; for a given diameter and variable porosity and for a given porosity and variable pore diameter (up and down)

$d = 6$ nm	Λ_{pores} (nm)			$\Delta\Lambda_{MCRT}$ (nm)
ϕ (%)	Eq. (9.3)	Eq. (9.18)	MCRT	
3.14	127.4	123.4	125.1	1.0
5.50	72.3	68.6	67.3	0.7
7.63	52.4	48.4	48.9	0.4
10.18	39.3	35.3	35.5	0.3
15.27	26.2	22.2	22.2	0.1
20.36	19.6	15.6	15.6	0.1
28.27	14.1	10.1	10.1	0.1
30.54	13.1	9.1	9.2	0.1
33.08	12.1	8.1	8.3	0.1
38.17	10.5	6.5	6.6	0.1
$\phi = 5\%$	Λ_{pores} (nm)			$\Delta\Lambda_{MCRT}$ (nm)
d (nm)	Eq. (9.3)	Eq. (9.18)	MCRT	
3.0	40.0	38.0	37.5	0.6
5.0	66.7	63.4	61.4	0.7
8.0	106.7	101.4	99.9	0.7
12.4	165.3	157.0	157.0	1.7
16.8	224.5	216.0	216.0	1.9
21.2	282.9	268.8	271.5	1.0
26.7	356.5	338.7	337.1	0.1
31.7	422.7	401.6	400.5	0.7

9.3.6.1 Influence of the Temperature

Different kinds of nanospheres organizations can be considered to model the nanoporous materials. Among them, a random distribution should be the most representative of the real material. However, in some calculations aligned pores can be modeled. In such a case, geometrical considerations can be put forward to understand the scattering mechanisms that rule phonon transport. This approach can be also considered to describe phononic structures.

Temperature dependence is first studied for Si and Ge in the case of thin films. In order to assess the reliability of MC simulations, calculations have been done for samples with and without pores. In each case, the film is bounded by hot and cold walls, with a temperature difference $\Delta T = 2$ K. The film thickness is increased as mean temperature decrease in order to match the variation of the phonon MFP in bulk. TC is calculated from phonon heat flux in z direction, assuming Fourier behavior, when steady state is reached. The porosity is set to a constant value of $\phi = 5\%$ with randomly distributed spheres ($R_s = 7.62$ nm, $N_s = 486$). In Fig. 9.3,

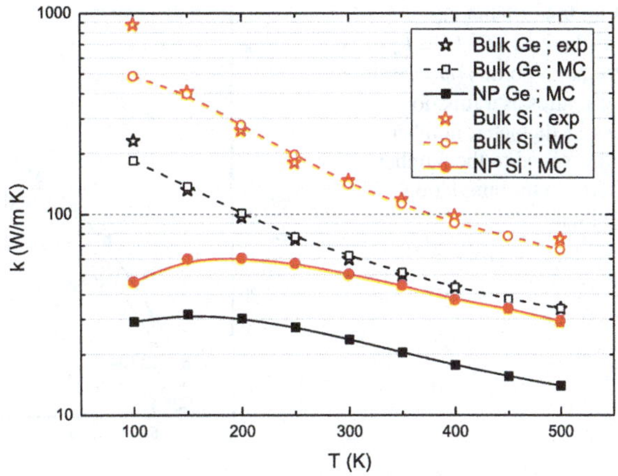

Fig. 9.3 Si and Ge TC; bulk and nano-porous ($\phi = 5\%$, mono-dispersion, random organization of the pores) structures variations with the temperature

our results and bulk experimental values are plotted. For "bulk" test case, a very good agreement is achieved for both Si and Ge until the temperature is lower than 150 K. Beyond this temperature, ballistic behavior of phonons requires films with very large thicknesses to retrieve the bulk thermal conductivity. In such a case the computational cost increases quickly for a given accuracy. TC of Si and Ge nano-porous devices (empty and filled squares) are, as expected, below the "bulk" ones. With the chosen porosity TC drop-off is stronger at low temperatures where phonon MFP is larger. This reduction of k is due to the introduction of a new scattering process related to the pore occurrence.

9.3.6.2 Influence of the Pore Size

Size of the pore is an important parameter for phonon transport in porous materials. Its impact on TC can be theoretically predicted using Eqs. (9.6) and (9.7). TC varies with the number of pores N_s as well as with the pore diameter d. Presently mono-dispersion is considered. For a given porosity (with pores of radius R), the product of the number of pores by the volume of the pores is constant so that the number of pores scales as $1/R^3$. The surface of the pores scales as R^2 so that the scattering surface of the pores is the number of pores times the surface of pores and scales as $1/R$. Thus, the smaller the pores are, the higher the scattering surface becomes. That picture is in agreement with numerical results obtained for nano-porous Si and Ge thin films ($L_z = 2\ \mu$m and $\phi = 5\%$) with sphere diameters varying from 3 to 30 nm, see Fig. 9.4. Furthermore, this effect prevails over the porosity value itself. It can be noted that even with a 5% porosity, a 90% reduction of the TC is achievable when pore radius is very small. As it has been observed in the previous study on temperature impact, the reduction is more effective on Si than on Ge, as phonon average MFP in silicon is larger than in Ge. For very small diameters, our results are coherent with those obtained by MD of Lee [30] who found for porous Si ($d = 2.26$ nm; $\phi = 7\%$) a TC

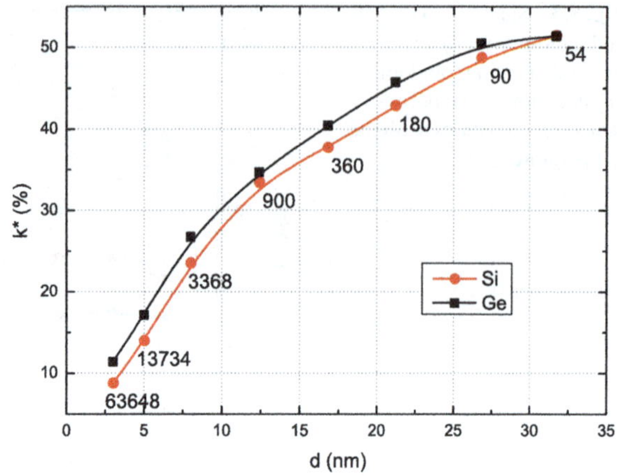

Fig. 9.4 Si and Ge normalized TC ($k^* = k_{np}/k_b$) for nano-porous ($\phi = 5\%$) structures as a function of pore's diameter; numbers correspond to the number of pores in the simulation domain

of 7.1 W m^{-1}K^{-1} when we found 12.6 W m^{-1}K^{-1} for $d = 3$ nm and $\phi = 5\%$. Yet, it shall be the lower diameter limit of these direct MC calculations as simulation time exponentially increases when trajectories and collisions between phonons and a huge number of pores need to be considered at each time step. In such a case EMC modeling is much more efficient.

As silicon and germanium have the same behavior regarding porosity dependency, only Si will be considered hereafter.

9.3.6.3 Influence of the Porosity

The porosity is a more adequate physical parameter to use when dealing nano-porous media. In Fig. 9.5 is plotted the influence of porosity in Si nano-porous thin films. For these calculations, 486 spheres are set in 3 dimensional cubic cells, that is a 27 spheres per cell organized pattern. The lowest porosity is obtained with 4 nm radius spheres, whereas the largest porosity is achieved for radius equal to 15 nm. It shall be noticed that (i) even at low porosity, the reduction effect is not negligible, (ii) the more the porosity increases, the greater the reduction is. In the same graph, the TC obtained from models discussed in Sect. 9.2 are also plotted. Except the Eucken's model, which neither takes into account the pore size nor the MFP, our results are in pretty good agreement with theoretical predictions of nano-porous media with spherical inclusions.

9.3.6.4 Influence of the MFP

As mentioned in Sect. 9.2, MFP directly affect TC as a result of phonon-pore scattering. This phenomenon is appraised in Fig. 9.6 that summarizes into a unique plot several of the previously discussed calculations of TC. It can be clearly seen that the

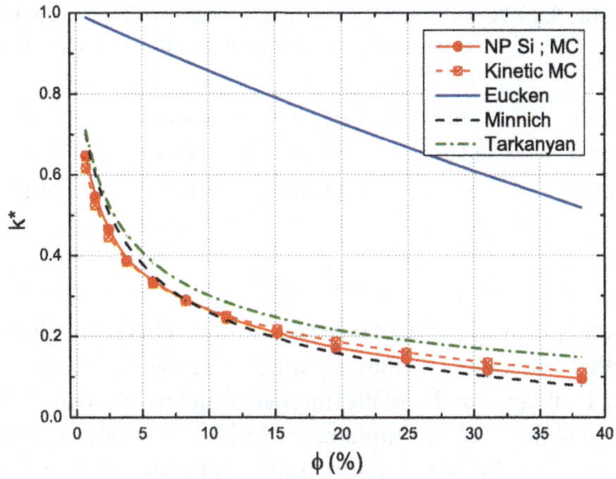

Fig. 9.5 Si normalized TC ($k^* = k_{np}/k_b$) in nano-porous structures at 300 K as a function of the porosity (*lower axis*) and of the diameter (*upper axis*); MC and models: Eucken Eq. (9.2); Minnich Eq. (9.6); Tarkahanyan Eq. (9.7)

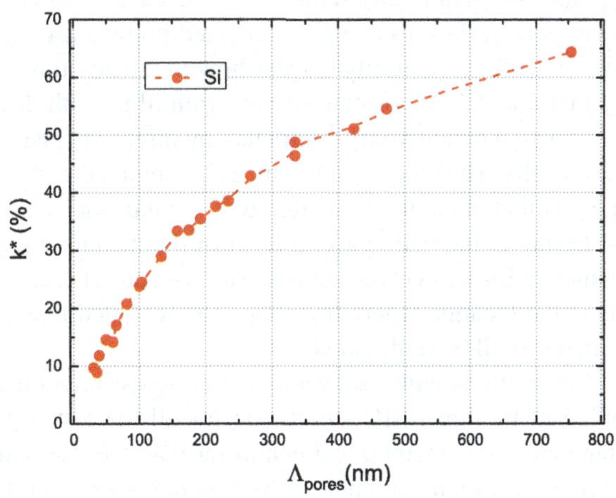

Fig. 9.6 Normalized TC ($k^* = k_{np}/k_b$) versus Λ_{pores} for nano-porous silicon at 300 K; data are taken from the previously discussed MC simulations

TC increases when Λ_{pores} becomes larger. MFP is the key parameter that gives the limit of the current MC simulations reliability. Indeed, when Λ_{pores} is smallest than $\simeq$5 nm several issues must be considered. First, phonon dispersion properties must be modified as in extremely thin nanostructures to take into account supplementary modes. Second, the phonon wavelength can be of the same order as the MFP and wave effect (coherence) should be considered. Last, on a numerical standpoint, time to reach the convergence of the simulations becomes very large and makes such technique less valuable.

Knowing that several parameters related to the pores (porosity, diameter, number of pores, distribution, MFP), we propose to study conductive transport in nano-porous devices from a more tractable model. The latter one relies on using an equivalent MC model, for which phonons scattering due to spherical pores is considered as occurring in an homogeneous slab with adjusted collisional parameters. In such a framework, we define a relaxation time associated to the diffusion with the

Table 9.2 TC k (W m^{-1}K^{-1}) of nano-porous Si; comparison between reference MC calculations and EMC simulations using four different techniques to model scattering by a porosity

ϕ [%]	Λ_{pores} (nm)	k	k_{iso}	k_+	k_-	k_R
3.80	236	62.34	80.04	119.5	62.28	70.10
8.25	134	42.40	59.6	97.8	42.7	48.7
15.00	82	27.46	41.1	78.47	28.21	32.82

nano-pores $\tau_{pores}(\omega, p)$ (see (Eq. 9.15)). This relaxation time is independent from the other processes, so that a phonon can undergo both a 3-phonons process (Normal or Umklapp) and a scattering event due to "porosity." The corresponding scattering probability can be implemented before the others scattering process. This process conserves the overall momentum, and adds a resistive contribution to heat transport. Technically, a difficulty remains. In the case of an equivalent medium, there is no more pore surface to compute reflected phonon propagation direction after scattering. Hereafter, we briefly discuss how to remedy this issue.

In Table 9.2, TC of nano-porous thin films with different porosities are given. k is the actual conductivity computed, by direct MC, with spherical pores in the simulation cells. The others values are TC, computed with EMC, that differ according to the way that phonons are scattered. k_{iso} stands for isotropic scattering, k_+ for forward scattering in the half-space in front of the incoming direction, and k_- for backward scattered. Last, k_R corresponds to a case where a fictitious normal surface is randomly set at the location where the phonon is scattered in order to define a hemisphere in which it shall be re-directed.

Calculations with backward scattering assumption provide TC close to the one obtained by direct MC simulation for all the tested porosities. From a numerical standpoint, this method is much faster than the one with real spherical pores. It has been test for mono and poly-dispersed pores with similar accuracy.

To conclude on this section, it can be stated that MC modeling approach is a versatile tool, which can handle several kinds of nanostructured porous materials. As long as incoherent interactions between phonons and pores exist this tool provides reliable outputs on thermal properties which can be compared to experimental data.

9.4 MD for Nano-Porous Material's Thermal Properties

9.4.1 Principle and Methodology

Molecular Dynamics (MD) is a numerical technique which aims to model the motion of each atom of a system with N atoms using the classical second Newton's law:

$$m_i \mathbf{a_i} = \sum_{j=1}^{N} \mathbf{F_{ij}} \qquad \text{with } i \neq j\,, \tag{9.19}$$

The forces and the individual positions and velocities for each atoms are computed at each time step Δt (usually of the order of the fs). Knowing these quantities during time evolution allows to compute many different properties as viscosity, chemical reaction velocity, rupture strength, or thermal conductivity. The forces applied on the atoms are calculated with

$$\mathbf{F_{ij}} = -\nabla V_{ij}. \tag{9.20}$$

In this equation, V is the interatomic potential between two particles i and j. The interatomic potentials are obtained either by experimental or theoretical data and can be classified in three categories: empirical, semi-empirical, and ab-initio. There are several types of interatomic potentials: 2-bodies (pair potential, like Lennard-Jones one), 3-bodies or many body potentials, each of them including or not Coulomb forces. The choice of the potential is really crucial to correctly model the system's behavior. For a same material, there are often several potentials which can represent more or less accurately the evolution of the system. Thus, MD users should be able to select the good potential according to the types of material, the sought physical properties, the needed precision and the computational cost. For instance, in the case of Silicon, one can choose the Stillinger-Weber (SW) [69] potential which is quite slow but accurate, or the Tersoff potential [70] to save some computational time if one does not need an exact modeling.

To compute the thermal conductivity k of a material with MD, there are several methods: Equilibrium Molecular Dynamics (EMD), Non-Equilibrium Molecular Dynamics (NEMD), and Approach to Equilibrium Molecular Dynamics (AEMD). The first one is based on the Green–Kubo formula which links the thermal conductivity to the flux fluctuations correlations [71]:

$$k_{xy} = \frac{\mathcal{V}}{k_B T^2} \int_0^\infty \left\langle J_x(0) J_y(t) \right\rangle dt, \tag{9.21}$$

$\mathcal{V}$ is the volume, x and y denote the directions. Thus, this relation directly provides the thermal conductivity tensor when steady state is reached. With NEMD, a temperature gradient is applied and k is obtained along a direction from the Fourier's law:

$$\mathbf{J} = -k\nabla T. \tag{9.22}$$

The temperature and the heat flux are computed from the atomic properties by, respectively:

$$T = \left\langle \sum_{i\in\mathcal{V}} \frac{m_i v_i^2}{3 k_B N} \right\rangle, \tag{9.23}$$

$$\mathbf{J} = \frac{d}{dt}\left(\sum_i \left(V(\mathbf{r_i}) + \frac{1}{2}m_i v_i^2\right)\mathbf{r_i}\right). \tag{9.24}$$

Thus, MD allows to represent the thermal transport by just modeling the motion of the atoms. The thermal transport is sometimes described in terms of transfer of vibrational energy between atoms. These vibrations have properties directly linked to phonons ones. Therefore, it is possible to get information about phonons as the Density of States, the participation ratio or the accumulative thermal conductivity with a MD simulation.

The systems simulated with MD are often small because of the computational cost. Indeed, a simulation box with dimensions of the order of few nanometers usually contains hundreds of thousands of atoms. It would be extremely expensive in terms of computational cost to model the dynamic of a system with micro-metric dimensions during several nanoseconds. In order to get the thermal conductivity of a macroscopic system, periodic boundary conditions can be applied in the different directions. With EMD, these conditions are not problematic as long as the dimensions of the periodic box are large enough to collect statistical data on the atoms. But with NEMD the dimension in the direction along which k is computed is limited to the distance between the two thermostats imposing the temperature gradient. To retrieve the thermal conductivity of an infinite size system, simulations are repeated with different sizes and k is extrapolated from a relation $1/k = f(1/L)$. Despite this, NEMD is sometimes faster than EMD because the latter needs a lot of data to reduce the statistical uncertainty. This requires to model the system during a long time (hundreds of ns). Concerning AEMD, the method is based on the same formula as EMD (Eq. 9.21), but half of the system is initialized at temperature T_1 and the other half is set to T_2. Then the thermal conductivity is computed while the system spontaneously evolves toward the thermodynamic equilibrium.

In MD, we must impose a thermodynamic ensemble to the system to study the dynamics. These ensembles indicate which thermodynamic variables are conserved during the simulation. For example, among the most used, one can choose the NVT ensemble (number of atoms, volume and temperature are conserved), the NPT (atoms, pressure and temperature) or the NVE (E for energy). The thermostats in the NEMD method are maintained at their temperature with the NVT ensemble. EMD are usually performed under NVE ensemble.

9.4.2 Nano-Porosity Modeling

Modeling a crystalline semiconductor with nano-pores with MD is a straight forward procedure. The crystalline structure is rapidly created by a simple program to generate a structure as it is the repetition of a periodic unit cell containing few atoms. In the case of Silicon, the crystalline structure is the same as diamond with a lattice constant $a_0 = 5.43$ Å. Then a nano-pore can be created by deleting atoms in specific regions of the system. Obviously, the size of the pore is limited to the size of the simulation box. Numerically, the absence of atoms is automatically managed by a good interatomic potential which can describe also the relaxation and reconstruction of the free surfaces.

The experimental elaboration of nano-pores in crystalline semiconductors is sometimes accompanied by an amorphization and/or an oxidation around the void. With MD, it is possible to take into account the effect of this layer on the thermal transport. A box of amorphous Silicon (a-Si) can be created (Fig. 9.7a) by simulating quenching and annealing of crystalline Si (c-Si) [72]. Then we have to define the regions where the amorphous shell surrounding the pore will be localized. The Si

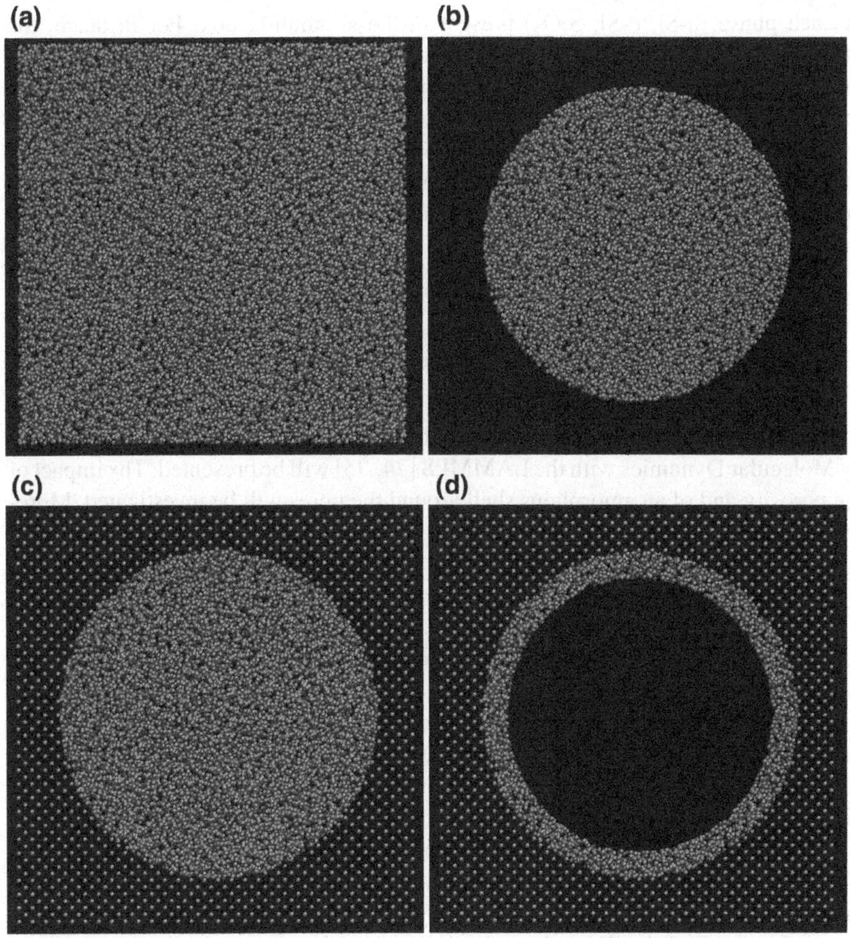

Fig. 9.7 Building of a nano-porous Silicon system with amorphous shell step by step. **a** Creation of the amorphous Silicon matrix. **b** Keep useful regions of a-Si. **c** Filling the space around the a-Si nanoinclusion with c-Si. **d** Creation of the nano-pore by deleting atoms

atoms which are outside this zone are deleted (Fig. 9.7b) and automatically replaced by other Si atoms which fulfill the crystalline arrangement (Fig. 9.7c). Then it only remains to erase the atoms in the pore region to get the nano-porous Si crystalline structure with an amorphous shell around the void (Fig. 9.7d). Finally, a last command deletes the atoms of a-Si and c-Si which are too close from each other (this can occur at the interface between a-Si and c-Si when the latter is created) in order to reproduce well the interfacial energy between the two phases. The same protocol can be accomplished with amorphous SiO_2 instead of a-Si to represent the oxidation. It may even extend to other materials in which an amorphization or an oxidation occurs when nano-pores are created.

When modeling such a system with MD, the interatomic potential must be carefully chosen. It must be able to correctly describe the interaction between atoms of each phase (a-Si, c-Si, SiO_2) present in the simulation box. For instance, the Stillinger-Weber potential has been adapted by Vink et al. [73] to model both a-Si and c-Si. In their work, they fitted the empirical coefficients of the SW potential for Si in order to respect both the a-Si and the c-Si behaviors. Thus, nanocomposites which contain amorphous and crystalline Silicon can be modeled with this potential and its particular coefficients. When Oxygen is present in the system (in the SiO_2), a potential which accurately describes interactions between O and Si is preferred, like the SiO Tersoff potential.

9.4.3 Nano-Porous Silicon Thermal Conductivity

In this section, the thermal conductivity of various nano-porous systems computed by Molecular Dynamics with the LAMMPS [74, 75] will be presented. The impact of the porosity and of an amorphous shell around the pores will be investigated. Moreover, a comparison of the phonon Density of State (DOS) of several configurations will be presented to give physical insights.

9.4.3.1 Crystalline Silicon with Spherical Nano-Pores

The thermal conductivity of Silicon with spherical nano-pores was calculated with the EMD technique. The dimension of the cubic box is $L_x = L_y = L_z = 10a_0 =$ 5.43 nm and a spherical void is created at the middle of the simulation box. The diameter of the pore is set between 0.5 and 6.5 nm. That means that the pores may be interconnected when the diameter is more than the box dimensions. The thermal conductivity of nano-porous crystalline Silicon is depicted in Fig. 9.8 as a function of the pore radius (lower axis) and the porosity (upper axis). The figure shows a drastic decreasing trend of the thermal conductivity when the porosity/radius increases, as it has been reported previously [34, 36, 76]. Moreover, a small fraction of voids as 10% can reduce the thermal conductivity of one order of magnitude (11 W/m.K instead of about 150 W/m.K for the bulk Si). This low value is due to the important

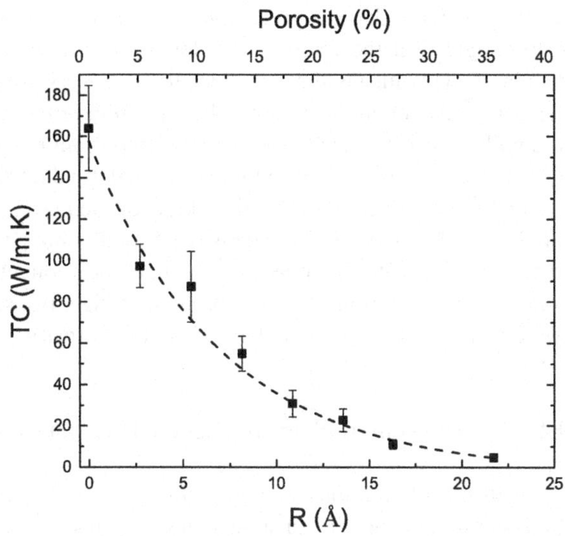

Fig. 9.8 Thermal conductivity of spherical nano-porous crystalline Silicon as a function of the pore radius and the porosity

scattering of the phonons on the free surface of the pores. Indeed, the total pore scattering surface of a system with N_p pores is determined by:

$$S = 4N_p\pi R^2 \tag{9.25}$$

The main parameter which controls the modification of thermal transport due to nano-porosity is the ratio of the scattering surface to the volume of material [34]. When this ratio increases, more phonon scattering occurs and the thermal transport is reduced. Assuming that L_x, L_y, and L_z are the dimension of the periodic box containing one pore, it can be written as

$$\frac{S}{\mathcal{V}} = \frac{4\pi R^2}{L_xL_yL_z(1-\phi)}. \tag{9.26}$$

From the definition of the porosity, we know that

$$L_xL_yL_z = \frac{\frac{4}{3}\pi R^3}{\phi}. \tag{9.27}$$

Then, replacing this term, the Eq. 9.26 becomes

$$\frac{S}{\mathcal{V}} = \frac{1}{R}\frac{3\phi}{1-\phi}. \tag{9.28}$$

The last term $3\phi/(1-\phi)$ tends to infinity when the porosity increases $(0 \leq \phi \leq 1)$. Thus, with a fixed radius of pores, increasing their number, we can obtain greater porosity which induces a great ratio $S/\mathcal{V}$ and so a low thermal conductivity. That seems logic. But the fraction $1/R$ tends to infinity when the radius of the pores decreases. This means that, in order to get a very weak thermal conductivity, the material must contain a great number (per unit volume) N_p of pores which must be as small as possible. Thus one keeps an important porosity while R stays small and $S/\mathcal{V}$ is large. This explains why the thermal conductivity is so low for our system, even with moderated porosity: the radius of the pore is less than 5 nm and the periodicity is 5.43 nm, so the pore density per unit volume is high and this has an impact on the phonon scattering rate. To conclude, the parameter $S/\mathcal{V}$ controls the thermal conductivity evolution of nano-porous materials because it takes into account the two quantities (N_p and R) hidden behind the porosity.

9.4.3.2 Amorphous Phase Impact on Thermal Conductivity

Geometric configurations studied in this section are globally the same as in the previous paragraph but with an amorphous silicon shell added around the pore with the method described in Sect. 9.4.2 in order to represent the amorphization which can occur during the nano-pores elaboration. Two distinct cases are investigated. First, the shell thickness is kept at a constant value $e = 0.5$ nm while the radius of the pore varies. Then, R is fixed at $2a_0 \simeq 1$ nm and e is the studied parameter. The thermal conductivities of the previous section are plotted again on Fig. 9.9 with double log axes, along with the new values for systems with the constant thickness shell for comparison.

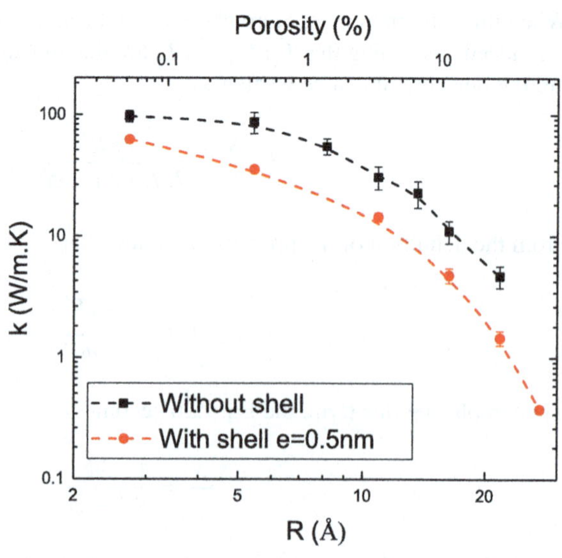

Fig. 9.9 Comparison of thermal conductivity versus radius of the pore for configurations with and without amorphous Silicon shell around the pore ($e = 0.5$ nm)

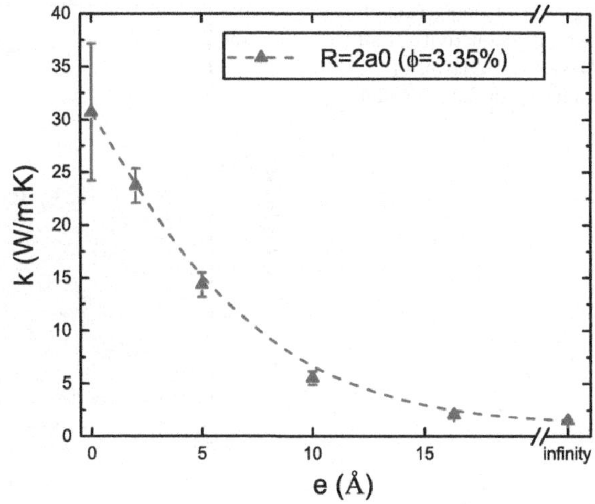

Fig. 9.10 Thermal conductivity versus amorphous shell thickness for constant *R*

The new curve is systematically below the former. Therefore, the amorphous shell contributes to the reduction of the thermal conductivity, as observed in references [31, 77]. Moreover, the TC reaches values lower than the amorphous limit (about 1.5–2 W/m.K) when *R* becomes greater than 2 nm, while a crystalline phase is still present in the system. Thus, the electron transport remains possible despite the amorphization around the pores. One can notice that the relative differences between same systems with and without amorphous shell is globally growing with porosity. This means that the reduction of the TC due to the amorphization increases when *R* increases. This can be explain in terms of amorphous fraction: a shell surrounding a wide sphere has a bigger volume than a shell with same thickness around a smaller sphere. So, for the systems compared in Fig. 9.9, the volume of amorphous Silicon is greater when the radius of the pore is large.

Now we comment on the impact of the amorphous shell thickness *e* on TC (Fig. 9.10). In our study, *e* varies from 0 to infinity while R is set to $2a_0 \simeq 1$ nm. In fact, $e = 0$ represents a simple nano-porous c-Si medium, and $e = \infty$ corresponds to a full amorphous Silicon matrix with nano-pores. The thermal conductivity obviously decreases when *e* increases, with the same exponential-like trend as in the Fig. 9.8. Thus, a small thickness shell of 1 nm is sufficient to reduce the TC as low as 5 W/m.K. Finally, the last configuration ($e = \infty$) has a TC lower than the amorphous limit, so nano-porosity can also influence thermal transport in an amorphous medium, as it was found by Coquil et al. [78].

Comparing same configurations of nano-porous Silicon with and without amorphous shell, it is possible to understand which is the main parameter universally controlling the thermal behavior of these systems. It was said in the previous Sect. 9.4.3.1 that for Si with spherical nano-pores the reduction of the TC was driven by the ratio of the scattering surface (i.e., the external surface of the pore) to the volume of material. This property can be extend to the case of an amorphous shell

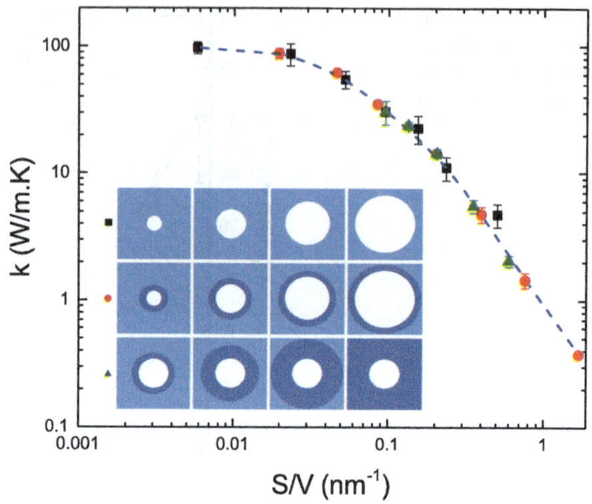

Fig. 9.11 Comparison of thermal conductivity of all the systems in terms of scattering surface to volume ratio

surrounding the pore. For this, it is necessary to define the "crystalline interface" as the surface separating c-Si to other phases (a-Si or void). For instance, the system in Fig. 9.7d have a crystalline interface of $4\pi(R+e)^2$. Thus, one takes into account equally the void and the amorphous phase for phonon scattering and limitation of thermal transport. Then, plotting the TC versus the ratio of crystalline interface to the material volume, results for the different configurations merge on the same curve (Fig. 9.11). It means that this parameter effectively controls the thermal conductivity for isotropic nano-porous systems. Moreover, it proves that a-Si is as much efficient as void to lower the TC at the nanoscale. The relatively high TC of c-Si is due to the extremely regular crystalline arrangement of its atoms. Thus, disturbing this regular atomic arrangement with a little void or adding a-Si, the TC decreases rapidly of one order of magnitude.

9.4.3.3 Phonon Density of States

The phonon Density Of State (DOS) of each system described before has been computed with Molecular Dynamics simulation. The *specden* tool included in the *VMD Signal Processing Plugin Package* was used to extract the DOS from the time-dependent coordinates of atoms. The most relevant DOS for comparison have been chosen and plotted in Figs. 9.12 and 9.13. The former shows the difference in the phonon frequency distribution due to the variation of the amorphous shell thickness. The main effects of this shell are (i) the strong reduction in the last peak (around 15 THz), corresponding to the TO mode, (ii) the global red shift which can clearly be observed at low frequencies and (iii) the smoothing of the curve, especially between 10 and 14 THz, for the LA and LO phonons. Finally, all these modifications due

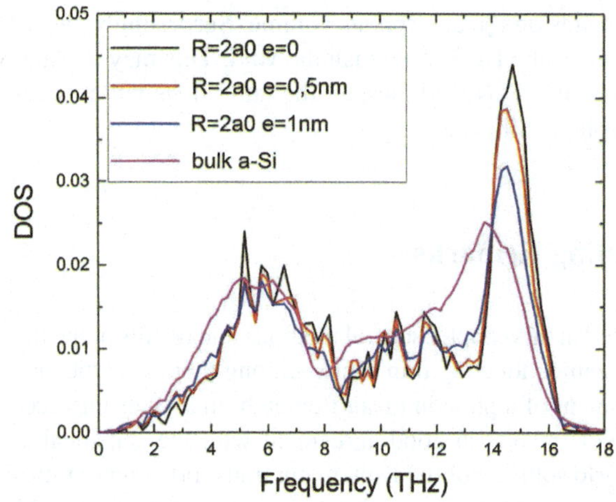

Fig. 9.12 Phonon DOS for systems with an amorphous shell of variable thickness around the pore ($R = 2a_0$)

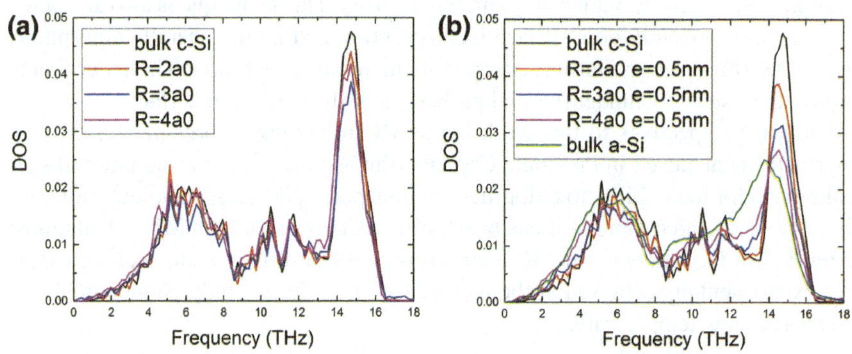

Fig. 9.13 Comparison of phonon DOS for systems with and without amorphous shell. **a** Without amorphous shell. **b** With an amorphous shell of constant thickness $e = 0.5$ nm

to the amorphous shell around the pores bring the DOS of the system closer to the amorphous Silicon one.

In the Fig. 9.13a, DOS of configurations without amorphous shell are compared for several pore radii. The difference between them is minimal. Only the TO peak is a little affected and the precision of the computations does not allow to tell whether the diminution of this peak is more marked for great porosity or not. So we can globally conclude that the nano-pores have a negligible influence on the phonon DOS. On the other hand, Fig. 9.13b shows the comparison of same systems but with an amorphous shell of thickness $e = 1$ nm. Here the DOS diverges a lot from the bulk c-Si one. The three effects previously listed for Fig. 9.12 are also observed here and the DOS approaches the a-Si one when R increases. Gathering the observations made for constant R and constant e, one can conclude that the modification of the

DOS mainly depends on the amorphous volume. Surprisingly, the a-Si phase, even if it is small, changes the DOS more than the void. This may explain why the TC is lowered by a-Si as much as void, despite the latter does not conduct heat at all, in contrast to amorphous Silicon.

9.5 Concluding Remarks

We have presented in this chapter several methods and results of the thermal transport in nano-porous semiconducting thin films. Among them analytic approaches which lie on the assessment of a phonon mean free path, that takes into account scattering with pores, provide results in good agreement with the statistical approach based on the Monte Carlo solution of the Boltzmann transport equation. Besides, effective Monte Carlo simulation tool recovers thermal conductivity of actual MC calculations with a good accuracy and a reduced computational cost. Using atomistic approaches such as Molecular Dynamic we demonstrate that TC of nano-porous materials can be appraised correctly, but for very small pore sizes. This technique is also an appropriate method to study pore surface reconstruction and more generally amorphous layers that often occurs during mesoporous material elaboration. The latter amorphous phase strongly hinder thermal properties of the studied materials.

Current perspectives to this work is the MC modeling of organized porosities such as those achieved in Phononic Crystals (PnCs) that attract the attention of several groups for their ability to tailor thermal transport in devices. Nanoinclusions are also a close research topic as it has been pointed out here through MD calculations. Extension of the latters to the MC methods is an interesting issue to deal with samples experimentally achieved at the mesoscales, in order to study phonon coherent transport at low temperature.

Appendix: Dispersion Properties and Relaxation Times

Parabolic fits $\omega = c_p K^2 + v_p K$ are used to describe the dispersion properties of Si and Ge along [100] axis, for LA and TA polarizations, as suggested by Pop et al. [62]. Parameters c_p and v_p are detailed in Table 9.3.

Relaxation-time constants are set to fit the experimental curve that gives the bulk material conductivity as a function of the temperature $k = f(T)$. Each scattering process has its own expression which depends of the frequency and the temperature [61]. Table 9.4 lists them for normal, umklapp and impurities scattering using the previous quadratic fits

Table 9.3 Data used to fit Si and Ge dispersion curves

	Units	Si	Ge
c_{TA}	$\times 10^{-7}$ ($m^2 s^{-1}$)	−2.28	−1.13
v_{TA}	$\times 10^{3}$ (m s^{-1})	5.24	2.60
c_{LA}	$\times 10^{-7}$ ($m^2 s^{-1}$)	−2.22	−1.5
v_{LA}	$\times 10^{3}$ (m s^{-1})	9.26	5.63
ω_{12}	$\times 10^{13}$ (rad s^{-2})	2.42	1.23

Table 9.4 Relaxation time parameters for Si and Ge

	Units	Si	Ge
B_{TN}	$\times 10^{-13}$ (s K^{-3})	9.0	30.0
B_{TU}	$\times 10^{-18}$ (s)	1.9	1.5
B_{L}	$\times 10^{-24}$ (K^{-4})	1.3	2.3
B_{I}	$\times 10^{-45}$ (s^{3})	1.2	24.0

$$
\begin{aligned}
&\tau_I{}^{-1} = B_I\omega^4, \tau_{N_{LA}}{}^{-1} = B_L\omega^2T^3, \tau_{U_{LA}}{}^{-1} = B_L\omega^2T^3 \\
&\tau_{N_{TA}}{}^{-1} = \begin{cases} B_{TN}\omega T^4 \text{ if } \omega < \omega_{12}, \\ 0 \text{ else} \end{cases} \\
&\tau_{U_{TA}}{}^{-1} = \begin{cases} 0 \text{ if } \omega < \omega_{12}, \\ B_{TU}\dfrac{\omega^2}{\sinh\frac{\hbar\omega}{k_BT}} \text{ else} \end{cases}
\end{aligned}
\tag{9.29}
$$

ω_{12} is the frequency which corresponds to half of the Brillouin zone for the TA mode.

References

1. Canham LT (1990) Silicon quantum wire array fabrication by electrochemical and chemical dissolution of wafers. Appl Phys Lett 57(10):1046
2. Cullis AG, Canham LT, Calcott PDJ (1997) The structural and luminescence properties of porous silicon. J Appl Phys 82(3):909
3. Granitzer Petra, Rumpf Klemens (2010) Porous SiliconA Versatile Host Material. Mater. 3(2):943–998
4. Buriak JM (2006) High surface area silicon materials: fundamentals and new technology. Philos Trans R Soc A 364(1838):217–225
5. Kelly A (2006) Why engineer porous materials? Philos Trans R Soc A 364(1838):5–14
6. Stewart MP, Buriak JM (2000) Chemical and biological applications of porous silicon technology. Adv Mater 12(12):859–869
7. Cahill DG, Ford WK, Goodson KE, Mahan GD, Majumdar A, Maris HJ, Merlin R, Phillpot SR (2003) Nanoscale thermal transport. J Appl Phys 93(2):793
8. Benedetto G, Boarino L, Spagnolo R (1997) Evaluation of thermal conductivity of porous silicon layers by a photoacoustic method. Appl Phys A-Mater Sci Process 64(2):155–159
9. Gesele G, Linsmeier J, Drach V, Fricke J, Arens-Fischer R (1997) Temperature-dependent thermal conductivity of porous silicon. J Appl Phys D: Appl Phys 30(21):2911

10. Ziman (1972) Electrons and phonons: the theory of transport phenomena in solids. Clarendon press
11. Chao Y (2011) 1.16 optical properties of nanostructured silicon. In: Comprehensive nanoscience and technology, vol 1. Elsevier edition
12. Olsson RH III, El-Kady I (2009) Microfabricated phononic crystal devices and applications. Meas Sci Technol 20(1):012002
13. Hopkins PE, Reinke CM, Su MF, Olsson RH, Shaner EA, Leseman ZC, Serrano JR, Phinney LM, El-Kady I (2011) Reduction in the thermal conductivity of single crystalline silicon by phononic crystal patterning. Nano Lett 11(1):107–112
14. Xiong Z, Zhao F, Yang J, Hu X (2010) Comparison of optical absorption in Si nanowire and nanoporous Si structures for photovoltaic applications. Appl Phys Lett 96(18):181903
15. Maldovan M (2013) Narrow low-frequency spectrum and heat management by thermocrystals. Phys Rev Lett 110(2):025902
16. Yang R, Chen G, Dresselhaus M (2005) Thermal conductivity of simple and tubular nanowire composites in the longitudinal direction. Phys Rev B 72(12):125418
17. Minnich A, Chen G (2007) Modified effective medium formulation for the thermal conductivity of nanocomposites. Appl Phys Lett 91(7):073105
18. Jeng M-S, Yang R, Song D, Chen G (2008) Modeling the thermal conductivity and phonon transport in nanoparticle composites using Monte Carlo simulation. J Heat Transf 130(4):042410
19. Chung JD, Kaviany M (2000) Effects of phonon pore scattering and pore randomness on effective conductivity of porous silicon. Int J Heat Mass Transf 43(4):521–538
20. Randrianalisoa J, Baillis D (2008) Monte Carlo simulation of cross-plane thermal conductivity of nanostructured porous silicon films. J Appl Phys 103(5):053502
21. Prasher R (2006) Thermal conductivity of composites of aligned nanoscale and microscale wires and pores. J Appl Phys 100(3):034307
22. Prasher R (2006) Transverse thermal conductivity of porous materials made from aligned nano- and microcylindrical pores. J Appl Phys 100(6):064302
23. Hao Q, Chen G, Jeng M-S (2009) Frequency-dependent Monte Carlo simulations of phonon transport in two-dimensional porous silicon with aligned pores. J Appl Phys 106(11):114321
24. Romano G, Carlo A, Grossman JC (2012) Mesoscale modeling of phononic thermal conductivity of porous Si: interplay between porosity, morphology and surface roughness. J Comput Electron 11(1):8–13
25. Hsieh T-Y, Lin H, Hsieh T-J, Huang J-C (2012) Thermal conductivity modeling of periodic porous silicon with aligned cylindrical pores. J Appl Phys 111(12):124329
26. Reinke CM, Su MF, Davis BL, Kim B, Hussein MI, Leseman ZC, Olsson-III RH, El-Kady I (2011) Thermal conductivity prediction of nanoscale phononic crystal slabs using a hybrid lattice dynamics-continuum mechanics technique. AIP Adv 1(4):041403
27. Dechaumphai E, Chen R (2012) Thermal transport in phononic crystals: the role of zone folding effect. J Appl Phys 111(7):073508
28. Hopkins PE, Rakich PT, Olsson RH, El-kady IF, Phinney LM (2009) Origin of reduction in phonon thermal conductivity of microporous solids. Appl Phys Lett 95(16):161902
29. Jain A, Yu Y-J, McGaughey AJH (2013) Phonon transport in periodic silicon nanoporous films with feature sizes greater than 100 nm. Phys Rev B 87(19):195301
30. Lee J-H, Grossman JC, Reed J, Galli G (2007) Lattice thermal conductivity of nanoporous Si: molecular dynamics study. Appl Phys Lett 91(22):223110
31. He Y, Donadio D, Lee J-H, Grossman JC, Galli G (2011) Thermal transport in nanoporous silicon: interplay between disorder at mesoscopic and atomic scales. ACS Nano 5(3):1839–1844
32. Fang J, Pilon L (2011) Scaling laws for thermal conductivity of crystalline nanoporous silicon based on molecular dynamics simulations. J Appl Phys 110(6):064305
33. Şopu D, Kotakoski J, Albe K (2011) Finite-size effects in the phonon density of states of nanostructured germanium: a comparative study of nanoparticles, nanocrystals, nanoglasses, and bulk phases. Phys Rev B 83(24):245416

34. Jean V, Fumeron S, Termentzidis K, Tutashkonko S, Lacroix D (2014) Monte Carlo simulations of phonon transport in nanoporous silicon and germanium. J Appl Phys 115(2):024304
35. Isaiev M, Tutashkonko S, Jean V, Termentzidis K, Nychyporuk T, Andrusenko D, Marty O, Burbelo RM, Lacroix D, Lysenko V (2014) Thermal conductivity of meso-porous germanium. Appl Phys Lett 105(3):031912
36. Verdier M, Termentzidis K, Lacroix D (2016) Crystalline-amorphous silicon nanocomposites: nano-pores and nano-inclusions impact on the thermal conductivity. J Appl Phys 119(17):175104
37. Xia R, Wu RN, Liu YL, Sun XY (2015) The role of computer simulation in nanoporous metals—a review. Materials 8(8):5060
38. Dettori R, Melis C, Cartoixà X, Rurali R, Colombo Luciano (2015) Model for thermal conductivity in nanoporous silicon from atomistic simulations. Phys Rev B 91:054305
39. Guo R, Huang B (2014) Thermal transport in nanoporous Si: anisotropy and junction effects. Int J Heat Mass Transf 77:131–139
40. Wen YW, Liu HJ, Pan L, Tan XJ, Lv HY, Shi J, Tang XF (2013) Reducing the thermal conductivity of silicon by nanostructure patterning. Appl Phys A 110(1):93–98
41. Fang J, Pilon L (2012) Tuning thermal conductivity of nanoporous crystalline silicon by surface passivation: a molecular dynamics study. Appl Phys Lett 101(1)
42. Fang J, Pilon L (2011) Scaling laws for thermal conductivity of crystalline nanoporous silicon based on molecular dynamics simulations. J Appl Phys 110(6)
43. Lee J-H, Galli GA, Grossman JC (2008) Nanoporous Si as an efficient thermoelectric material. Nano Lett 8(11):3750–3754. PMID: 18947211
44. Lee J-H, Grossman JC, Reed J, Galli G (2007) Lattice thermal conductivity of nanoporous Si: molecular dynamics study. Appl Phys Lett 91(22)
45. Nan C-W (1993) Physics of inhomogeneous inorganic materials. Prog Mater Sci 37(1):1–116
46. Maxwell JC (1904) A treatise on electricity and magnetism, vol 1, 3rd edn. Clarendon Press, Oxford
47. Rayleigh JWL (1892) On the influence of obstacles arranged in rectangular order upon the properties of a medium 34:481–502
48. Kandula M (2011) On the effective thermal conductivity of porous packed beds with uniform spherical particles. J Porous Media 14(10):919–926
49. Hasselman DPH, Johnson LF (1987) Effective thermal conductivity of composites with interfacial thermal barrier resistance. J Compos Mater 21:508–515
50. Nan C-W, Birringer R, Clarke DR, Gleiter H (1997) Effective thermal conductivity of particulate composites with interfacial thermal resistance. J Appl Phys 81(10):6692
51. Chen G, Borca-Tascuic D, Yang RG (2004) Encyclopedia of nanoscience and nanotechnology, vol 7. H. S. Nalwa, American Scientific Publisher edition
52. Zhou J, Li X, Chen G, Yang R (2010) Semiclassical model for thermoelectric transport in nanocomposites. Phys Rev B 82(11):115308
53. Tarkhanyan RH, Niarchos DG (2013) Reduction in lattice thermal conductivity of porous materials due toinhomogeneous porosity. Int J Therm Sci 67:107–112
54. Luo T, Chen G (2013) Nanoscale heat transfer from computation to experiment. Phys Chem Chem Phys 15(10):3389
55. Volz S, Lemonnier D, Saulnier J-B (2001) Calmped nanowire thermal conductivity based on the phonon transport equation. Microscale Thermophys Eng 5(3):191–207
56. Mazumder S, Majumdar A (2001) Monte Carlo study of phonon transport in solid thin films including dispersion and polarization. J Heat Transf 123(4):749
57. Lacroix D, Joulain K, Lemonnier D (2005) Monte Carlo transient phonon transport in silicon and germanium at nanoscales. Phys Rev B 72(6)
58. Wong BT, Francoeur M, Meng P (2011) A Monte Carlo simulation for phonon transport within silicon structures at nanoscales with heat generation. Int J Heat Mass Transf 54(9–10):1825–1838
59. Praud J-PM, Hadjiconstantinou NG (2011) Efficient simulation of multidimensional phonon transport using energy-based variance-reduced Monte Carlo formulations. Phys Rev B 84(20):205331

60. Neil W, Mermin DN (1976) Solid state physics. Saunders College, Philadelphia
61. Holland M (1963) Analysis of lattice thermal conductivity. Phys Rev 132(6):2461–2471
62. Pop E, Dutton RW, Goodson KE (2004) Analytic band Monte Carlo model for electron transport in Si including acoustic and optical phonon dispersion. J Appl Phys 96(9):4998
63. Lacroix D, Traore I, Fumeron S, Jeandel G (2008) Phonon transport in silicon, influence of the dispersion properties choice on the description of the anharmonic resistive mechanisms. Eur Phys J B 67(1):15–25
64. Singh BK, Roy MK, Menon VJ, Sood KC (2003) Dispersive effects and correction term in two-mode phonon conduction model for Ge. J Phys Chem Solids 64(12):2369–2377
65. Chen Y, Li D, Lukes JR, Majumdar A (2005) Monte Carlo simulation of silicon nanowire thermal conductivity. J Heat Transf 127(10):1129
66. Liu L-C, Huang M-J (2010) Thermal conductivity modeling of micro- and nanoporous silicon. Int J Therm Sci 49(9):1547–1554
67. Lacroix D, Joulain K, Terris D, Lemonnier D (2006) Monte Carlo simulation of phonon confinement in silicon nanostructures: application to the determination of the thermal conductivity of silicon nanowires. Appl Phys Lett 89(10):103104
68. Chen G (1998) Thermal conductivity and ballistic-phonon transport in the cross-plane direction of superlattices. Phys Rev B 57(23):14958–14973
69. Stillinger FH, Weber TA (1985) Computer simulation of local order in condensed phases of silicon. Phys Rev B 31(8):5262–5271
70. Tersoff J (1988) New empirical approach for the structure and energy of covalent systems. Phys Rev B 37(12):6991–7000
71. Evans DJ, Morriss GP (2007) Statistical mechanics of nonequilibrium liquids, 2nd edn. ANU E Press, Canberra
72. France-Lanord A, Blandre E, Albaret T, Merabia S, Lacroix D, Termentzidis K (2014) Atomistic amorphous/crystalline interface modelling for superlattices and core/shell nanowires. J Phys Condens Matter 26(5):055011
73. Vink RLC, Barkema GT, van der Weg WF, Mousseau N (2001) Fitting the Stillinger-Weber potential to amorphous silicon. J Non-Cryst Solids 282(2–3):248–255
74. http://lammps.sandia.gov/
75. Plimpton S (1995) Fast parallel algorithms for short-range molecular dynamics. J Comput Phys 117(1):1–19
76. Yang L, Yang N, Li B (2014) Extreme low thermal conductivity in nanoscale 3D Si phononic crystal with spherical pores. Nano Lett 14(4):1734–1738
77. Blandre E, Chaput L, Merabia S, Lacroix D, Termentzidis K (2015) Modeling the reduction of thermal conductivity in core/shell and diameter-modulated silicon nanowires. Phys Rev B 91(11)
78. Coquil T, Fang J, Pilon L (2011) Molecular dynamics study of the thermal conductivity of amorphous nanoporous silica. Int J Heat Mass Transf 54(21–22):4540–4548

Chapter 10
Scale Transition for Mass Transport in Porous Structures

Gavriil Gavriil and Frank A. Coutelieris

Nomenclature

Latin Letters

A	Pre-exponential coefficient [$m^3\ s^{-1}\ kg^{-1}$]
a_i	"Mix Jannaf" method coefficients [–]
a_{pol}	Polarizability [$Å^3$]
B	Body force vector [–]
b_{ij}	Adsorbed species stoichiometric coefficient [–]
c_{ij}	Bulk species stoichiometric coefficient [–]
c	Total molar concentration [$mol\ m^{-3}$]
C_A	Concentration of passive solute [$mol\ m^{-3}$]
C_i	Molar concentration of the *i*th component [$mol\ m^{-3}$]
C_F	Quadratic drag factor [–]
c_p	Effective thermal conductivity [$W\ m^{-1}K^{-1}$]
c_{ps}	Thermal conductivity of the solid [$W\ m^{-1}\ K^{-1}$]
c_{pr}	Thermal conductivity of the pores [$W\ m^{-1}\ K^{-1}$]
D	Diffusivity [$m^2\ s^{-1}$]
D_i	Diffusivity of *i*th species in the mixture [$m^2\ s^{-1}$]
$D_{i,eff}$	Effective diffusion coefficients of species *i* [$m^2\ s^{-1}$]
$D_{i,eff}^T$	Soret diffusion coefficients of species *i* [$m^2\ s^{-1}$]
D_{ij}	Maxwell–Stefan-diffusion coefficient [–]
E_a	Activation energy [$J\ mol^{-1}$]
h	Enthalpy [$J\ kg^{-1}$]
h_i	Enthalpy of the *i*th species in the system [$J\ kg^{-1}$]
h_o	Total enthalpy [$J\ kg^{-1}$]
i*	Internal energy as a function of the state variables ρ and T [$m^2\ kg\ s^{-2}$]
J_i	Diffusion flux of the *i*th species [$kg\ m^{-2}s^{-1}$]

G. Gavriil · F.A. Coutelieris (✉)
Department of Environmental & Natural Resources Management, University of Patras, Seferi 2, 30100 Agrinio, Greece
e-mail: fcoutelieris@upatras.gr

P. Bettotti (ed.), *Submicron Porous Materials*,
DOI 10.1007/978-3-319-53035-2_10

$J_{i,n}$	Flux balance of the *i*th species [kg $m^{-2}s^{-1}$]
k	Thermal conductivity [W m^{-1} K^{-1}]
k_c	Reaction rate constant [m^3 s^{-1} $^{kg-1}$]
k_{eff}	Effective thermal conductivity [W m^{-1} K^{-1}] [
k_{fj}	Forward reaction rate [–]
k_{rj}	Reverse reaction rate [–]
k_{ov}	Overall reaction rate [m s^{-1}]
k_{ov_mt}	Overall mass transport coefficient define by (Eq. 10.62) [m s^{-1}]
L	Characteristic length [m]
MW_i	Molecular Weight of species i [kg]
n	Vector normal to the surface [–]
n_1, n_2, n_3	Coefficient in chemical reaction (36) [–]
n_s	Number of spheres [–]
n_c	Number of components [–]
N_A	Molar flux on *i*th surface [mol m^{-2} s^{-1}]
N_{ads}	Total number of adsorbed species [–]
N_b	Total number of bulk (deposited) species [–]
N_g	Total number of gas phase species [–]
N_r	Total number of chemical reactions [–]
N_s	Total number of chemical species [–]
P	Pressure [atm]
Pe	Peclet number [–]
q	Heat flux [J m^{-2} s^{-1}]
q_R	Radiative heat flux [J m^{-2} s^{-1}]
R	Universal gas constant, R = 8.314 [J mol^{-1} K^{-1}]
$\dot{R}_i$	Production–destruction rate [–]
r	Radius of sphere [m]
s	Source or Sink term [mg/s]
S_{inlet}	Inlet surface [m^2]
S_{outlet}	Outlet surface [m^2]
$S_{adsorbing\_surface}$	Total adsorbing surface area [m^2]
S_h	Heat source or sink term [kJ/s]
S_{Mx}, S_{My}, S_{Mz}	Rate of increase of momentum due to sources in the *i*th direction [kg m/s]
Sh_O	Sherwood number [–]
$\frac{S}{V}$	Surface to volume ratio [m^{-1}]
t	Time [s]
T	Temperature [K]
T^*	Dimensionless temperature [–]
u	Velocity [ms^{-1}]
u_i	Velocity of species i [ms^{-1}]
u_x	Velocity in the x direction [ms^{-1}]
u_y	Velocity in the y direction [ms^{-1}]
u_z	Velocity in the z direction [ms^{-1}]

v'_{ij},	Forward stoichiometric coefficients for the *i*th species in the *j*th reaction
v''_{ij}	Backward stoichiometric coefficients for the *i*th species in the *j*th reaction
V	Volume [m^3]
V_{sph}	Sphere occupied volume [m^3]
V_{total}	Total volume in porous media [m^3]
V_{void}	Void volume in porous media [m^3]
x	Molar fraction
x_i	Mass fraction for species i
X_i	Surface site fractions [–]
Y_i	Mass fraction of *i*th species [–]
$Y_{\rho,i}$	Species mass fraction in the pore fluid [–]
Y_i^W	Gas-phase mass fractions adjacent to the wall [–]

Greek Letters

α	Average pore size [m]
α_i	Activity of species i in a mixture [–]
δ	Diffusion length scale [m]
	Porosity [–]
ε^*	Characteristic energy [K]
κ^*	Boltzmann constant [m^2 kg s^{-2} K^{-1}]
	Permeability [m^2]
Λ_i	Chemical symbol for species i [–]
λ	Adsorption efficiency [–]
μ	Dynamic viscosity [kg m^{-1} s^{-1}]
μ_i	Dynamic viscosity for the *i*th species [kg m^{-1} s^{-1}]
μ^*	Chemical potential [kJ]
ρ	Density [kg m^{-3}]
ρ_w	Gas phase density [kg m^{-3}]
ρ_s	Surface site density [kg m^{-3}]
σ_i	Collision diameter [Å]
τ	Shear stress tensor [N m^{-2}]
τ_{ij}	Components of the viscous stress tensor, where i = x, y, z and j = x, y, z
τ'	Tortuosity [–]
φ	Parameter defined by (Eq. 10.64) [–]
$\Phi_{i,j}$	Dimensionless quantity defined by (Eq. 10.19) [–]
ω	Parameter defined by (Eq. 10.65) [–]
ω_i	Volumetric production/consumption rate of species in the mixture [mol m^{-3} s^{-1}]
Ω_μ	Collision integral

Subscripts

eff Effective [–]
macro Macroscopic [–]
meso Mesoscopic [–]
micro Microscopic [–]
sph Spheres [–]

10.1 Introduction

Porous materials, either artificial or natural, are used widely in the industrial or scientific applications, technological developments, as well as they can be found in the natural environment. For example, porous materials are used for catalytic pellets, chalk, underground reservoirs, etc.

However, mass transport phenomena in porous media is an engineering field that requires a lot of research, because of the great industrial and technological interest, in order to be properly understood. As porous materials are characterized by high complexity in their invisible internal structure, modeling with the use of a CFD simulator is a useful tool for understanding transport processes. Computational fluid dynamics (CFD) is the use of computer-based simulation to analyze complex systems involving fluid flow, heat transfer, and associated phenomena such as chemical reactions, using mathematical models. The development and application of CFD have recently increased and become a powerful tool in the design and analysis stages of several engineering and industrial systems and processes [44].

When mathematics are used to model transport processes, the problem of scales arises. The major reason that scale transition must be investigated is to obtain scale dependence of the results occurring by a transport process [40]. It is well established that small-scale heterogeneities can have a significant impact on macroscopic results; therefore, the average effects of small-scale heterogeneities in large-scale numerical grids must be accounted. Actually, how to upscale the reactive transport parameters from a small lab-scale to a large field-scale for modeling is one of the most important issues in applications of the reactive transport modeling to contaminant remediation and related risk assessments [14, 56].

Upscaling is the process of replacing a heterogeneous domain by a homogeneous one, which reproduces the equivalent response with the same boundary conditions imposed [14, 41]. This process inevitably has to maintain parameters known in a small scale, by spatial averaging to obtain values for larger scales when the same processes considered. There are various upscaling methods (see the review article by Cushman et al. [14] and the references therein), including volume averaging (Whitaker, Diffusion and dispersion in porous media [45, 49], homogenization [24], renormalization [26, 33], fractals [29, 31], ensemble averaging [18], Eulerian perturbation [15], spectral integral approach [21, 39], Lagrangian perturbation schemes [39], central limit approach [53], fast Fourier transform and Green's function method [25], continuous time random walk [8, 34], effective medium theory [20], and indicator geostatistics-based approach [16]. A brief description of the major upscaling methods is given below.

The volume averaging method is in many aspects similar to the asymptotic homogenization, although it employs different mathematical formulation. A comparison of the two approaches can be found in the paper by Bourgeat et al. [9]. This method is used by numerous authors for modeling transfer phenomena in porous media (see for instance [3, 5]. In recent years, the formalism of the volume averaging method has been substantially developed [52] and applied to various problems including multiphacial phenomena [37]. The derivation of the equations, describing the process at a larger scale, is performed by taking the volume average of the governing equations from a smaller scale. Various levels of averaging are possible. The local volume averaging is used for passing from the pore scale to the local (Darcy) scale [50]. The resulting equations can be successively upscaled using the large-scale averaging technique [37]. The large-scale equations are written in terms of averaged (macroscopic) variables. The local variables are represented as a sum of the averaged value and the spatial deviation. The averaged large-scale equations should be completed by the relations between the averaged values and their local scale deviations. These relations are given by differential equations, which constitute the closure problems. The solution of the closure problem allows, obtaining the large-scale effective parameters of the system from the smaller scale to use for the large one.

Homogenization is an asymptotic method for upscaling parameters of periodic media [19]. The basic idea for homogenization is as follows. Consider a porous medium with microscopic structure characterized by period ℓ and a macroscopic structure with the characteristic length of L. A spatial scale parameter can be defined as $\varepsilon = \ell/L$. Instead of working on one function (i.e., velocity field) u, a family of functions u^{ε} is considered to find the limit of u^{ε} as $\varepsilon \rightarrow 0$. The limit is considered as the result of upscaling procedure for homogenization, which consists of finding differential equations whose solutions can satisfy the above limit. The asymptotic process $\varepsilon \rightarrow 0$ stands for the transition from microscopic to macroscopic scale. Hence, the upscaling process of homogenization is completed by making the microscopic scale approximate to zero [24].

Renormalization approach is also called the renormalization group that has been initially developed by Wilson [48], to study critical phenomena in phase transition. For example, this method is used to upscale permeability in fractured porous media [26]. In renormalization, an overall behavior for a certain small part (a unit cell) of a fractured porous medium is considered to be controlled by heterogeneity at a particular scale. This scale is treated as the scale of the initial stage (step 0) of the renormalization procedure. Then step-by-step transformation gives the effective properties at step, n + 1, resulting from the renormalization of a set of calculated properties at step n according to a probability distribution function of $f(n)$ for the property (for example, permeability). This method retrieves the effective parameters, by recurrent applications of renormalization group transformation from the basic cell to the entire domain of interest.

According to Noetinger and Estebenet [34], continuous time random walk (CTRW) begins with a finite-volume discretization of a fractured porous medium for small-scale diffusion equation. A particle is then released jumping from one grid block to adjacent one with transition probability proportional to the transmissivity. When the particle starts its random walk in the domain, its movement path can be defined by its coordinates on the computational grid and by a random function of time τ, whose value is 1 if the particle moves in fracture at time t, and 0 otherwise. The correlation function of τ is closely related to fluid pressures in fracture and in matrix, and can be calculated by Monte Carlo simulations. Finally, the upscaled mass transfer coefficient is computed using a suitable integration of the correlation function with respect to [7, 30, 42].

Over the last decade a number of powerful approaches have been developed to intelligently upscale transport phenomena in porous materials. Most of the work on upscaling methods has been done by Whitaker who is probably the pioneer in the field of transport phenomena in different scales in a porous medium. Whitaker and Quintard [36, 38, 51], examined the problems arising when the upscaling of transport phenomena occurring in a specific scale and in a porous material is needed. The analysis presented by them can easily be extended to include more than two regions and have shown how the method of large-scale averaging can be applied to systems in which the Darcy-scale properties vary continuously. By Darcy-scale is meant the length scale corresponding to the first averaging process. In a two-region model of a heterogeneous porous medium, the point equations are joined with boundary conditions to form local volume-averaged transport equations. In their study of incompressible flow [36], it has been showed that discrete and continuously varying Darcy-scale properties led to the same form of the large-scale transport equations. Under these circumstances, the choice between continuous models and discrete models is largely a matter of personal preference.

The main objective of this study is to develop a methodology for estimating the appropriate value for one parameter at some scale when its value at another specific scale is known. When this is assured for every parameter at any scale, a sufficient scale-up and/or scale-down procedure for the fundamental processes occurring in a porous medium, applied to many systems. The aim is to match all geometrical, flow and transport parameters in order to move from a small-scale model to a full-scale setup and vice versa. The results obtained by different scale approaches must be compared and analyzed to obtain a strategy (scale transition rule) toward a globally valid description to cope with the physicochemical and geometrical characteristics of a porous medium in scale transition.

10.2 Theory

Simulations have been performed through a commercially available CFD platform named CFD-ACE+. In the following part of this chapter, the simulation process is going to be analyzed because it is important to understand how each module of the above mentioned software treats the problems submitted to it.

10.2.1 Flow Module

For the flow simulation through the porous media, the "Flow Module" of the CFD-ACE+ has been used. Both, inlet and outlet flows, in and out of the porous medium, with the velocity magnitude set by the user as normal velocity, can be simulated to obtain velocity and pressure fields. The flow of the gas phase has been set as laminar flow due to low to medium Reynolds numbers. Using the following two laws, CFD-ACE+ develops the governing equations used for the Flow Module.

- The mass of a fluid is conserved, i.e., there is no loss or gain of mass in the system.
- The time rate of change of momentum equals the sum of the forces on the fluid (Newton's second law).

CFD-ACE+ uses an iterative method to solve the Navier–Stokes equations, from which their results show the satisfaction of the two laws mentioned above.

The main equation that must always be satisfied is the generic transport equation:

$$\frac{\partial(\rho c)}{\partial t} + \nabla(\rho u c) - \nabla(\mathcal{D}\rho\nabla c) = s \tag{10.1}$$

The terms that appear in this equation admit the following physical interpretation:

the rate-of-change term $\frac{\partial(\rho c)}{\partial t}$ is the net gain/loss of mass per unit volume and time,

the convective term $\nabla(\rho u c)$ is due to the downstream transport with velocity u,

the diffusive term $-\nabla(\mathcal{D}\rho\nabla c)$ is due to a nonuniform spatial distribution of *c*,

the source or sink term *s* combines all other effects that create or destroy mass.

For the time being, we assume that the parameters ρ, u, D, and *s* are known. In real life applications, they may depend on the concentration *c* and/or other variables. In particular, conservation laws of the form of the generic transport equation constitute the Navier–Stokes equations, in which the conserved variables are the mass, momentum, and total energy. The simplest component of this PDE system is the continuity equation:

$$\frac{\partial \rho}{\partial t} + \nabla(\rho u) = 0 \tag{10.2}$$

The first term on the left-hand side is the time rate of change of the density while the second term describes the net mass flow across the control volume's boundaries (convective term).

The equations used by the flow module are presented and described next. As Newton's second law states, "*the alteration of motion is ever proportional to the motive force impressed; and is made in the direction of the right line in which that force is impressed*" [32]. This means that if a force generates a motion, a double force will generate double the motion, a triple force triples the motion, whether that forces are impressed altogether and at once, or gradually and successively. And this motion (being always directed the same way with the generating force), if the body moved before, is added to or subtracted from the former motion, according as they directly conspire with or are directly contrary to each other; or obliquely joined, when they are oblique, so as to produce a new motion compounded from the determination of both.

The x-component of the momentum equation is found by setting the rate of change of x-momentum of the fluid particle equal to the total force in the x direction on the element due to surface stresses plus the rate of increase of x-momentum due to sources:

$$\frac{\partial(\rho u_x)}{\partial t} + \nabla(\rho u_x) = \frac{\partial(-P + \tau_{xx})}{\partial x} + \frac{\partial \tau_{yx}}{\partial y} + \frac{\partial \tau_{zx}}{\partial z} + S_{Mx} \tag{10.3}$$

Similar equations can be written for the y- and z- components of the momentum equation:

$$\frac{\partial(\rho u_y)}{\partial t} + \nabla(\rho u_y) = \frac{\partial(-P + \tau_{xy})}{\partial x} + \frac{\partial \tau_{yy}}{\partial y} + \frac{\partial \tau_{zy}}{\partial z} + S_{My} \tag{10.4}$$

$$\frac{\partial(\rho u_z)}{\partial t} + \nabla(\rho u_z) = \frac{\partial(-P + \tau_{xz})}{\partial x} + \frac{\partial \tau_{yz}}{\partial y} + \frac{\partial \tau_{zz}}{\partial z} + S_{Mz} \tag{10.5}$$

where p is the static pressure, τ_{ij} is the viscous stress tensor and S_M is the rate of increase of x, y, and z momentums due to sources. The momentum equations, given above, contain as unknowns the viscous stress components τ_{ij}, therefore a model must be provided to define the viscous stresses.

In Newtonian flows, the viscous stresses are proportional to the deformation rates of the fluid element. The nine viscous stress components (of which six are independent for isotropic fluids) can be related to velocity gradients to produce the following shear stress terms:

$$\tau_{xx} = 2\mu\frac{\partial u_x}{\partial x} - \frac{2}{3}\mu(\nabla u) \tag{10.6}$$

$$\tau_{yy} = 2\mu\frac{\partial u_y}{\partial y} - \frac{2}{3}\mu(\nabla u) \tag{10.7}$$

$$\tau_{zz} = 2\mu\frac{\partial u_z}{\partial z} - \frac{2}{3}\mu(\nabla u) \tag{10.8}$$

$$\tau_{xy} = \tau_{yx} = \mu(\frac{\partial u_x}{\partial y} + \frac{\partial u_y}{\partial x}) \tag{10.9}$$

$$\tau_{xz} = \tau_{zx} = \mu\left(\frac{\partial u_x}{\partial z} + \frac{\partial u_z}{\partial x}\right) \tag{10.10}$$

$$\tau_{zy} = \tau_{yz} = \mu(\frac{\partial u_y}{\partial z} + \frac{\partial u_z}{\partial y}) \tag{10.11}$$

The substitution of the above shear stress terms (Eqs. 10.6–10.11) into the equations of momentum (Eqs. 10.3–10.5) yields the Navier–Stokes equations:

$$\frac{\partial(\rho u_x)}{\partial t} + \nabla(\rho u_x) = \frac{\partial P}{\partial x} + \nabla(\mu \nabla u_x) + S_{Mx} \tag{10.12}$$

$$\frac{\partial(\rho u_y)}{\partial t} + \nabla\left(\rho u_y\right) = \frac{\partial P}{\partial y} + \nabla\left(\mu \nabla u_y\right) + S_{My} \tag{10.13}$$

$$\frac{\partial(\rho u_z)}{\partial t} + \nabla(\rho u_z) = \frac{\partial P}{\partial z} + \nabla(\mu \nabla u_z) + S_{Mz} \tag{10.14}$$

The “Mix Kinetic Theory” option that uses the kinetic theory of gasses for the calculation of viscosity of the whole mixture has been used. For monatomic gas, the viscosity is defined as:

$$\mu_i = 2.6693{*}10^{-5}\frac{\sqrt{MW_i T}}{\sigma_i^2 \Omega_\mu} \tag{10.15}$$

where μ_i is the dynamic viscosity of species i, MW_i is the molecular weight of species i, T is the temperature, σ_i is the characteristic diameter of the molecule and Ω_μ is the collision integral. The collision theory gives that:

$$\Omega_\mu = \frac{1.16145}{T^{*0.14874}} + \frac{0.52487}{\exp(0.77320T^*)} + \frac{2.16178}{\exp(2.43787T^*)} \tag{10.16}$$

where T^* is the dimensionless temperature given as:

$$T^* = \frac{\kappa^* T}{\varepsilon^*} \tag{10.17}$$

where ε^* is the characteristic energy and κ^* is the Boltzmann constant.

These equations are used for calculating the viscosity of each gas in the mixture. To calculate the mixtures viscosity using the previous equations, the following equation is also applied:

$$\mu_{mix} = \sum_{i=1}^{N_S} \frac{x_i \mu_i}{\sum_j x_j \Phi_{ij}} \tag{10.18}$$

where x_i, x_j are mass fraction of species i and species j, μ_i is the viscosity of species i (Eq. 10.15) and $\Phi_{i,j}$ is a dimensionless quantity given by:

$$\Phi_{i,j} = \frac{1}{\sqrt{8}} \left(1 + \frac{\mathrm{MWi}}{\mathrm{MWj}}\right)^{-1/2} \left[16 + \left(\frac{\mu_i}{\mu_j}\right)^{1/2} \left(\frac{\mathrm{MWi}}{\mathrm{MWj}}\right)^{1/4}\right] \tag{10.19}$$

10.2.2 Heat Transfer Module

The heat transfer analysis has been performed with the use of the Heat Transfer Module, which is an internal part of the CFD-ACE-SOLVER. The use of the heat transfer module implies the solution of the total enthalpy form of the energy equation. Many types of heat transfer analysis can be performed with this module, from basic conduction/convection to complex radiation. Heat transfer analysis can be performed in stand-alone mode (pure heat transfer analysis) or coupled with other modules (such as the flow, mixing, etc.) for a multi-physics simulation.

Heat transfer processes are computed by solving the equation for the conservation of energy. This equation can take several forms and CFD-ACE+ numerically solves the energy equation in the form known as the total enthalpy equation. This form is fully conservative and is given in following Eq. 10.20:

$$\frac{\partial(\rho h_o)}{\partial x} + \nabla(\rho u h_0) = \nabla\left(k_{eff} \nabla T\right) + \frac{\partial P}{\partial t} + \left[\frac{\partial(u_x \tau_{xx})}{\partial x} + \frac{\partial(u_x \tau_{xy})}{\partial y} + \frac{\partial(u_x \tau_{xz})}{\partial z}\right]$$
$$+ \left[\frac{\partial(u_y \tau_{xy})}{\partial x} + \frac{\partial(u_y \tau_{yy})}{\partial y} + \frac{\partial(u_y \tau_{zy})}{\partial z}\right] + \left[\frac{\partial(u_z \tau_{xz})}{\partial x} + \frac{\partial(u_z \tau_{yz})}{\partial y} + \frac{\partial(u_z \tau_{zz})}{\partial z}\right] + S_h \tag{10.20}$$

where h_0 is the total enthalpy and is defined as:

$$h_0 = i^* + \frac{P}{\rho} + \frac{1}{2}(u_x^2 + u_y^2 + u_z^2) \tag{10.21}$$

By i* is denoted the internal energy as a function of the state variables ρ and T while k_{eff} represents the effective thermal conductivity of the material.

Under laminar flow conditions, the effective thermal conductivity is the thermal conductivity of the fluid, k. Static pressure is defined as p and as τ_{ij} is referred to the viscous stress tensor. S_h contains terms for additional sources due to reactions, body forces, etc. The method used to evaluate the specific heat and conductivity properties for the use in the simulation is the "MIX_JANNAF" method [54], which uses the input data to create fitting curves of the following form:

$$\frac{C_p}{R} = a_1 + a_2 T + a_3 T^2 + a_4 T^3 + a_5 T^4 \tag{10.22}$$

$$\frac{h}{RT} = a_1 + \frac{a_2 T}{2} + \frac{a_3 T^2}{3} + \frac{a_4 T^3}{4} + \frac{a_5 T^4}{5} + \frac{a_6}{T} \tag{10.23}$$

The coefficients used for the simulation can be found in [22].

Chemistry Module

The chemistry module by CFD-ACE+ has been used to solve mixing and reacting flow problems.

The Species Mass Fraction approach is used and it requires the solution of a transport equation for every species in the system. This approach is required for:

- Multicomponent diffusion problems
- Surface reaction problems
- A multistep finite rate gas-phase reaction.

For the mass diffusion, the "multi-component diffusion" option is selected therefore the solver calculates the diffusion of the mixture instead of reading a specific constant value imposed by the user. When species diffuse at different rates, their mass fractions do not automatically add up to unity and some corrections have to be invoked to guarantee the species conservation. To treat this problem, the "Stefan-Maxwell" option is selected. This option enforces the species conservation by employing the Stefan-Maxwell equations. This is the most rigorous of all the approaches, even if it is more computationally expensive [54]. The Stefan–Maxwell equations are:

$$\frac{\nabla \mu_i^*}{RT} = \nabla \ln \alpha_i = \sum_{j=1 j \neq i}^{n_c} \frac{x_i x_j}{\mathfrak{D}_{ij}}(u_j - u_i) = \sum_{j=1 j \neq i}^{n_c} \frac{c_i c_j}{c^w \mathfrak{D}_{ij}}\left(\frac{J_j}{c_j} - \frac{J_i}{c_i}\right) \tag{10.24}$$

where x is the molar fraction, μ* is the chemical potential, α_i is the activity (i.e., a measure of the effective concentration of a species in a mixture), i, j are indexes for component i and j, n_c is the number of components, D_{ij} is the Maxwell–Stefan-diffusion coefficient, u_i is the velocity of component i, c_i is the molar concentration of component i, c is the total molar concentration and J_i is the mass flux of component i.

The chemical rate used by the chemistry module can be expressed in a general way, for a system of N_r chemical reactions, involving N_s species, as:

$$\sum_{i=1}^{N_s} \nu_{ij}'\Lambda_i = \sum_{i=1}^{N_s} \nu_{ij}''\Lambda_i, j=1, \ldots, N_r \tag{10.25}$$

where Λ_i are chemical symbol for species i and v'_{ij}, ν''_{ij} are the forward and backward stoichiometric coefficients for the *i*th species in the *j*th reaction, respectively. As the stoichiometric coefficients are integers for elementary reactions are normally 0, 1, or 2.

The following equation (Eq. 10.26) can be used to write the equation (Eq. 10.25) in a more compact way:

$$V_{ij} = v_{ij}'' - V_{ij}' \tag{10.26}$$

$$\sum_{i=1}^{N_s} \nu_{ij}\Lambda_i = 0, j=1, \ldots, N_r \tag{10.27}$$

For this simulation, the surface reaction option has been used. The decane oxidization is placed onto the surface of the particles for all scales. (Details about the problem formulation can be found in the following sections). The surface reaction provides a boundary condition for the mass fractions of species in the fluid, rather than a source term in then transport equations.

The general form of the surface reaction considered in CFD-ACE+ is:

$$\sum_{i=1}^{N_g} a_{ij}^{'} A_i + \sum_{i=1}^{N_{ads}} b_{ij}^{'} B_{(s)} + \sum_{i=1}^{N_b} c_{ij}^{'} C_{(b)} = \sum_{i=1}^{N_g} a_{ij}^{''} A_i + \sum_{i=1}^{N_{ads}} b_{ij}^{''} B_{(s)} + \sum_{i=1}^{N_b} c_{ij}^{''} C_{(b)} \tag{10.28}$$

where a_{ij} are the gas species stoichiometric coefficients, b_{ij} are the adsorbed species stoichiometric coefficients, c_{ij} are the bulk species stoichiometric coefficients, N_g is the total number of gas-phase species, N_{ads} is the total number of adsorbed species and N_b is the total number of bulk (deposited) species.

For the above reaction the surface reaction rate can be expressed as:

$$\dot{S}_j = k_{fj} \prod_{i=1}^{N_g} A_{iw}^{a_{ij}^{'}} \prod_{i=1}^{N_s} B_{(s)}^{b_{ij}^{'}} - k_{rj} \prod_{i=1}^{N_g} A_{iw}^{a_{ij}^{''}} \prod_{i=1}^{N_s} B_{(s)}^{b_{ij}^{''}} \tag{10.29}$$

where k_{fj} is the forward and k_{rj} is the reverse rate, respectively.

As seen from the above expression, the surface reaction rate is assumed to be independent of the concentration of the bulk species. The gas-phase concentrations at the surface are expressed as:

$$A_{iw} = \frac{\rho_w Y_i^w}{m_i}, B_{i(s)} = \rho_s X_i \tag{10.30}$$

where ρ_w is the gas-phase mass density, ρ_s is the surface site density, Y_i^w is the gas-phase mass fractions adjacent to the wall and X_i is the surface site fractions.

The mass flux of reacting species to the surface or in the bulk for species produced by the reaction equals the rate at which the species are consumed or produced respectively by the reaction on the surface.

The species flux balance at the reacting surface yields:

$$j_{i,n} = M_i \sum_j \left(a_{ij}'' - a_{ij}'\right) \dot{S}_j, i = 1, \ldots, N_g \tag{10.31}$$

$$\frac{dB_i(s)}{dt} = \sum_j \left(b_{ij}'' - b_{ij}'\right) \dot{S}_j, i = 1, \ldots, N_s \tag{10.32}$$

where, the left-hand side of (Eq. 10.31) is the diffusive flux of species i normal to the surface. The right-hand side of (Eq. 10.31) is the production rate of species i per unit area of surface, on mass basis. Equations (10.31) and (10.32) are solved by coupled Newton–Raphson iterations.

The reaction (mass) flux can be computed by using two different approaches, namely the sticking coefficient method and the general rate method. The sticking coefficient method evaluates the production rate based on sticking probability and precursor thermal flux, while the finite-rate chemistry uses the kinetic expression (Eq. 10.29) to evaluate the reaction rate. In this study, the finite rate has been used for the simulations since we are interested in calculating the production/consumption of species throughout the porous media rather than studying the surface coverage phenomena.

For practical applications, the most useful macroscopic quantity is the adsorption efficiency, λ, defined by the ratio of the solute adsorption rate divided by the rate of the upstream influx [12]:

$$\lambda = u \frac{\iint_{S_{inlet}} c_A u n dS - \iint_{S_{outlet}} c_A u n dS}{\iint_{S_{inlet}} c_A u n dS} \tag{10.33}$$

where $\iint_{S_{inlet}} c_A u n dS$ is the total mass entered the volume of interest through the inlet surface, $\iint_{S_{outlet}} c_A u n dS$ is the total mass exit the volume of interest through the outlet surface, c_A is the concentration of passive solute, u is the velocity in the medium and n is the vector normal to the surface.

Porous media module

For the simulation of the macroscopic case the porous media module has been used. The porous media feature allows to model fluid flow in a volume representing a material consisting of a rigid solid matrix with interconnected void. The interconnectedness of the void (the pores) allows the flow of one or more fluids through the material. In the simplest situation (single-phase flow), the void is saturated by a single fluid. (In two-phase flow, a liquid and a gas can share the void space).

In most porous media, the distribution of pores with respect to shape and size is irregular. On the pore scale (the microscopic scale), the flow quantities (velocity, pressure, etc.) will clearly be irregular. But for most applications, the quantities of interest are space-averaged (macroscopic) in nature (measured over areas that cross many pores). Experiments have shown that such macroscopic quantities change in a regular manner with respect to space and time and are therefore amenable to numerical modeling. Flow through porous media plays an increasingly important role in a multitude of engineering applications and the porous media module in CFD-ACE+ can be applied to many such problems [54].

In the next section, the treatment of the conservation equations by porous media module will be described. The mass conservation equation is as follows:

$$\frac{\partial(\varepsilon\rho)}{\partial t} + \nabla(\varepsilon\rho u) = 0 \tag{10.34}$$

where ε is the porosity of the medium.

Equation (10.25) is actually the same equation as (Eq. 10.2) with the addition of the parameter ε. The porosity of the medium represents the volume occupied by the pores to the total volume of the porous solid.

The equation for momentum conservation within the porous media is written as:

$$\frac{\partial(\varepsilon\rho u)}{\partial t} + \nabla(\varepsilon\rho u) = -\varepsilon\nabla P + \nabla(\varepsilon\tau) + \varepsilon B - \frac{\varepsilon^2\mu}{k}u - \frac{\varepsilon^2 C_f\rho}{\sqrt{k}}u \tag{10.35}$$

where C_F is a quadratic drag factor, τ is the shear stress tensor, B is the body force vector, u is the fluid velocity and κ is the permeability of the porous matrix.

It is important to note here that permeability is a measure of the ability of a porous material to allow fluids to pass through it while the last two terms in the equation represent an additional drag force imposed by the pore walls on the fluid within the pores, and usually results in a significant pressure drop across the porous solid. In a purely fluid region, the standard Navier–Stokes equation is recovered.

In terms of boundary conditions, no special treatment to the momentum and pressure correction equations is required at fluid-porous solid interfaces [54].

The energy conservation equation within the porous regions is written as:

$$\frac{\partial(\varepsilon\rho h)}{\partial t} + \nabla(\varepsilon\rho u h) = \nabla q + \varepsilon\tau\colon \nabla u + \varepsilon\frac{dP}{dt} \tag{10.36}$$

where h is the total enthalpy of the fluid. The heat flux, q, is comprised of contributions due to thermal conduction, thermal radiation, and species diffusion, and is written as:

$$q = c_p \nabla T + q_R + \sum_{i=1}^{N_g} J_i h_i \tag{10.37}$$

where N_g is the total number of gas phase species in the system, h_i are the enthalpies of gas species in the system, J_i are the diffusion fluxes of gas species in the system, T is the bulk temperature and q_R is the radiative heat flux.

The thermal conductivity of the porous medium, c_p, is an effective thermal conductivity of the pores and solid regions in combination and may be written as:

$$c_p = -2c_{ps} + \frac{1}{\frac{\varepsilon}{2c_{ps} + c_{pr}} + \frac{1-\varepsilon}{3c_{ps}}} \tag{10.38}$$

where c_{ps} and c_{pr} are the thermal conductivities of the solid and pores, respectively.

The mass conservation equations for individual gas phase species (i.e., for i = 1, …, N_g) is written as:

$$\frac{\partial(\varepsilon \rho Y_i)}{\partial t} + \nabla(\varepsilon \rho u Y_i) = \nabla J_i + \omega_i \tag{10.39}$$

where Y_i are the mass fractions of the ith species and ω_i are the production rates of the ith species in the gas phase. When heterogeneous surface reactions are treated within the porous solid, this production rate term (ω_i) will be modified so that the species are produced in a volumetric sense, having the basic physics governing their production [54].

The species diffusion flux may be written as:

$$J_i = \rho D_{i,eff} \nabla Y_i + \rho D^T_{i,eff} \frac{\nabla T}{T} + \frac{\rho Y_i}{MWi} D_{i,eff} \nabla MWi - M \sum_j D_{j,eff} \nabla Y_j - \nabla MWi \sum_j D_{j,eff} Y_j - \rho Yi \sum_j D^T_{jj,eff} \frac{\nabla T}{T} \tag{10.40}$$

where MW_i is the molecular weight of the mixture.

The first term represents Fickian diffusion due to concentration gradients, and the second term represents Soret (or thermo) diffusion. The last four terms are correction terms necessary to satisfy the "Stefan-Maxwell" equations for multi-component species systems. In this case, $D_{i,eff}$ and $D^T_{i,eff}$ are the effective mass diffusion and Soret diffusion coefficients of species i, within the porous medium, respectively, and depend on the porosity of the medium.

The following expression of Dagan model [54] has been provided for the effective diffusion coefficient through porous media:

$$D_{j,eff} = MAX(0, \frac{3\varepsilon - 1}{2\varepsilon})D_i \tag{10.41}$$

An alternate correction is the Bruggeman type:

$$D_{j,eff} = \varepsilon^{\tau'} D_j \tag{10.42}$$

where τ' is the tortuosity of the porous media.

Tortuosity is the degree of interconnection resulting in paths through the solid, which are some percentage longer than a direct path. The value of tortuosity is obtained by using a method proposed in [43]. The real tortuosity can be obtained by averaging over all possible flow paths in porous media. For simplicity though, they made a review of the available formulations for the scaled diffusion coefficient with tortuosity containing all the available relations for tortuosity with porosity.

Such relations are presented next, for mono-sized spheres [11]:

$$\tau'^2 = \varepsilon^{-\frac{1}{2}} \tag{10.43}$$

while for overlapping spheres [2] it is:

$$\tau'^2 = 1 - \frac{\ln(\varepsilon)}{2} \tag{10.44}$$

and for heterogeneous catalyst [6] it is:

$$\tau'^2 = \frac{\varepsilon}{(1 - (1 - \varepsilon)^{1/3})} \tag{10.45}$$

The preferred relation for the case studied is the relation of Beekman [6] (Eq. 10.45) because as it is mentioned above the surface where the reaction occurs is platinum catalyzing the reaction.

The cell sizes used in performing CFD analysis are typically much larger than the characteristic pore dimensions in a porous media. This makes it impossible to grid individual surfaces adjoining the pores. Therefore, modeling the heterogeneous chemical reactions occurring within catalyst-lined pores requires a volume-averaged implementation of the net effect of heterogeneous reactions occurring on the catalyst surfaces (ESI CFD Inc., [54].

Surface reaction within porous media is treated by performing a reaction–diffusion balance at the surfaces of the catalysts. The balance equation is given by:

$$\rho D_i \nabla Y_i = R_i' \tag{10.46}$$

where R_i' is the production–destruction rate per unit area of the ith species due to chemical reactions.

In discrete form, (Eq. 10.46) may be rewritten as:

$$\rho D_i \frac{Y_i - Y_{\rho,i}}{\delta} = R_i \tag{10.47}$$

where $Y_{\rho,i}$ is the ith species mass fraction in the pore fluid, Y_i is the mass fraction at the pore fluid/catalyst interface and δ is the diffusion length scale.

When considering surface reactions on a well-defined flat substrate, such as in chemical vapor deposition applications, this length scale is extracted from the computational grid. For the case of porous media, this length scale cannot be obtained from the grid since the region under consideration is already much smaller than the size of a typical control volume. This length scale can be computed using various correlations available in the literature. The simplest approach is to assume that this length scale is equal to the volume to surface ratio of the porous media in question [54].

The volumetric production/destruction rate, (in kmolm^{-3} s^{-1}) of species, i, due to heterogeneous surface reaction, can be obtained from the surface flux in equation:

$$\dot{\omega} = R_i \left[\frac{S}{V}\right]_{eff} = \rho D_i \frac{Y_i - Y_{\rho,i}}{\delta} \left[\frac{S}{V}\right]_{eff} \tag{10.48}$$

where $\left[\frac{S}{V}\right]_{eff}$ is the effective surface to volume ratio of the catalyst.

10.3 Problem Statement

The main objective of this study is the development of a new method for upscaling a process occurring in a porous medium. For the sake of application, we consider a porous medium consisting of a solid phase, being distributed within a defined volume of interest and of given specific geometrical characteristics, such as porosity, average pore size, surface over volume ratio, tortuosity, etc. The void space (porous) is occupied by a Newtonian liquid or gas mixture that flows under laminar flow conditions and contains a solute, which can react heterogeneously on the solid/liquid or solid/gas interfaces. A simple, fast hydrocarbon reaction takes place on the catalytic gas/solid interfaces at relatively low temperature. This reaction process can adequately represent a sorption mechanism under the pressure and temperature conditions specified by the phenomena that take place.

The general overall reaction representing the oxidation of a conventional hydrocarbon fuel is:

$$\text{Fuel} + n_1 O_2 \rightarrow \text{n}_2\text{C}O_2 + \text{n}_3 H_2\text{O} \tag{10.49}$$

where n_1, n_2, and n_3 are coefficients determined by the choice of fuel. This global reaction is often a convenient way of approximating the effects of many elementary reactions, which actually occur. Its rate must therefore represent an appropriate average of all of the individual reaction rates involved. The reaction rate for the above reaction is usually expressed as:

$$k_{ov} = Ae^{\left(\frac{-Ea}{RT}\right)}[Fuel]^n[Oxidizer]^b \tag{10.50}$$

In this study, the mixture is assumed to be decane rich, therefore the decane oxidization reaction is assumed to occur whenever decane ($C_{10}H_{22}$) and molecular oxygen (O_2) reach the gas/solid interface through advection and/or diffusion. The parameters used for the calculation of the viscosity have been taken from literature data and are presented in Table 10.1.

The overall reaction is as follows:

$$C_{10}H_{22} + 15.5O_2 \rightarrow 11H_2O + 10CO_2 \tag{10.51}$$

It should be noted here that the use of the decane as well as of its oxidization reaction are options taken just for presentation purposes, not affecting the mathematics and the transport itself, as next presented. Obviously, any other species and reaction could be used instead.

The oxidation mechanism generally involves many reaction steps, making it difficult to find the exact expression of the reaction rate in widely various operating conditions, especially various temperatures. According to specific kinetic studies for the oxidation of other hydrocarbons [35, 47], the reaction rate is expected to depend on both the reactants and the eventual presence of inhibiting species that influence the reaction by their competitive adsorption on the catalytic reactive sites. The functionality of such dependence is usually quite complex and might depend strictly on the rate-determining steps of the overall reaction mechanism. Thus, the reaction rate is given as:

$$r_{C_{10}H_{22}} = -k_c T C_{C_{10}H_{22}} = -Ae^{-\frac{E_a}{RT}} C_{C_{10}H_{22}} \tag{10.52}$$

where k_c is the reaction rate constant, T is the temperature, $C_{C_{10}H_{22}}$ is the decane concentration, E_α is the activation energy, and R is the universal gas constant. For the current simulations, the heterogeneous reaction characteristics were as follows [10]:

Table 10.1 Values used for the molecular properties of Decane ($C_{10}H_{22}$)

Parameter	Symbol	Values	Units	References
Characteristic energy (Lennard-Jones)	ε^*	311.0	K	[4]
Collision diameter (Lennard-Jones)	σ_i	8.2	Å	[4]
Polarizability	a_{pol}	19.10	$Å^3$	[28]

Table 10.2 Inlet fuel mixture composition

Species	$C_{10}H_{22}$	CO_2	O_2	H_2O	NO	N_2	Total
Mass fraction	0.0002	–	0.05	0.1	0.0002	0.8496	1.0

$E_a = 226.783$ kJ/mole with a pre-exponential factor, $A = 1.66 \times 10^{24} m^3/s$ kg. Furtherin this analysis, the reaction is set to temperatures from 473 K up to 479, with the oxygen concentration of the mixture being much higher than that of decane. Under these circumstances, it is reasonable to consider this reaction, of first order (Eq. 10.52), as the oxygen concentration is a constant in the Arrhenius pre-exponential factor

$$r_{C_{10}H_{22}} = -k_c(T)C_{C_{10}H_{22}} = -A\exp\left(-\frac{E_c}{RT}\right)C_{C_{10}H_{22}} \tag{10.53}$$

The values of the composition for the inlet mixture used as boundary conditions set for the simulations are presented in Table 10.2.

The scale of the transport problem could be crucial in obtaining accurate results of macroscopic interest, such as adsorption efficiency. When microscopic geometry is used to model the processes described above, all the local heterogeneities have been taken into account, and thus significantly affect the results. As the final results are of macroscopic interest, it is crucial to develop an accurate, general mechanism for scale transition. ThIs study focuses on the exact matching of geometrical, flow, and transport characteristics when scale transition is required. For this reason, a typical porous model of mesoscale is presented, and the results obtained were subsequently scaled-up (macroscale) and scaled-down (microscale).

To define the 3-D domain and solve the set of partial differential equations analyzed above the CFD-GEOM was used. For the microscopic scale the geometry used is a unit cell, consisting of two concentric spheres (Fig. 10.1), while for the mesoscopic scale, an assemblage of space filling unit cells has been chosen (Fig. 10.2). For the macroscopic scale a structured domain was created (Fig. 10.3). For each scale, a different set of geometry files with different geometrical characteristics was made.

10.3.1 Microscopic Model

For the microscopic scale, the geometry used is a unit cell or also known as sphere-in-cell. It is consisted by two concentric spheres with their diameters set to result in a specific ratio describing the solid volume fraction of the represented solid. The idea behind the unit cell approach is that it represents the whole porous medium in a much smaller scale [13, 46]. Therefore, all the processes occurring in the unit cell represent the processes found in the real scale of the porous medium.

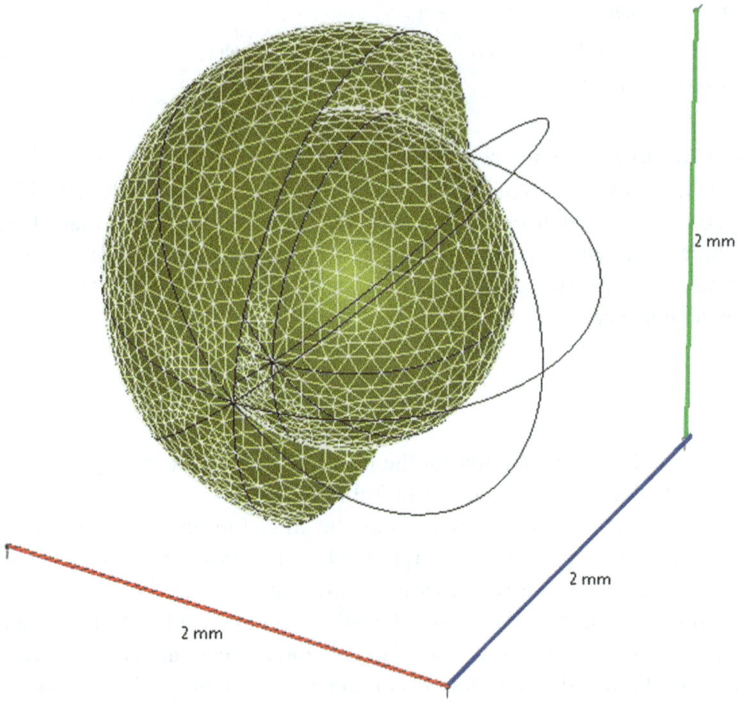

Fig. 10.1 Microscopic geometry for porosity 0.7

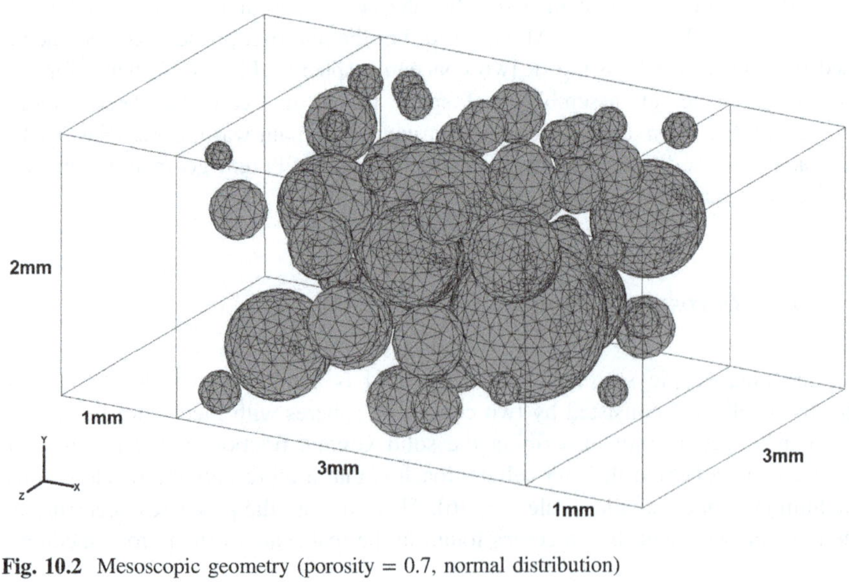

Fig. 10.2 Mesoscopic geometry (porosity = 0.7, normal distribution)

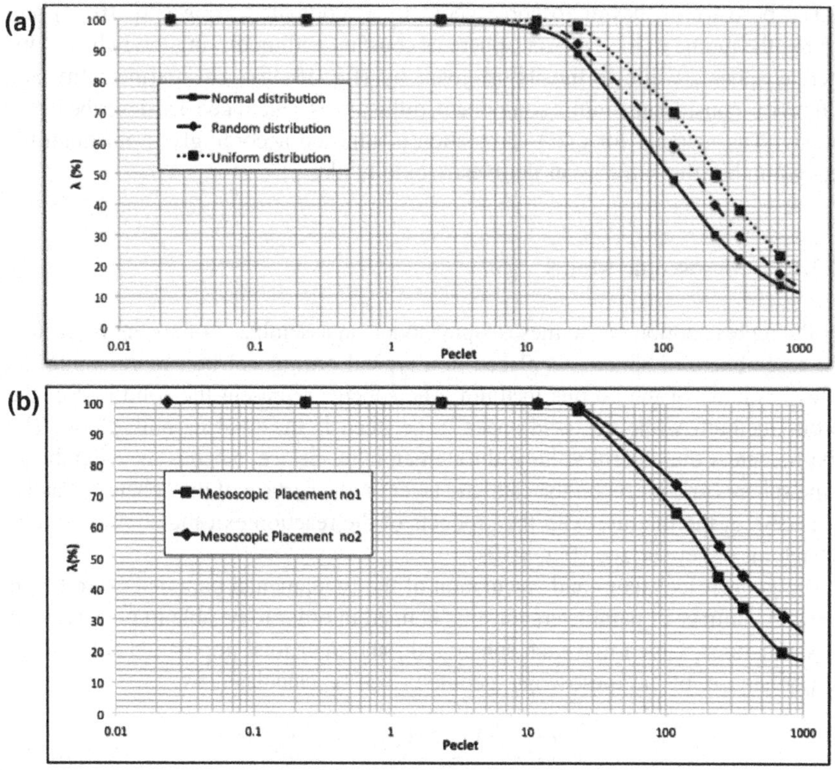

Fig. 10.3 Geometry dependence tests: **a** distribution, **b** placement (ε = 0.7)

In the present study this model is studied to provide a comparative scale for the outcomes. A representation of the unit cell can be seen in Fig. 10.1, where a typical geometry is presented for porosity 0.7. The inner sphere (smaller one) represents the surface (interface) where the catalytic reaction occurs. The outer sphere represents the control volume where the fuel is found. The hidden surface on the left is the inlet surface from which the fuel mixture enters the unit cell, and the surface on the right is the outlet surface of the medium.

All of the transport equations described above were solved by the simulator for the unit cell taking into account numerous assumptions. Regarding the boundary conditions, the inlet velocity of the mixture varied from 0.0001 to 7 m/s and fixed pressure is applied to the outlet surface in order to have pressure gradient and hence flow. The inner sphere is the catalytic surface where the oxidization reaction happens and it is assumed to have zero mass flux on it. Furthermore, the fuel mixture composition is the same as for all of the cases and it is presented in (Table 10.2). The inlet temperature is set to 473 K, while zero heat flux is set to the outlet.

The porosity of the microscopic geometry is computed using (Eq. 10.54) preserving the radius of the external sphere as constant and modifying the radius of the internal sphere, while tortuosity is given by (Eq. 10.45). The permeability and diffusivity considering multi-component diffusion is calculated as described previously. The velocity for every simulation is adjusted accordingly to maintain the Peclet number the same as in the mesoscopic model.

10.3.1.1 Mesoscopic Model

For the representation of the mesoscopic model, an assemblage of sphere is created. The assemblage of spheres is placed in a typical orthogonal box representing the control volume of the porous medium. The spheres represent the solid part of the porous medium while the void space in between is the fluid or bulk phase. The mixture enters the control volume through an inlet surface, reaches the assemblage of the spheres and reacts on the catalytic layer on the surface of the spheres. Finally, the excess of the mixture with the products of the reaction exit the control volume through an outlet surface (Fig. 10.2).

The geometrical setup and calculation of the fundamental parameters are found below. A number of parameters must be defined in order to be able to proceed with the simulations. Given that each sphere radii and their total population are known, the porosity, ε, of the mesoscopic geometry can be calculated by:

$$\varepsilon = \frac{V_{void}}{V_{total}} = \frac{V_{total} - V_{sph}}{V_{total}} = 1 - \frac{V_{sph}}{V_{total}} \tag{10.54}$$

where V_{void} represents the void volume of the porous media, V_{total} is the total volume of the medium, and V_{sph} is the volume occupied by the spheres.

The geometry of the 3 × 2 × 3 mesoscale porous medium with a typical porosity of 0,7 and normal distribution of sphere is depicted in Fig. 10.2. The exterior walls of the control volume are hidden so the assemblage of spheres can be viewed. In order to eliminate the end effects error two extra domains have been added. The domains located attached to the inlet and outlet surfaces allow the flow to deploy before entering and exiting the porous medium volume where the Peclet number will be as required by the model.

Sphere radii and sphere coordinates in the control volume are calculated by an algorithm implemented in Fortran 90/95 whose steps are described below:

Step (1) Incorporating a random number generator, the position of the sphere's center is generated, always set in a box with specific dimensions (3 × 2 × 3 mm).

Step (2) Using a random number generator along with the appropriate routine, a radius is generated, forced to follow a certain distribution (i.e., normal, random or uniform).

Table 10.3 Geometrical characteristics for mesoscopic porous medium (normal distribution)

Porosity [–]	Permeability [m^2]	Tortuosity [–]	Average pore size [m]	S/V [m^{-1}]
0.6	8.91E-10	1.50	0.00027	8874.31
0.7	8.23E-10	1.45	0.00038	8594.38
0.8	7.40E-10	1.38	0.00036	8244.43
0.9	1.32E-09	1.30	0.0011	9098.30

Step (3) The void space surrounding each sphere is checked for availability. If it is free (not occupied by another sphere), the radius value is accepted; otherwise step 2 is repeated until an acceptable value is produced.

Step (4) The sphere is posed.

Step (5) Steps 1–4 are repeated until the volume of the positioned spheres satisfies the predefined porosity value.

The geometrical characteristics of the mesoscopic geometries (assemblage of spheres) used in the simulations are presented in Table 10.3. It is necessary to mention that the average pore size and the surface to volume ratio, S/V, are manually calculated for the mesoscopic geometries and are also used for the macroscale model described in the following section.

A real porous medium has an unlimited combination of the structures of its porous. Hence, it is necessary to examine whether the sphere radii and their locations in the porous medium has a significant effect on the mass transfer occurring. In order to examine this problem, all the necessary distribution tests were performed. As the sphere radii are variable, three different types of radius distribution have been used. The radii distributions examined are, the normal distribution, a random distribution, and a uniform radii distribution were all spheres has the same radius. The effect of the three different distributions, on the mass transfer phenomenon for various Peclet numbers in a porous medium, with porosity 0.7, is presented in Fig. 10.3a.

As it is presented on the chart, the different type of radius distribution does not affect the overall mass transfer for the area of low and moderated Peclet numbers. However, for Peclet numbers over 100 a discrepancy is observed between the plotted values. By definition higher *Pe* numbers equal to higher convection. Hence, high velocities correlate to higher probabilities for the soluble to escape from the porous medium. Under these conditions, the radii distribution can affect the solution significantly as it defines the pore size.

Moreover, as in a real porous medium, porous can be placed with an unlimited combination of placements in space. Hence, even if the spheres follow a certain distribution it is important to study the effect of the different pore placement in the porous medium. The effect of different sphere placement in the porous volume on mass transfer occurring in a porous medium is presented in Fig. 10.3b for constant porosity (0.7) and for various *Pe* numbers. The behavior is very similar to the one seen in Fig. 10.3a. It can be seen that for low *Pe* numbers no deviation appears, while a small deviation can be seen for the higher number.

As a result, since neither the distribution tests nor the placement tests, critically affect the transport processes when the regime is not strongly convectional, only the normal distribution has been considered for the detailed simulations.

10.4 Macroscopic Model

In order to scale-up the mesoscopic model presented in the above section, a bigger scale geometry was build representing the scale of the macroscopic model. The mesoscopic scale was built in the scale of centimeters. A representation of the macroscopic geometry is presented in Fig. 10.4. Similarly, to the mesoscopic case, an extra volume has been added attached to the inlet and outlet surfaces respectively in order to allow to the flow to fully develop. At this scale, a porous medium is represented by a box where the pores and the solid cannot be defined in detail but only by assumptions and mathematical equations. Specifically, it is assumed that both pores and voids exist in the whole control volume, while all quantities are calculated by spatial averaging. Therefore, the transport equations used to solve the model should and are modified accordingly.

The boundary conditions applied regarding the macroscopic model are presented below:

- Flow: Normal velocity varying between 10^{-6} to 7 m/s at inlet, fixed pressure of 1 atm to outlet surface, non-slip condition to walls, Darcy's law for porous region.
- Heat transfer: 473 K at inlet, zero heat flux $(n\nabla T = 0)$ everywhere else.
- Mass transport: Constant fuel mixture at inlet (Table 10.2), zero mass flux $(n\nabla C_i = 0)$ to walls and outlet, decane oxidization on catalytic solid surfaces assuming zero accumulation.

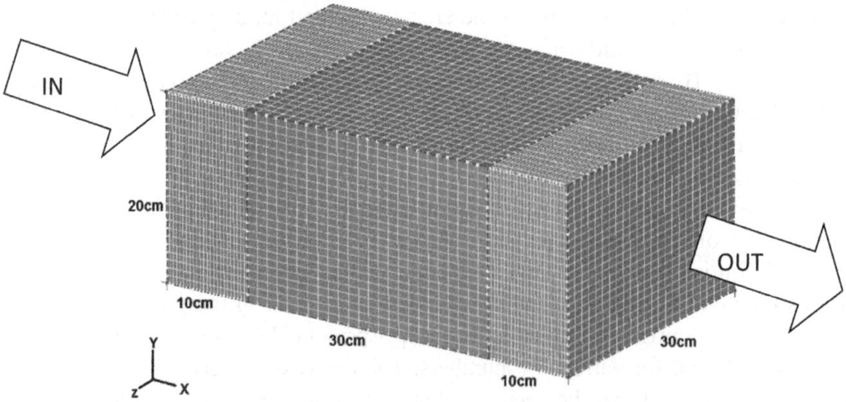

Fig. 10.4 Macroscopic geometry (ε = 0.7)

The parameters used in the simulation of the macroscopic model cannot be measured manually since the pores and geometrical characteristics do not physically exist. These crucial parameters such as porosity, tortuosity and permeability can be defined from the same porosity mesoscale solution (Table 10.3). Prior to analyzing the adjustment of the parameters it is necessary to mention that similar flow characteristics such as *Pe* number, should be preserved in order to allow a direct comparison of the model results between the three scales (meso-, macro- and micro-). For the mesoscopic scale, the Peclet number (Pe_{meso}) is given as:

$$Pe_{meso} = \frac{u_{meso} L_{meso}}{D} \tag{10.55}$$

where u_{meso}, and L_{meso}, are the inlet velocity and characteristic length for the mesoscopic scale, respectively, and D is the diffusion coefficient, calculated by Stefan–Maxwell equations and considered practically constant. Similarly, the *Pe* number for the macroscopic scale (Pe_{macro}) can be computed as:

$$Pe_{macro} = \frac{u_{macro} L_{macro}}{D} \tag{10.56}$$

As it has been previously mentioned in order to maintain the same flow characteristics, the above dimensionless number should be equal to:

$$u_{macro} = \frac{u_{meso} L_{meso}}{L_{macro}} \frac{1}{\varepsilon} \tag{10.57}$$

Given that u_{meso}, L_{meso} and L_{macro} are already known, u_{macro} is easily computed. Additionally, the average pore size of the macroscopic geometry (α_{macro}) is given by:

$$\alpha_{macro} = \frac{L_{macro}}{L_{meso}} a_{meso} \tag{10.58}$$

where α_{meso} is the average pore size of the mesoscopic geometry, calculated manually for each different mesoscopic geometry. Specifically, a pore size is considered to be, the distance between each two neighboring spheres. The distance is measured by subtracting the summation of the two radii from the center-to-center distance. Accordingly, the average value of these distances, for all the combinations of neighboring spheres, constituting the porous structure, is the average pore size for each case.

To maintain the constant porosity and the sphere number during scale-up, the average pore size should be modified proportionally to the rate of the scale transition (L_{macro}/L_{meso}), since otherwise it is assumed that the hypothetical spheres representing the solid in the macroscopic model are squeezed, reducing the pore spaces between them.

The surface to volume ratio, $\left(\frac{S}{V}\right)_{macro}$, can be computed for the macroscopic simulations as:

$$\left(\frac{S}{V}\right)_{macro} = \frac{L_{meso}}{L_{macro}} \left(\frac{S}{V}\right)_{meso} \tag{10.59}$$

where $\left(\frac{S}{V}\right)_{meso}$ is the surface to volume ratio of the mesoscopic scale considering normal distribution of sphere radii. It is calculated by the summation of the spheres' surface area divided by their volume:

$$\left(\frac{S}{V}\right)_{meso} = \frac{\sum_{i=1}^{n_s} S_{meso,i}}{\sum_{i=1}^{n_s} V_{meso,i}} \tag{10.60}$$

where n_s is the number of spheres.

10.5 Numerics

While the micro- and mesoscopic geometries are discretized in space by an unstructured grid, the simpler macroscopic porous medium, is discretized by a non-homogeneously structured grid. In order to solve the nonlinear-coupled conservation equations described previously, the finite-volume method was adopted allowing the SIMPLEC (Semi-Implicit Method for Pressure-Linked Equations Consistent) algorithm to calculate the pressure. An HP Compaq 6000 Pro MT PC with the following characteristics was used: Intel® Core ™2 Duo CPU E7500 at 2.93 GHz and 2 GB RAM memory. Given the required accuracy of at least 10^{-4} for all the involved quantities, the convergence time and the maximum iterations differ for each scale.

For the mesoscopic geometry, the convergence time is significantly higher than that of the macroscopic simulations due to its complex geometry. The convergence time varies between 160 to 300 min, while for the macroscopic and microscopic simulations it ranges from 5 to 60 and 10 to 30 min, respectively. Additionally, the required iterations for convergence are 300, 2000 and 4000 for the meso-, macro- and macroscopic simulations, respectively. To ensure that the simulation outcomes are not affected by the discretization used, grid independency tests were performed for each scale and the number of cells used was 217000 for the meso-, 35000 for the macro-, and 110000 for the microscale geometry.

10.6 Validation

In order to allow direct comparisons between the results, it is necessary to transform the adsorption efficiency to a dimensionless number. The most frequently used dimensionless number for this case is the overall Sherwood number, which represents the overall mass transfer from bulk to solid phase [27, 55]. The overall Sherwood number, Sh_O, is defined as:

$$Sh_o = \frac{Lk_{Ov\_mt}}{D} \tag{10.61}$$

where k_{ov_mt} is the overall mass transport coefficient, defined by the relation:

$$k_{ov\_mt} \Delta c S_{adsorbing\_surface} = \iint_{S_{adsorbing\_surface}} [-N_A]_{on\_the\_surface} dS_{adsorbing\_surface} \tag{10.62}$$

where $[N_A]_{on\_the\_surface}$ is the component of the molar flux which is vertical to the collector surface, $S_{adsorbing\_surface}$ is the total adsorbing surface, Δc is the concentration difference between the characteristic concentration far from the adsorbing surfaces and the average concentration on the adsorbing surface. By using simple mathematical manipulations, (Eq. 10.61) becomes:

$$\lambda = \frac{\varphi \omega}{Pe} Sh_o \tag{10.63}$$

where,

$$0 \le \varphi = \frac{\Delta c}{C_{inlet}} \le 1 \tag{10.64}$$

and

$$0 \le \omega = \frac{S_{adsorbing\_surface}}{S_{inlet}} \tag{10.65}$$

Figures 10.5, 10.6 and 10.7 present the validation for the three models with typical results found in the relative literature. Specifically, the comparison of simulation results for the microscale model with those published in [27] is presented in Fig. 10.5, and the validation for the case of sphere assemblage (mesoscale) was made against the results of [17] and is depicted in Fig. 10.6. Finally, the results for the macroscopic scale are compared with the results of [1] in Fig. 10.7. In each case, it can be observed that the qualitative behavior of the adsorption efficiency is in good accordance with the existing literature data. The fairly high discrepancy in terms of quantitative comparison observed in all scales,

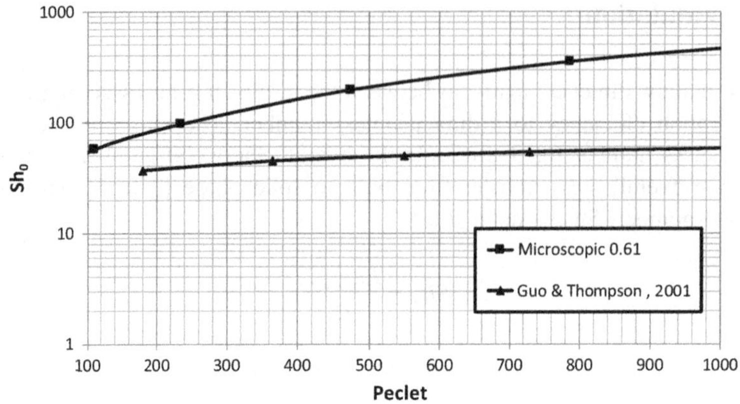

Fig. 10.5 Validation of microscopic scale results for low Peclet numbers with results presented in (Guo and Thompson 2001)

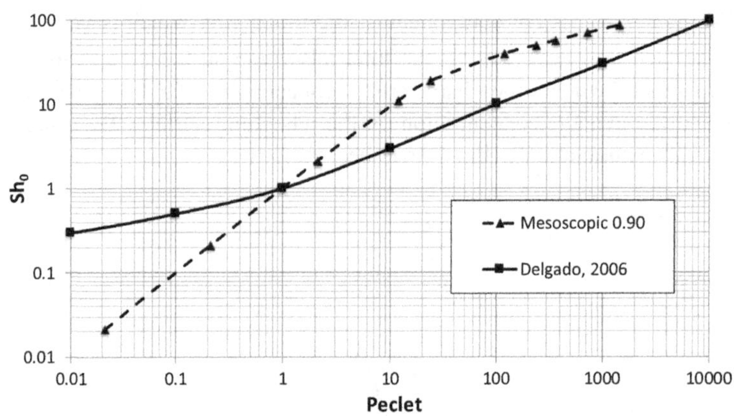

Fig. 10.6 Validation of mesoscopic scale results with those presented in [17]

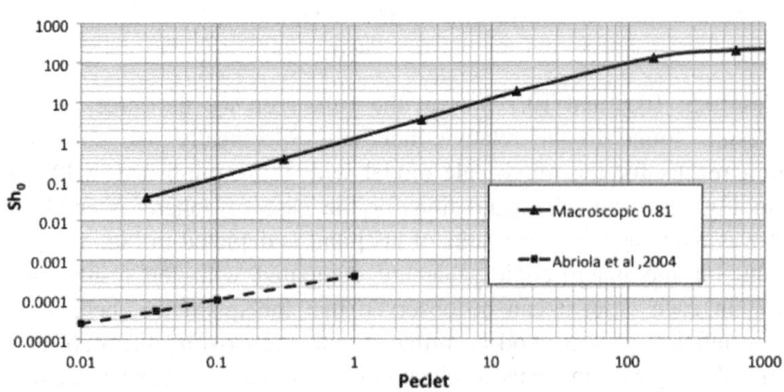

Fig. 10.7 Validation of macroscopic scale results with results presented in [1]

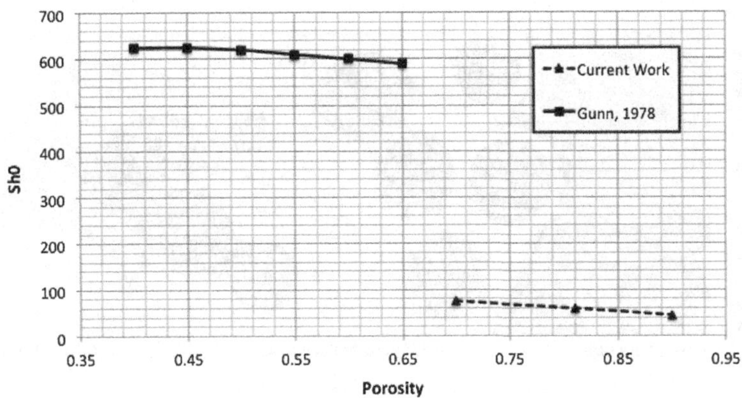

Fig. 10.8 Validation of porosity effect with the behavior presented by Gunn [23]

can be attributed to the arbitrary values of the parameters ϕ and ω used here, as well as the different surface reactions and/or sorption mechanisms considered in these works. Unfortunately, in order to validate the results of this work, a direct comparison of experimental or theoretical results of the reaction of decane oxidation in porous materials would be necessary. However, neither experimental or theoretical results have been found as a reaction like this is not frequently used when studying mass transfer in a porous medium. To further ensure the validity of the model, the effect of porosity on the dimensionless number known as overall Sherwood number is presented in Fig. 10.8, where the results of the model of the mesoscopic scale are compared to those published by [23]. As expected, the results comparison shows that Sh_O number decreases as the porosity value increases. A general trend is observed here, i.e., that adsorption efficiency is found to decrease with Peclet number and porosity. This trend also fully compatible with numerous published results by [1, 17, 23, 27] regarding all the scales considered in this work.

10.7 Results and Discussion

The mass fraction of decane oxidization is presented in Fig. 10.9 for all scales. As it has been mentioned previously, in order to be able to compare these results, the post processing of the simulations was carried out keeping both porosity and Peclet number constant (porosity = 0.7 and Pe = 50) for all scales. In Fig. 10.9, it is clearly observed that the mass fraction gradients are toward the solid mass, wherever it has been explicitly defined, i.e., mesoscopic (Fig. 10.9a) and microscopic (Fig. 10.9b) scale, and it is proportional to the flow for the macroscopic case (Fig. 10.9c) where the exact location of the solid phase is uncertain. Furthermore, decane consumption increases slightly with scale since the available adsorbing mass also increases with it. By conserving constant porosity ensures that the solid volume

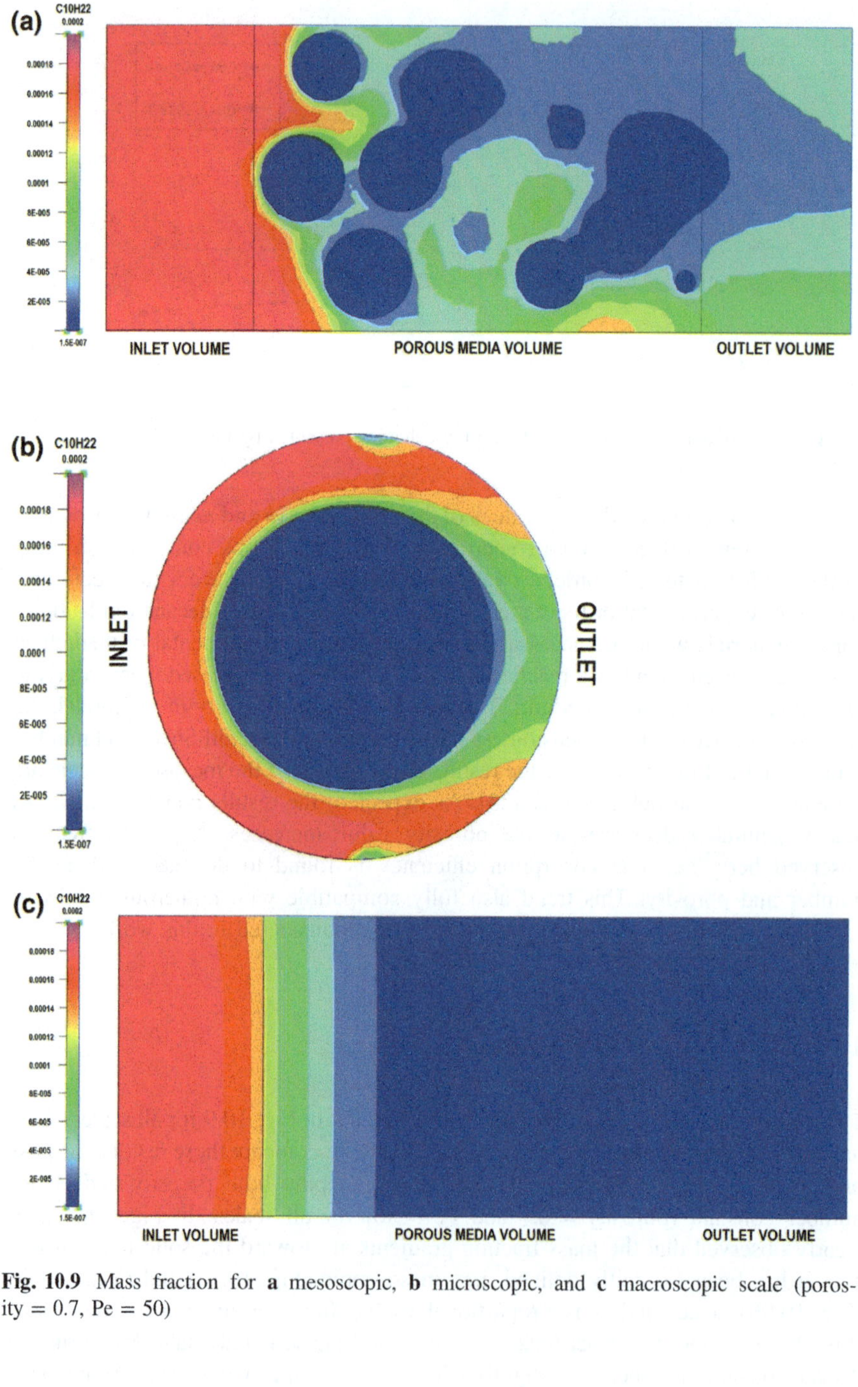

Fig. 10.9 Mass fraction for **a** mesoscopic, **b** microscopic, and **c** macroscopic scale (porosity = 0.7, Pe = 50)

fraction is the same in all cases. However, the adsorbing surface (i.e., the adsorbing mass itself) is not the same and follows the S/V trend (see discussion).

In Figs. 10.10, 10.11, 10.12 the adsorption efficiency, λ, as a function of the Peclet number is depicted for various porosity values. Figure 10.10 represents the mesoscopic geometry while Fig. 10.11 the macroscopic geometry, and Fig. 10.12 the microscopic geometry. It can be observed that as the Peclet number increases the adsorption efficiency value decreases, for all porosities and all geometrical scales, since the flow becomes more convective and as a result the species are forced to exit the porous media before participating in the oxidization reaction. By definition low *Pe* number, equals to a more dominant the diffusion regime than a convective one. This phenomenon allows the reactant species to approach the catalytic surface where the reaction happens to equalize the concentration gradient of the reacted species (Fick's law). It is important to mention that for a high reaction rate constant such as the one used in the present simulations, λ reaches the maximum value of 100%. The effect of porosity on λ values is also presented in the

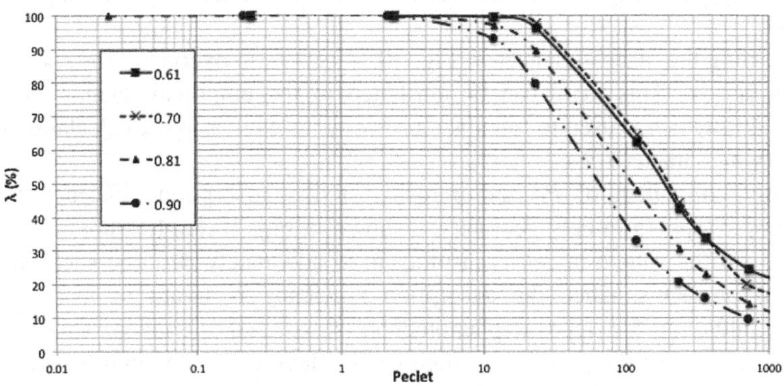

Fig. 10.10 Adsorption efficiency of the mesoscopic geometry as a function of Pe for various ε

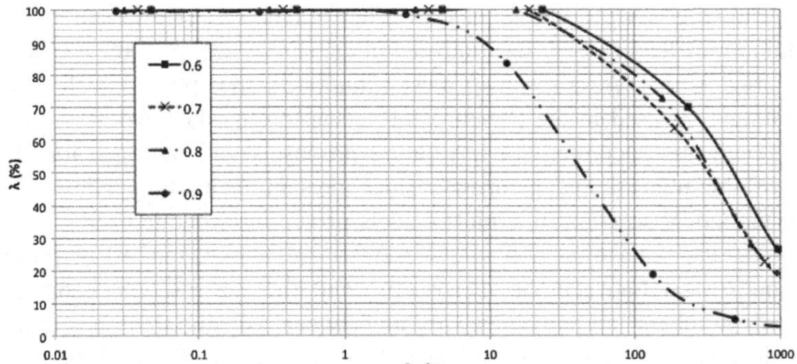

Fig. 10.11 Adsorption efficiency of the macroscopic geometry as a function of Pe for various ε

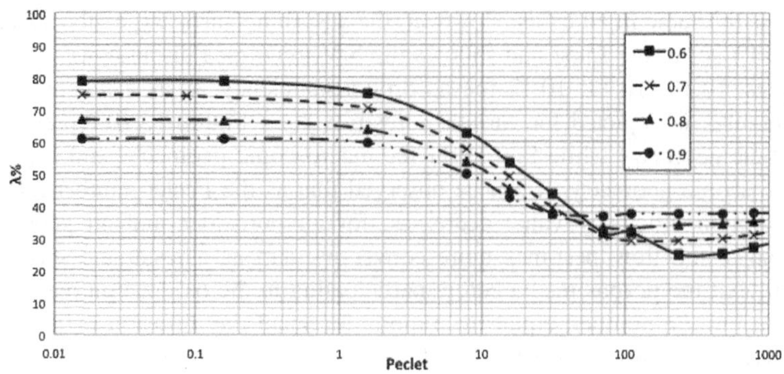

Fig. 10.12 Adsorption efficiency of the microscopic geometry as a function of Pe for various ε

following figures. As porosity increases, the empty space (void volume) increases and consequently, the solid volume, which represents the catalytic surface for the heterogeneous reaction of decane oxidation (Eq. 10.51), decreases, leading to lower λ values.

Regarding the macroscopic case, Fig. 10.11 shows an overlap of the $\varepsilon = 0.7$ and $\varepsilon = 0.8$ curves. This trend can be attributed to the empirical relation given by (Eq. 10.45), which is used to calculate tortuosity, as well as to the calculation of the geometry's characteristics.

Finally, the effect of *Pe* and porosity on the adsorption efficiency for the microscopic (sphere-in-cell) geometry is shown in Fig. 10.12. It is noteworthy that for $Pe > 50$ the sphere-in-cell is insufficient to adequately describe such physicochemical problems, since the adsorption efficiency is found to increase with porosity, although the opposite phenomenon is expected to happen. This behavior can be attributed to the presence of high local velocities located in the contact area of the inlet boundary condition with the outlet boundary condition (i.e., at the area of the two—inlet and outlet—adjacent boundary cells). Conceptually, unit cell approach suffers by intrinsic disadvantages in the adequate representation of a porous structure since it is neither space filling, nor energy conservative and does not take into account the interference of other grains, etc. (Kuwabara 1959). All these limitations lead to high local advection causing gradient to the species mass fraction (Fig. 10.13). As a result of these limitations, the sphere-in-cell geometry will no longer be used in the present study because the other two geometries are more efficient and accurate. Nor any future proposals will be made for the correction of the microscopic geometry, since present technology allows for the detailed numerical solution of similar models providing accurate results.

Applying the scale transition rule presented previously, Fig. 10.14 presents a direct comparison between the mesoscopic and the macroscopic geometry for different *Pe* numbers and porosities, varying between 0.60 and 0.90. It is found that all porosity values follow the same general trend. In these graphs, it is critical to

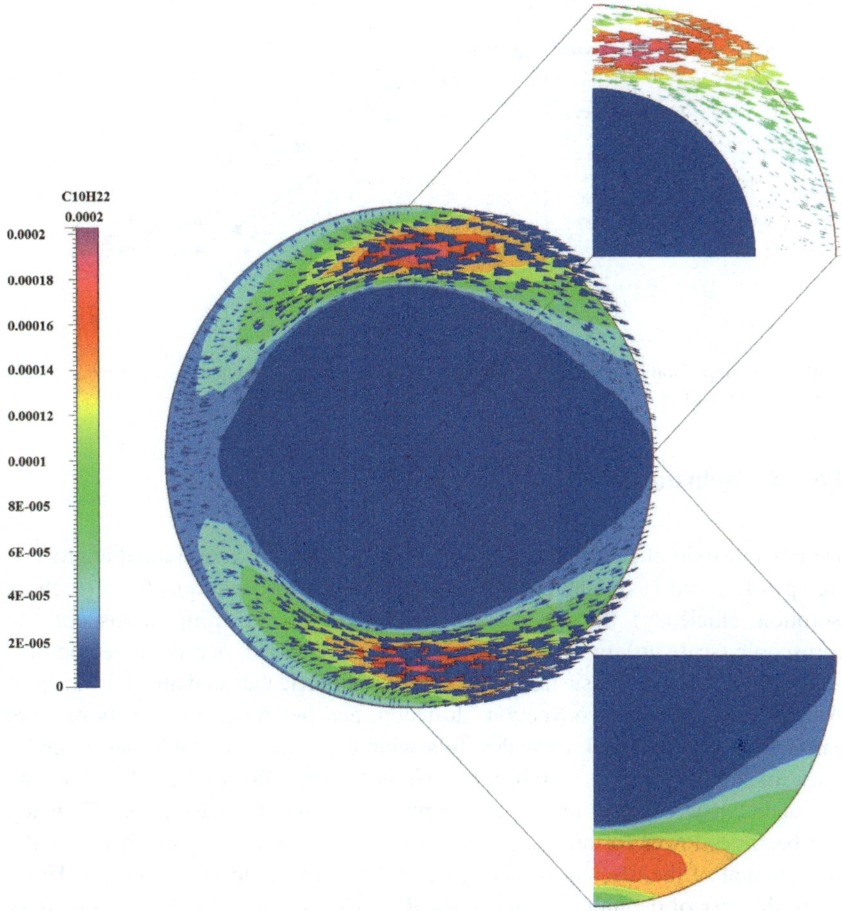

Fig. 10.13 Discontinuities in the flow field for the microscopic geometry

note the presence of a deviation bar that has been obtained from the distribution dependency test (Fig. 10.3a). For the simulations with low Peclet numbers (diffusive regimes), the mesoscopic and macroscopic models predict the same λ. The difference found between the two geometries for intermediate and high Pe values ($50 < Pe < 1000$) can be explained by the more detailed consideration and solution of the mesoscopic models. Furthermore, it can be observed that the decane consumption increases slightly with the scale, since the available adsorbing mass and reactive surface also increases with the scale. While constant porosity ensures that the solid volume is the same in all cases, the adsorbing surface (i.e., the adsorbing mass itself) is not the same and follows the S/V trend.

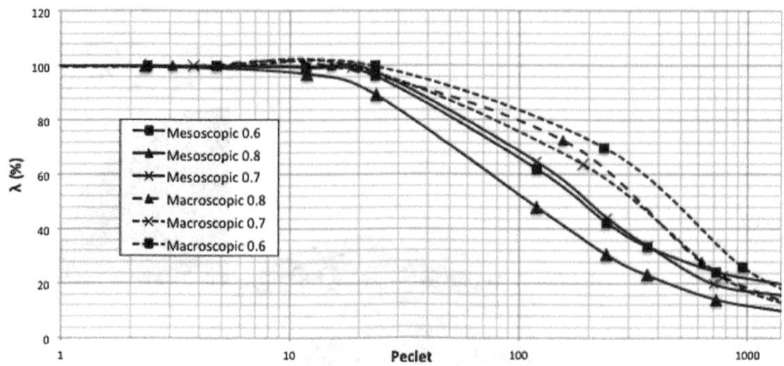

Fig. 10.14 Comparison between mesoscopic and macroscopic geometry for various Peclet numbers and porosities

10.8 Conclusions

The mass transport phenomena occurring in a porous media were studied in order to develop a fast and easy method of matching the macroscopic quantities (such as adsorption efficiency) during a scale transition process from mesoscopic to macroscopic (scale-up), or to microscopic (scale-down) and vice versa geometries. In order to be able to transit from one scale to another, the fundamental transport processes (laminar flow, convection, diffusion, and heterogeneous reaction) were detailed described for the three scales, following the same flow conditions given by the use of a dimensionless number, known as Peclet. After the simulation of the three scales, it was found that as *Pe* number increases, the adsorption efficiency value becomes lower for all scales, meaning that as the convection increases the lower amount of the reactant reaches the catalytic surface in order to react. However, in the case of the unit cell model, this behavior was found to be accurate only for relatively low to medium *Pe* numbers ($Pe < 50$). For *Pe* numbers over 50, the velocity is found to reach extremely high values at the two points where the inlet surface meets the outlet surface, leading to discontinuities. Therefore, it is proved that the sphere-in-cell geometry is insufficient to accurate simulate transport phenomena for high Peclet numbers. Furthermore, the same effect of porosity on adsorption efficiency, was found for all scales, as it is observed that the λ value increases, as porosity decreases. This phenomenon can be explained because as porosity decrease the catalytic surface, where the oxidization reaction takes place also increases. Hence, if the surface increases then the available surface for a reaction increases too. Finally a method of matching the geometrical parameters when scale transition occurs is proposed, underlying the necessary steps that should be followed. As the first step the calculation of the geometrical characteristics of the detailed geometry is a necessity as it will be the base where the transition will be relied on. Next, the inlet mixture flow needs to be adjusted in order to preserve the Pe number for all of the scales. Since the diffusion coefficients are practically constant, this adjustment can be made only by adapting the velocity to the necessary value. Finally, the last two most important adjustments are required to be made in order to have an identical characterization of the geometry without any discrepancies. The characteristic ratio, S/V of the porous material, is a measurement value of the available surface where the reaction may occur. By correctly adjusting this value, the geometry will have the same amount of reaction if it is simulated in any of the needed scales. Last but not least the average pore size needs to be adjusted proportionally to the transition ratio. If the average pore size is not adjusted correctly, it could lead to a discrepancy of the geometry as the hypothetical pores would have been squeezed or extended in order to preserve that lower or higher void space, respectively.

References

1. Abriola LM, Bradford SA, Lang J, Gaither CL (2004) Volatilization of binary nonaqueous phase liquid mixtures in unsaturated porous media. Vadose Zone Journal 3(2):645–655
2. Akanni KA, Evans JW, Abramson IS (1987) Effective transport coefficients in heterogeneous media. Chem Eng Sci 42(8):1945–1954
3. Bachmat Y, Bear J (1990) Introducing to modeling of transport phenomena in porous media. Vol 4 of series: "Theory and applications of transport in porous media". Kluwer Academic Publishers, Dordrecht, The Netherlands
4. Battin LF, Fournet R, Glaude PA, Judenherc B, Warth V (2000) Modeling of the gas-phase oxidation of n-decane from 550 to 1600 K. Proc Combust Inst 28(2):1597–1605
5. Bear J (1972) Dynamics of fluids in porous media. American Elsevier Publishing Co., New York
6. Beekman JW (1990) Mathematical description of heterogeneous materials. Chem Eng Sci 45 (8):2603–2610
7. Berkowitz B, Scher H (1998) Theory of anomalous chemical transport in random fracture networks. Phys Rev E 57(5):58–69
8. Berkowitz B, Cortis A, Dentz M, Scher H (2003) Modelling non-Fickian transport in geological formations as a continuous time random walk. Rev Geophys 44(2):1–49
9. Bourgeat A, Quintard M, Whitaker S (1988) Comparison between homogenization theory and volume averaging method with closure problem. Comptes Rendus de l'Académie des Sciences, Série 2(306):463–466
10. Brenner G (2009) CFD in process engineering, vol 100(341). In: Ernst HH, Egon K (eds) 100 Volumes of notes on numerical fluid mechanics. Springer, Berlin
11. Bruggemann DA (1935) Berechnung verschiedener physicalischer konstate von heterogen substanzen. Annalen der Physik 4, 167:636–664
12. Coutelieris FA, Delgado JMPQ (2012) Advanced structured materials. Springer
13. Cunningham MA (1910) On the velocity of steady fall of spherical particles through fluid medium. Proc R Soc A 83(563):357–365
14. Cushman JH, Bennethum LS, Hu BX (2002) A primer on upscaling tools for porous media. Adv Water Resour 25(8–12):1043–1067
15. Dagan G (1989) Flow and transport in porous formations. Springer, New York
16. Dai Z, Ritzi RW, Huang C, Rubin YN, Dominic DF (2004) Transport in heterogeneous sediments with multimodal conductivity and hierarchical organization across scales. J Hydrol 294(1–3):66–86

17. Delgado JMPQ (2006) Mass transfer and dispersion around an active cylinder in cross flow and buried in a packed bed. Heat Mass Transf 42(12):1119–1128
18. DyKaar B, Kitanidis PK (1992) Determination of the effective hydravlic conductivity for heterogeneous porous media using a numerical spectral approach: 2. Results. Water Resour Res 28(4):1167–1178
19. Ene HI (1990) Application of the homogenization method to transport in porous media, Cushman JH (ed). Academic Press, London
20. Fokker PA (2001) General anisotropic effective medium theory for the effective permeability of heterogeneous reservoirs. Transp Porous Media 44(2):205–218
21. Gelhar LW, Axness CL (1983) Three dimensional stochastic analysis of microdispersion in aquifers. Water Resour Res 19F:161–180
22. Goos E, Burcat A, Ruscic B (2005) Extended third millennium ideal gas and condensed phase thermochemical database for combustion with updates from active thermochemical tables. Update of: "Third millennium ideal gas and condensed phase thermochemical database for combustion with updates from active thermochemical tables, A. Burcat and B. Ruscic, Report ANL 05/20 and TAE 960, Technion-IIT, Aerospace Engineering and Argonne National Laboratory, Chemistry Division, September 2005
23. Gunn DJ (1978) Transfer of heat or mass to particles in fixed and fluidized beds. Int J Heat Mass Transf 21(4):467–476
24. Horung U (1997) Homogenization and porous media. Springer, New York
25. Huang H, Hu BX (2001) Nonlocal nonreactive transport in heterogeneous porous media with rate-limited sorption and interregional diffusion. Water Resour Res 36(7):639–747
26. King PR (1989) The use of renormalization for calculationg effective permeability. Transp Porous Media 4(1):37–58
27. Kløv PE, Øren SPE, Stensen JÅ, Lerdahl TR, Berge LI, Bakke S, Boassen T, Statoil ASA, Virnovsky G, Rogaland Research, "Pore-to-Field Scale Modeling of WAG.", Society of Petroleum Engineers Inc., Copyright 2003
28. Miller TM (1998) Atomic and molecular polarizabilities. J Phys Chem A 102(14):2399–2407
29. Molz FJ, Rajaram H, Lu S (2004) Stochastic fractal-based models of heterogeneity in subsurface hydrology: origins, applications, limitations and future research questions. Rev Geophys 42(1)
30. Montroll EW, Scher H (1973) Random walks on lattices. IV. Continuous time random walks and influence of adsorbing boundaries. J Stat Phys 9(2):101–135
31. Neuman SP (1990) Universal scaling of hydravlic conductivities and dispersivities in geological media. Water Resour Res 26(8):1749–1758
32. Newton I (1687) The mathematical principles of natural philosophy, 2nd edn. London
33. Noetinger B (1994) The effective permeability of a heterogeneous porous media. Transp Porous Media 15(2):99–127
34. Noetinger B, Estebenet T (2000) Upscaling of double porosity fractured media using continuous time random walk methods. Transp Porous Media 39(3):315–337
35. Oh SH, Cavendbh JC (1981) Transients of monolithic catalytic converters. Response to step changes in feedstream temperature as related to controlling automobile emissions. Ind Eng Chem Prod Res Dev 21(1):29–37
36. Quintard, M., Whitaker, S., "Convection, dispersion, and interfacial transport of contaminants: homogeneous porous media.", Vol. 17(4), 1994: 221–239
37. Quintard M, Whitaker S (1988) Two-phase flow in heterogeneous porous media: The method of large scale averaging. Transp Porous Media 3(4):357–413
38. Quintard M, Whitaker S (1994) Transport in ordered and disordered porous media V: geometrical results for two-dimensional systems. Trasp Porous Media 15(2):183–196
39. Rajaram H (1997) Time and scale dependent effective retardation factors in heterogeneous aquifers. Adv Water Resour 20(4):317–330
40. Rizzi A, Nørstrud H, Siikonen T (2009) CFD developments in the Northern European countries, vol 100. In: Egon HH, Ernst K (eds) 100 Volumes of 'Notes on numerical fluid mechanics. Springer, Berlin

41. Rubin Y (2003) Applied stochastic hydrogeology. Oxford University Press, New York
42. Salamon P, Fernàndez-Garcia D, Gómez-Hernández JJ (2006) Modeling mass transfer processes using random walk particle tracking. Water Resour Res 42(11)
43. Shen L, Chen Z (2007) Critical review of the impact of tortuosity on diffusion. Chem Eng Sci 62(14):3748–3755
44. Silbert EL, Ertas D, Gary SG, Thomas CH, Dov L, Steven JP (2001) Granular flow down an inclined plane: Bagnold scaling and rheology. Phys Rev E 64(5):64–78
45. Slattery JC (1967) Flow of viscoelastic fluids through porous media. Am Inst Chem Eng 13 (6):1066–1071
46. Uchida S (1954) Viscous flow in multiparticle systems: slow viscous flow through a mass of particles, pp 197–201, Motai T (ed). Int Sci Technol Univ Tokyo 46(2):1194–1195(in Japanese)
47. Voltz SE, Morgan CR, Llederrnan D, Jacob SM (1973) Kinetic study of carbon monoxide and propylene oxidation on planitnum cataysts. Ind Eng Chem Prod Res Dev 12(4):294–301
48. Whilson KG (1971) Renormalization group and critical phenomena. Phys Rev B 4(9):583–600
49. Whitaker S (1967) Diffusion and dispersion in porous media. Am Inst Chem Eng 13(3):420–427
50. Whitaker S (1986) Flow in porous media i: a theoreticale derivation of Darcy's law. Transp Porous Media 1(1):3–25
51. Whitaker S (1993) The closure problem for two phase flow in homogeneous porous media. Chem Eng Sci 49(5):765–780
52. Whitaker S (1999) The method of volume averaging, vol 13. Springer, Netherlands
53. Bhattacharya RN, Gupta UK (199) Application of central limit theorems to solute dispersion in saturated porous media: From kinetic to darcy scale. In: Cushman JH (ed) Dynamic of fluid in hierarchical Porous media. Academic Press, New York, pp 61–96
54. ESI CFD Inc., CFD-ACE+ Modules_Manual, vol 2. ESI-group, Huntsville
55. Yadav PK (2011) On the slow viscous flow through a swarm of solid spherical particles covered by porous shells. Appl Math 11(2):112–121
56. Zhou Q, Molz FJ, Liu HH, Zhang Y, Bodvarsson GS (2007) Field-scale effective matrix diffusion coefficient for fractured rock: results from literature survey. J Contam Hydrol 93(1–4):161–187

Chapter 11
Positron Beam-Based Ortho-Positronium Porosimetry

S. Mariazzi

11.1 Positron

The positron (e^+) is the anti-particle of the electron. Its existence was predicted by Dirac in 1930 [1]. Few years later it was observed by Anderson [2–4]. Finally, the possibility of a bound state of a positron with an electron, the positronium (Ps), was demonstrated by Deutsch in 1951 [5, 6].

In vacuum, the positron is a stable particle with a mean lifetime of more than 2×10^{21} yr. [7]. Within experimental limits, its mass is the same as the electron and it has the same intrinsic angular momentum 1/2 but opposite charge and magnetic moment [7]. The stability of e^+ in vacuum and the possibility to slow it down in solids allow one to transport positron beams for use in various applications [8]. The use of variable energy slow-positron beams has provided the opportunity to obtain unique information on a wide variety of issues in solid-state physics. In particular positron and positronium are highly sensitive to open volumes with dimensions ranging from mono-vacancies (positrons) to pores of several tens of nm (positronium). Thus, in the last decades, positron beams have been extensively used to study the open volume structures and related mechanical, optical and electrical properties of several materials (see for example [9]).

Positrons can be produced by nuclear decay reactions (i.e., $^{22}Na \rightarrow {}^{22}Ne + \beta^+$ with a lifetime $\tau_{1/2} = 2.6$ years) or by stopping high-energy photons (with $E >> 1.22$ MeV) in a dense material through $e^+ - e^-$ pair formation [10]. Positrons produced by radioactive decay have an energy distribution ranging from 0 to a fraction of MeV. In order to obtain a monochromatic positron beam, it is necessary to "moderate" the positrons by using a crystalline material, placed close to the radioactive source. Positrons implanted in a solid lose rapidly their kinetic

S. Mariazzi (✉)
Stefan Meyer Institute for Subatomic Physics, Austrian Academy of Sciences, 3 Boltzmanngasse, 1090 Vienna, Austria
e-mail: Sebastiano.Mariazzi@cern.ch

P. Bettotti (ed.), *Submicron Porous Materials*,
DOI 10.1007/978-3-319-53035-2_11

energy (see the following paragraph) and, in some crystalline materials like solid neon or tungsten, a fraction of thermalized or quasi-thermalized positrons can be re-emitted from the surface into the vacuum. The result of moderation is the emission of low-energy positrons (few eV) preferentially perpendicular to the surface of the moderator and with a narrow energy distribution [7]. Using a system of electrostatic or magnetic lenses and apertures, the moderated positrons can be extracted and directly transported to a target. In alternative, cooling and storage of positrons is made possible by using a buffer gas combined to electrical and magnetic fields [11] which ensure transversal and radial confinement of e^+ . This procedure allows the storage of a large number of positrons (up to 10^8 – 10^9 positrons) and the production of bunches with a well-defined geometry [12–14]. Such bunches can be subsequently implanted on a target with a time spread of few nanoseconds [12, 15].

The positron implantation profile obeys well to the so-called Markovian function [16] with the following relationship between the mean positron implantation depth ($\bar{z}$) and the positron implantation energy E Eq. (11.1):

$$\bar{z} = \frac{40}{\rho} E^n \tag{11.1}$$

where ρ, is the density of the material (g/cm^3) and n is a phenomenological parameter with value around 1.6 [17]. The depth and the energy are expressed in units of nm and keV, respectively. As the energy of slow-positron beams is typically tunable from few eV up to tens of keV, one can investigate target from the surface down to several micrometers of depth, depending on the sample density.

11.2 Interaction of Positron with Solids and Ps Formation

When energetic positrons are implanted in a solid, they can experience several different processes that are schematically summarized in Fig. 11.1.

A fraction of positrons is backscattered and does not enter the solid (case A in Fig. 11.1). This amount depends on the energy of the incident particles, on their angle with respect to the surface of the sample and on the target atomic number [18, 19].

Positrons entering the material begin to lose their kinetic energy by several interactions with the host. At the end of this stopping process, e^+ remains with just their thermal energy in a delocalized Bloch state and diffuse around the medium (see for example [17]) (case B in the Fig. 11.1). The positrons implanted at low energy (typically less than 1 keV) can reach the surface before thermal equilibrium. In this case, they can be re-emitted in vacuum as epithermal positrons [20] (case C in the Fig. 11.1). For thermalized positrons, there are several possible annihilation channels. They can diffuse in bulk material, remaining in a delocalized state until they annihilate with electrons (case D in the Fig. 11.1). In this annihilation process two γ rays are emitted Eq. (11.2):

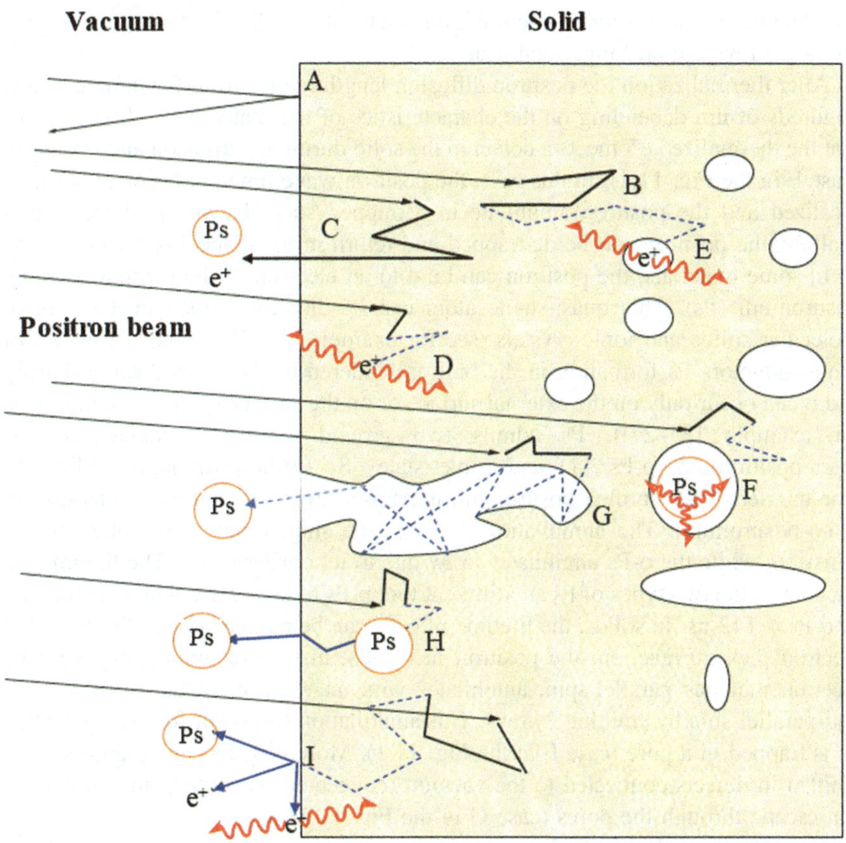

Fig. 11.1 Schematic representation of the processes that can involve the positrons interacting with a solid: **A**: backscattering of energetic positrons **B**: stopping, thermalization (continuous line) and diffusion (dashed line) of the implanted positron **C**: partial thermalization and reemission of epithermal positrons as positron or as positronium **D**: annihilation of a delocalized positron in bulk **E**: positron trapping in a defect and annihilation in this state **F**: Positronium formation in open volume with nano size. **G**: positronium formation in defects connected to the vacuum and escape through connected porosities. **H**: positronium formation in bulk and emission from the surface **I**: positron back-diffusion to the surface and positron reemission (as positron or as positronium) or positron trapping in a surface localized state

$$e^+ + e^- \to 2\gamma \tag{11.2}$$

If the positron and electron, involved in the annihilation, are both at rest, the two photons are emitted collinearly in opposite direction. In this situation, the energy of each γ ray is $mc^2 = 511$ keV, due to energy conservation. Where m is the rest mass of the electron and c is the speed of light. Instead, if the positron–electron couple is not at rest, it has a momentum $P_{e^+e^-}$ with respect to the laboratory frame. This fact introduces a shift in the γ-rays energy given by the characteristic Doppler broadening of the positron–electron photopeak at 511 keV. Moreover, the two emitted γ rays are no longer collinear.

After thermalization the positron diffusion length ranges from few nm up to few hundreds of nm depending on the characteristics of the material [7]. It is possible that the thermalized e^+ meets a defect in the solid during its diffusion and is trapped (case E in the Fig. 11.1). In this case, the positron wave function becomes strongly localized and the positron annihilate in a trapped state. However, if the trap is shallow, the positron can be de-trapped and return in the delocalized state.

In some materials, the positron can bind to an electron of the medium forming positronium (Ps). This quasi-stable atom can be directly formed in the bulk of molecular solids and ionic crystals (see for example [21–25]), while in metals or semiconductors Ps formation in the bulk is hindered by the high-electron density and it can occur only on the external surface or on the surface of open volumes (see for example [25–27]). Ps admits two ground states: a singlet-state $^1S_0$ (para-positronium, p-Ps ↓↑) and triplet-state $^3S_1$ (ortho-positonium, o-Ps ↑↑). One-quarter of the formed positronium is para-positronium and three-quarters are ortho-positronium. The annihilation of Ps in the singlet state takes place by 2γ emission, while the o-Ps annihilates in 3γ due to its nonzero spin. The lifetimes in vacuum of the two types of Ps are different too: p-Ps lives 125 ps, while the lifetime of o-Ps is 142 ns. In solids, the lifetime of o-Ps can be reduced due to the so-called electron pick-off-reaction: the positron in the Ps, instead of annihilating with its electron that has parallel-spin, annihilates with an electron of the medium with anti-parallel-spin by emitting 2γ rays. This annihilation can occur for example when Ps is trapped in a pore (case F in the Fig. 11.1). Moreover, Ps can be produced or emitted in defects connected to the vacuum (connected porosities). In this case, it can escape through the pores (case G in the Fig. 11.1).

As told, in molecular solids and ionic crystals positronium can be formed directly in the bulk. In this materials, the Ps diffusion length ranges from few nm up to several tens of nm [21, 28]. If Ps is formed close to the surface a fraction can diffuse and reach the surface being emitted into vacuum (case H in the Fig. 11.1).

Another possible path of the thermalized positron is illustrated in case I of Fig. 11.1. At low implantation energy or, in general, if the positron diffusion length is long enough, the positron can back diffuse to the surface. At the surface, the positron has three possibilities. First, it can be reflected back by the surface potential into the lattice. Second, the positron can be trapped in a surface state. And third, the positron can be emitted in vacuum as positron or as positronium, depending on the surface characteristic of the material [27].

11.3 Ps as a Probe for Porosity: Observables

The different processes of Ps formation and interaction with a material described in the previous paragraph generate several observables that can be used to characterize the material itself. These observables are introduced in the present section.

11.3.1 o-Ps 3γ Annihilation

The first important observable related to Ps annihilation is the energy of the emitted gamma rays. When single positrons are implanted in the target, the energy of the gamma rays generated by their annihilations can be measured by using high purity Ge detectors and a spectrum of gamma energy distribution can be acquired collecting many events (see example reported in Fig. 11.2). Since the o-Ps annihilation occurs primarily into three γ photons, their average energy is less than 511 keV. Therefore, the presence of 3γ o-Ps annihilation affects the shape of the spectrum (Fig. 11.2) in the gamma energy region below the characteristic 511 keV peak.

In order to extract information about the Ps formation from the gamma distribution, the parameter R_F is introduced Eq. (11.3) (see also Fig. 11.2):

$$R_F = \frac{valley\,area}{peak\,area} = \frac{V_F}{P_F} \tag{11.3}$$

The "valley area" is typically chosen as the number of counts with gamma energy E_γ between 410 and 500 keV, whereas the "peak area" as the number of counts in the photopeak (511–4.25 keV < E_γ < 511 + 4.25 keV). This parameter only gives information about the amount of o-Ps annihilating via 3γ, but not about the total yield of o-Ps.

In the particular case of Ps formation at the surface the pick-off annihilation vanishes and all the formed o-Ps annihilate in 3γ rays. In this situation, it is possible to obtain the total Ps (either o-Ps or p-Ps) fraction F defined as the fraction of positrons that form Ps Eq. (11.4) [26, 29].

$$F = \left\{ 1 + \frac{P_1\left[R_1 - R_F(E)_{REF}\right]}{P_0\left[R_F(E)_{REF} - R_0\right]} \right\}^{-1} \tag{11.4}$$

With this definition the Ps fraction F ranges between 0, corresponding to a situation in which Ps is not formed, and 1, when all the positrons form positronium. The term "P" is the number of counts in the peak area while R is the parameter R_F of Eq. (11.3). The subscripts 0 and 1 correspond to the no Ps-formation situation

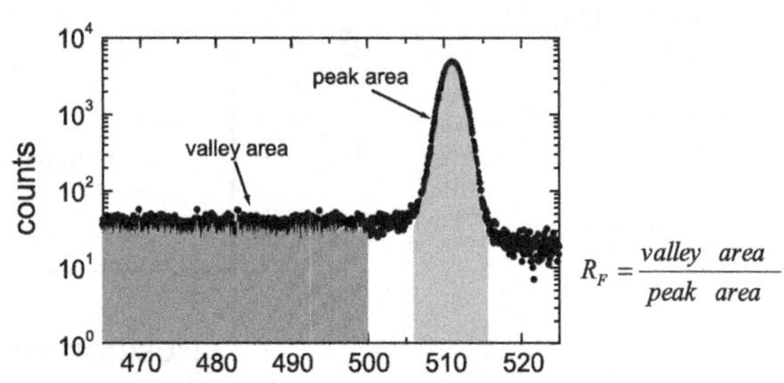

Fig. 11.2 Definition of R_F parameter. The measured curve shows the energy distribution of the photons emitted by the annihilation. The "valley area" is the region where the 3 gammas are counted

and 100% Ps formation situation, respectively. The calibration of P_1 and R_1 is performed measuring the peak and the valley areas on the surface of a Ge single crystal at 1000 K, where almost all the implanted positrons are expected to back diffuse and form Ps [26, 30]. The calibration of P_0 and R_0 is achieved measuring the peak and the valley areas in bulk of a Ge single crystal where no Ps is formed. $R_F(E)_{REF}$ represents the ratio in Eq. (11.3) calibrated so that $R_F(E)_{REF} = R_0$ in absence of Ps formation (i.e., for $E \to \infty$).

The parameter F gives information about the Ps yield and it can be used to characterize metallic and semiconductor surfaces [26, 30]. For insulating materials with mesoporous (2 < pore size < 50 nm) and microporous (pore size < 2 nm) structure, the parameter $F_{3\gamma} = \frac{3}{4}F$ (corresponding to the normalized fraction of implanted positron annihilating as o-Ps) is frequently used [31]. Furthermore, the parameter $R_{\%}$ defined as in Eq. (11.5)

$$R_{\%} = \frac{R_F(E)_{REF} - R_0}{R_1 - R_0} 100 \tag{11.5}$$

is sometimes used [32, 33].

As an example, a curve of $F_{3\gamma}$ versus positron implantation E measured in a nanoporous silicon sample [34] is shown in Fig. 11.3.

Figure 11.3 shows a decrease of the fraction of implanted positrons annihilating as o-Ps via 3 gammas starting from the surface of the target toward the silicon bulk. In general, the shape of the $F_{3\gamma}(E)$ curve is related both to the distribution of the open volumes in the target and the diffusion of o-Ps inside those volumes. As a consequence, information about the pore distribution and interconnectivity can be achieved by the study of this parameter (see paragraph 1.4.2).

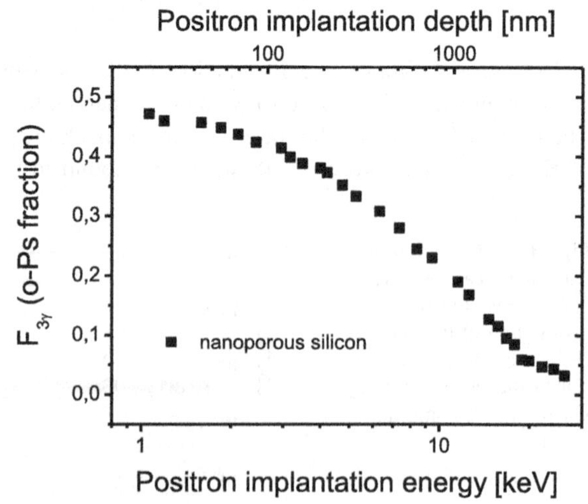

Fig. 11.3 Example of $F_{3\gamma}$ versus positron implantation E measured in a nanoporous silicon target. The reported positron implantation depth has been calculated using Eq. (11.1) and assuming for the target a density $\rho = 1.9$ g/cm^3 [34, 35]

11.3.2 o-Ps Lifetime

The second important observable related to o-Ps is the lifetime. In general, the annihilation rate λ of a positron delocalized in a considered material, corresponding to the inverse of its lifetime τ, is determined by the overlap of the positron density $\rho_+(\vec{r})$ with the electron density $\rho_-(\vec{r})$ [17] Eq. (11.6):

$$\lambda = \frac{1}{\tau} = \pi r_0^2 c \int \rho_+(\vec{r})\rho_-(\vec{r})\gamma[\rho_-(\vec{r})]d\vec{r} \tag{11.6}$$

where r_0 is the classical electron radius, c is the speed of light, $\vec{r}$ is the position vector. The term $\gamma[\rho_-(\vec{r})] = 1 + \Delta\rho_-/\rho_-$ is the so-called enhancement factor and it describes the increase $\Delta\rho_-$ of the electron density around the positron due to the Coulomb attraction between a positron and the surrounding electrons. Each material has a characteristic electron density and, therefore, a positron in each material has a characteristic lifetime. Positrons trapped in open-volume defects have a longer lifetime in the range of hundreds of ps due to the local reduction of the electron density [17]. Positron lifetime characterizes the type and the size of open-volume defects. From the analysis of a lifetime spectrum, one can extract two–four lifetimes with the relative intensity. The concentration of defects can be determined from the intensity of the positron lifetime [9].

In presence of Ps formation, its annihilation can be easily detected and distinguished by the other annihilation channels. Indeed, o-Ps self-annihilation and o-Ps pick-off annihilation give rise to long lifetimes ranging from ns up to 142 ns that are easily identifiable.

Measurements of a depth-resolved Positron Annihilation Lifetime Spectroscopy (PALS) are carried out by implanting in the investigated target a single positron at a time and measuring the delay between the instant the positron enters the sample and its annihilation. Therefore, in order to obtain a lifetime spectrum, it is necessary to use a start signal and stop signal. The stop signal is provided by the detection of a gamma photon generated by the positron annihilation. Two methods have been developed to determine the start signal: one exploits secondary electrons emitted from the solid surface at the positron impact [10]. In the second method, the positrons are bunched in time by an adequate apparatus, in this way the bunch arrival time at the sample surface is well determined (see for example [36]). An example of PALS spectrum measured at 5 keV positron implantation energy on nanoporous Si with oxidized surface and capped with 30 nm of TiO_2 is shown in Fig. 11.4a. In Fig. 11.4b, the single lifetimes and the relative intensities, extracted from lifetime spectra are reported, as measured at different positron implantation energies in the same sample. The long lifetime τ_3 corresponds to the annihilation of o-Ps inside the pores while τ_1 and τ_2 are ascribable, respectively, to: (1) p-Ps self-annihilation and positron annihilations in Si and (2) annihilation of positron in the layer of silica covering the open volumes [34].

Lifetime measurements are also possible with pulses containing a large number of positrons provided that the temporal width of the pulse is short compared to the

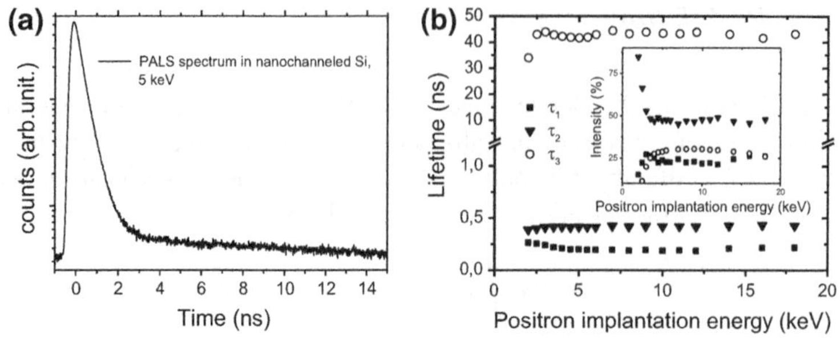

Fig. 11.4 **a** Lifetime spectrum measured at 5 keV positron implantation energy nanoporous Si with oxidized surface and capped with 30 nm of TiO_2. **b** Lifetimes and relative intensities (Inset) extracted from the lifetime spectra measured at different positron implantation energies in the same sample

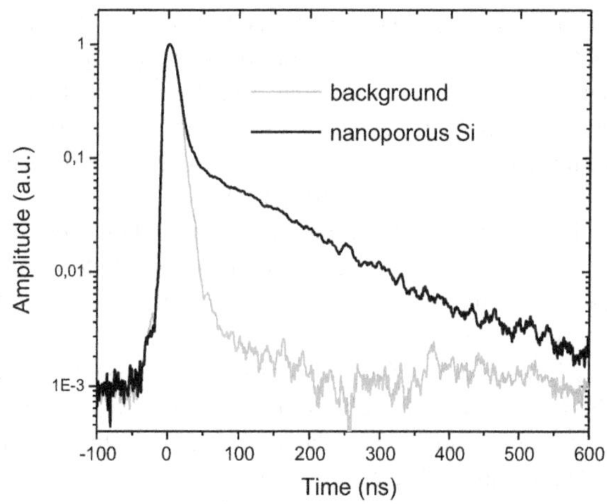

Fig. 11.5 Example of SSPALS spectra measured on the surface of a target without Ps formation (*gray curve*) and in a nanoporous silicon sample with Ps formation (*black curve*). In both cases positron implantation energy was 3.3 keV

investigated lifetimes. As pulses of few ns can be produced [12], o-Ps self annihilations and o-Ps quenching annihilations can be distinguished from faster annihilation channels. The gamma events generated by simultaneous annihilations of many positrons are collected by using fast detectors that do not show saturated signals caused by the intense gamma flux [37]. Such detectors allow one to record a lifetime distribution from a single shot of positrons. This variation of PALS spectroscopy is referred to as Single Shot Positron Annihilation Lifetime Spectroscopy (SSPALS) [38]. An example of SSPALS spectra is reported in Fig. 11.5.

The SSPALS obtained by implantation of positrons in the target without Ps formation shows a sharp peak corresponding to the fast 2 gamma annihilations of implanted e^+ . On the right side of the peak, the signal quickly decreases and reaches the noise level in ~100 ns. On the other hand, the lifetime spectrum measured in the nanoporous silicon sample shows a long tail from around 20 ns up to 600 ns generated by Ps formation and decay.

11.3.3 o-Ps Emission Kinetic Energy

Thanks to its relatively long lifetime, o-Ps formed in porous targets can diffuse along the pores and eventually be emitted into vacuum after reaching the surface (case G in Fig. 11.1). When formed in the bulk of an insulator, o-Ps is emitted into the pore with a typical energy of few eV [39]. During the diffusion in the pore, it loses energy and cools down through collisions with the walls. Finally, it can be emitted into vacuum with a reduced kinetic energy [40–45].

The o-Ps emission energy can be measured by using two different techniques: (i) o-Ps Time Of Flight (TOF) spectroscopy or (ii) Doppler broadening of a transition line. In technique (i) the o-Ps emission energy, perpendicular to the target surface, is measured detecting the time (t) between positron implantation in the target and o-Ps annihilation in flight at a given, well-known distance z_0 from the target surface [25–27, 35, 46–48]. The signal of positron arrival is usually derived by detecting secondary electrons emitted when e^+ is implanted in the investigated target. The gamma rays emitted by o-Ps self-annihilation are detected with scintillators (like NaI (Tl)) placed behind a thick gamma ray shield with a slit of few mm width. The center of the slit is located at the distance z_0 from the target surface which represents the flight path of o-Ps during its lifetime. More than 10^3 events are collected for each TOF spectrum (light gray in Fig. 11.6). The background originating from gammas due to prompt positron annihilation and scattered into the slit (dashed dark-gray line in Fig. 11.6) is recorded in a target without Ps formation and subtracted by the TOF spectrum. Finally, the effects of o-Ps finite lifetime and permanence in front of the slit is corrected multiplying by the factor $\frac{1}{t}e^{t/(142\,ns)}$ (black curve in Fig. 11.6) [25].

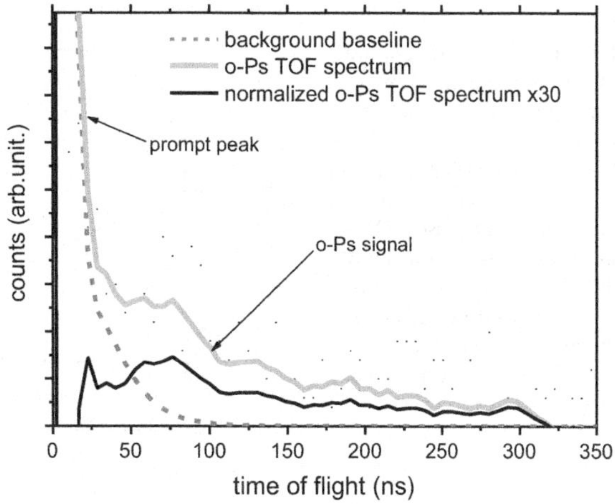

Fig. 11.6 Example of TOF spectrum measured in a nanoporous silicon after positron implantation with an energy of 7 keV (*light gray curve*) compared to a background curve (*dashed dark gray*) measured in a Si target without o-Ps formation at the same energy. The *black curve* is the normalized difference of the two previous curves (see text). The spectra were measured at a distance $z_0 = 1$ cm from the target

The measured time of flight t can finally be converted to the o-Ps kinetic energy along the direction perpendicular to the target $E_{\perp}$ by the relationship $E_{\perp} = m\frac{z_0^2}{t^2}$, where m is the positron rest mass.

In the alternative method of the Doppler broadening (ii), Ps is laser excited from the n = 1 ground state to n = 2 (or from n = 1 to n = 3). In this case, a linewidth is obtained by scanning the laser wavelength. When the o-Ps atom is at rest, the transition is expected to occur only when the laser wavelength is perfectly on resonance with the transition energy. On the other hand, if o-Ps has an energy distribution, a shift and Doppler broadening of the transition line are introduced. The measurement of the transition linewidth is usually performed by studying the effect of laser excitation on SSPALS spectra [45, 49]. Excited o-Ps atoms are selectively quenched by a magnetic field or photoionized with a second dedicated laser. Either of these processes results in a decrease of the o-Ps population decaying into three gamma rays. The fraction of excited o-Ps can be evaluated by analyzing the decrease in the area below the SSPALS spectra when quenching or photoionization are applied. In Fig. 11.7 a SSPALS spectrum given by self-annihilation of Ps emitted by a nanoporous silicon sample (laser OFF) is compared to a SSPALS spectrum obtained with the same parameters but after laser excitation to n = 3 and successive photoionization (lasers ON). The fraction S of excited and ionized o-Ps atoms can be evaluated from the areas f_{off} and f_{on} of the SSPALS spectra in a proper time window: $S = (f_{off} - f_{on})/f_{off}$ [50]. In the inset of Fig. 11.7, a scan of S parameter (in the present case calculated from 50 and 250 ns from the prompt peak) versus the wavelength of the excitation laser is reported.

If the laser bandwidth is negligible with respect to the Doppler profile, the width at half-maximum of the linewidth ($\Delta\lambda$) can be finally correlated to the standard deviation of the o-Ps velocity (σ_v) along the laser propagation axis through the relation Eq. (11.7):

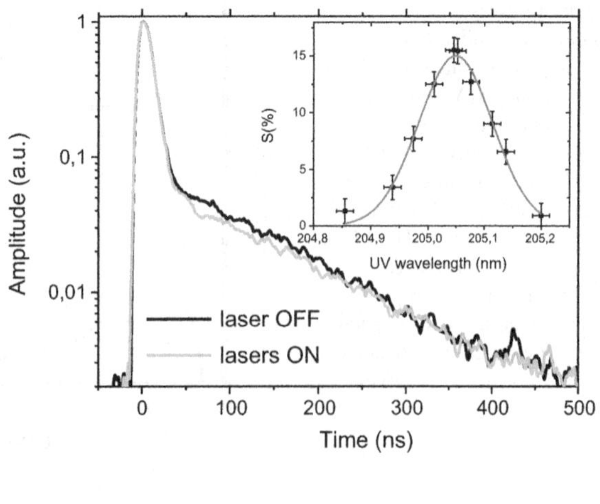

Fig. 11.7 SSPALS spectra of a Ps cloud emitted by a nanoporous silicon sample without laser excitation (laser OFF) and with laser excitation to n = 3 and subsequent photoionization (lasers ON). In the inset, the linewidth of the n = 1 → n = 3 transition of Ps obtained by scanning the excitation laser wavelength is reported (see text)

$$\Delta\lambda = \frac{2\lambda_0}{c}\sigma_v\sqrt{2ln2} \tag{11.7}$$

Where λ_0 is the wavelength for an atom at rest.

11.4 Ps as a Probe for Porosity: Information

11.4.1 Open Volume Dimension

As said in the previous paragraphs, the lifetime of o-Ps in vacuum is 142 ns, thus the corresponding annihilation rate is $\lambda_T = 1/142$ ns. In the bulk of material, the o-Ps lifetime is shortened by pick-off and the annihilation rate is commonly assumed to be equal to the spin-averaged rate $\lambda_A = (\lambda_S + 3\lambda_T)/4$ where $\lambda_S =$ 1/125 ps [51]. This is equivalent to the assumption that positron in o-Ps feels all the time the presence of one extra electron with inversely oriented spin at negligible distance. The average o-Ps annihilation rate in a pore is then written as the weighted sum of the bulk annihilation rate and the vacuum annihilation rate Eq. (11.8):

$$\langle\lambda\rangle = \frac{1}{\tau_{o-Ps}} = \frac{\lambda_S + 3\lambda_T}{4}P + (1-P)\lambda_T, \tag{11.8}$$

where the weight P represents the probability of superimposition of o-Ps wave function with the bulk.

This probability is usually estimated considering o-Ps as a quantum particle confined in a pore modeled as a potential quantum well. In the simplest model (Tao-Eldrup model), [52, 53], the pores are modeled as spherical, infinite potential wells of radius $R + \Delta$, where R is the radius of the electron-free volumes and Δ is a fitting parameter representing the thickness of the region where the o-Ps wave function is superimposed to the bulk. For many materials Δ results between 0.166 and 0.19 nm [51]. In this picture the overlap probability P is Eq. (11.9):

$$P = \int_R^{R+\Delta} \phi(r)r^2dr = 1 - \frac{R}{R+\Delta} + \frac{1}{2n\pi}\sin\left(\frac{2n\pi R}{R+\Delta}\right) \tag{11.9}$$

Where $\phi(r)$ is the o-Ps wave function in the potential well and n is the principal quantum number. By combining Eqs. (11.9) and (11.8), a relationship between the pore radius and the o-Ps lifetime (τ_{o-Ps}) in the pore can be found.

This model is usually considered valid for pores smaller than about 1 nm where the lowest energy level in the well is much higher than the thermal energy of o-Ps at a moderate temperature. In this range of dimension and temperature, the principal quantum number can be assumed to be $n = 1$.

In order to account for Ps in any pore size and at any temperature T, the previous model has been extended. To avoid complications introduced by the use of Bessel functions, a rectangular pore geometry has been used instead of a spherical one (Rectangular Tao-Eldrup model, RTE). o-Ps atoms are assumed to be in thermal equilibrium with the pore, implying that the occupation of the energy levels is governed by the Boltzmann distribution. With these assumptions the overlap probability P results as in Eq. (11.10) [54]:

$$P = 1 - F_i(a, \Delta, T) F_j(b, \Delta, T) F_k(c, \Delta, T) \tag{11.10}$$

where a, b, and c are the pore dimensions along the three axes and i, j, and k are the related principal quantum numbers. The terms $F_i(x, \Delta, T)$ are Eq. (11.11):

$$F_i(x, \Delta, T) = 1 - \frac{2\Delta}{x} + \sum_{ijk,\, ijk} \frac{\frac{1}{i\pi}\left(\sin\left(\frac{2i\pi\Delta}{x}\right)\right) e^{-\frac{\beta}{kT}\left(\frac{i^2}{x^2}\right)}}{\sum\limits_{ijk=1}^{\infty} e^{-\frac{\beta}{kT}\left(\frac{i^2}{x^2}\right)}} \tag{11.11}$$

By combining Eqs. (11.8), (11.10), and (11.11), the o-Ps annihilation rate as a function of the pore dimensions a, b, and c can be obtained. The limiting cases of thin one-dimensional (1D) pores, square infinitely long two-dimensional (2D) nanochannels, and three-dimensional (3D) cubic pores can be modeled by setting one or more of the $F_i(x, \Delta, T)$ terms equal to unity.

The assumption of a rectangular pore shape is clearly unphysical but the problem can be overcome determining the pore size in terms of the mean free path, $l = 4$ V/S, where V and S are the volume and the surface of the pore respectively [55].

In spite of some inaccuracies of these models (like the assumption of the value of λ_A as independent from the electron density of the medium, the assumption of complete thermalization of o-Ps in the pores and the difficulty in the determination of Δ), they have shown to be highly successful in the determination of the pore size in zeolite, zirconia, and silica (Fig. 11.8).

The results in the characterization of pore's dimension, achieved with beam-PALS using the models described above, are highly consistent with those obtained with other techniques like small angle neutron scattering (SANS), acoustic wave/gas absorption (SAW-BET) and ellipsometric porosimetry (EP) [54, 56]. The disadvantage that o-Ps porosimetry can be applied only in samples where Ps is formed and can reach the pores (see paragraph 1.2) is compensated by other advantages offered by this technique. In particular, PALS has been shown to be sensitive to pores much smaller than those observable with SANS and to closed pores or multilayer porous films that are not accessible to gas-absorption techniques [57]. Moreover, it can be performed at room temperature preserving the investigated sample from structural changes that can occur at cryogenic temperature like in the case of adsorption isotherms of liquid nitrogen at 77 K [58]. Thanks to its characteristics, beam-PALS has gained in the last two decades a widespread

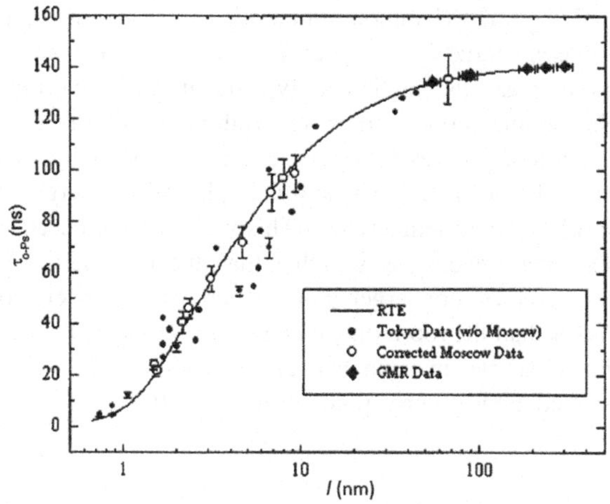

Fig. 11.8 Comparison of τ_{o-Ps}, versus l as predicted by RTE model with some experimental data measured in silica gel [54]

popularity and has become a technique frequently used to investigate the dimension of free volumes in porous materials. In particular, it has allowed to perform depth profile of pore size (see for example [57, 59]) and study the effects of preparation conditions on pore dimension (examples in [60, 61]) of microelectronic materials such as low dielectric constant (low-k) films. Moreover, beam-based PALS has allowed to study the depth-size distribution of pores in polymers [62], in barriers and permeable membranes [56] and to measure the size of the interstitial void spaces in densified SiO_2 glasses [63].

PALS is by far the most used and reliable positronic technique for the measurement of pore dimension but recently the measurement of o-Ps emission kinetic energy has demonstrated to be an alternative suitable technique to achieve information about pore dimension in presence of pores connected to the surface [45, 64]. When a positronium atom is confined in a pore, it can be simulated like a particle in a potential well and its minimum energy is determined by the dimension of the pore itself through the so-called quantum confinement [65]. For instance, a regular nanochannel can be approximated by a 3-D infinite potential well with a quadratic base of side a. Thus, its minimum energy E is given by Eq. (11.12):

$$E = \frac{\hbar^2 \pi^2}{2ma^2}, \tag{11.12}$$

where m is the positron rest mass and $\hbar$ the reduced Planck constant. If Ps is able to diffuse through the pore and reach the vacuum its kinetic energy can be measured by o-Ps TOF spectroscopy or Doppler broadening of a transition line (see Sect. 11.3.3). By increasing the positron implantation energy, it is possible to increase the number of interactions of Ps with the walls of the channel and hence the Ps emission energy from the target is expect to decrease progressively until the quantum confinement limit is eventually reached. By using this technique, Cassidy et al. [45] have studied porous silica films prepared using the cationic surfactant cethyl trimethylammonium chloride as pore generator and tetraethoxysilane as the mineral source for the silica structure. In these samples they observed a lower limit to the mean Ps kinetic energy associated with motion in the direction x, parallel to the sample surface, consistent with conversion of the confinement energy of Ps in 2.7-nm-size pores to kinetic energy in vacuum. For this sample with small pore size, the scan of the positron implantation energy has been performed with the target at room temperature. This was possible because the lower limit to the mean Ps kinetic energy was higher than the Ps kinetic energy corresponding to room temperature. For larger pores, where energy levels corresponding to temperatures lower than the room one, it is necessary to keep the target at cryogenic temperature. For example, in nanochannels with a size of 20 nm, the minimum Ps energy is around 1 meV corresponding to ~7 K [65].

11.4.2 Open Volume Distribution and Interconnectivity

While the pore size calibration with lifetime is quite universal in insulators, the interpretation of the related o-Ps intensity obtained with PALS measurements is not so straightforward [66–68]. There are several factors that can influence the relative intensity of Ps annihilating in the pores, like the dependency of the Ps fraction on the chemistry of the insulator (the o-Ps formation can be significantly altered by additives and free radicals) or the pore dimension (the distance among small pores is typically shorter than the one among large pores. Thus, the diffusion of o-Ps into the pores is favored in the first case).

Nevertheless, the intensity corresponding to the different o-Ps lifetimes has been successfully used to estimate the open volume distribution and in particular to get information about pore interconnection and the presence of densified layers [59, 69, 70].

As an example, in Ref. [69], low-k methylsilsesquioxane-like films were depth profiled with PALS by varying implantation energy from 1 to 8 keV. Effects of densification on the open volume distribution due to plasma treatment were studied. Authors observed that PALS spectra typically require at least two long-lived Ps: one component consistent with Ps annihilating with the vacuum lifetime of ~140 ns and a second component of Ps annihilating in the pores with a lifetime $\leq$45 ns. In the unprocessed porous film an intensity related to the ~140 ns lifetime higher than 8% were found at each positron implantation energy. This points out that Ps is able to back diffuse and reach the surface from high depth indicating that the pores are highly interconnected and open to the vacuum. On the opposite, in the samples processed with plasma no Ps is formed in the pores below 3 keV and only the lifetime component $\leq$45 ns is observed for implantation energy higher than 3 keV. This indicates that the plasma treatment induces the densification of a surface layer preventing Ps formation and out diffusion. However, at higher depth the structure of the pores remains unchanged as indicated by the lifetime that remains $\leq$45 ns like in the unprocessed film.

Similar information on the open volume distribution [71, 72] and pore interconnectivity [32, 64, 73] can be determined from the o-Ps 3γ annihilation. In particular, the diffusion length of Ps in methyl-silsesquioxane low-k films [32] and in porous silicon produced by electrochemical etching [34] has been investigated as a measurement of the pore interconnectivity. Such analysis was proposed for the first time by [32, 71] and is based on a model first used to study Ps formation and out diffusion in ice [31]. The idea is that the $F_{3\gamma}(E)$ curve measured in a porous film is given by the sum of two contributions: (i) a fraction of o-Ps annihilating by 3γ into pores *a(E)* and (ii) a fraction of o-Ps annihilating into 3γ after diffusion in the porosity and emitted into vacuum *b(E)*. If the porosity is roughly homogeneously depth distributed the term "*a(E)*" can be approximated to a constant [33, 34] and *b(E)* can be written as Eq. (11.13):

$$b(E) = \frac{F_1 + (F_2 - F_1)e^{-\frac{1}{2}\left(\frac{E}{E_1}\right)^{\beta}}}{1 + \left(\frac{E}{E_0}\right)^{1.6}}, \tag{11.13}$$

Where the numerator is a phenomenological profile for o-Ps formation. $F_1,\ F_2,\ \beta$ and E_1 are fitting parameters. In particular, F_2 is related to the fraction of formed o-Ps. The ratio $\frac{1}{\left[1+\left(\frac{E}{E_0}\right)^{1.6}\right]}$ represents the effective diffusion probability of o-Ps from connected pores into vacuum. $E_0 = \sqrt[1.6]{L_{Ps}\rho/40}$, where ρ is the density of the investigated sample and L_{Ps} is a measure of the o-Ps diffusion length [7, 31]. This length is similar to molecular or atomic diffusion in pores and, obviously, it depends on the connectivity of pores with vacuum.

It is important to emphasize that this model is based on the diffusion equation which assumes a constant velocity of o-Ps into the pores while its velocity is progressively reduced by interactions with the walls of the pore. Currently no models of o-Ps based on Boltzmann equation have been developed taking into account the energy loss of Ps hitting the nanochannels surface. In spite of this limitation, diffusion models were found to give reliable fits with constant o-Ps diffusion in ordered and disordered porous materials [45].

An example of the analysis procedure described above of a $F_{3\gamma}(E)$ curve measured in a porous silicon sample synthetized by electrochemical etching is reported in Fig. 11.9.

$F_{3\gamma}$ shows a maximum at around $E = 0.3$ keV. At lower positron implantation energy the number of detected 3γ events is reduced by the smaller positronium formation at the surface [31] and by the presence of fast o-Ps able to escape from the target region and annihilate far from the Ge detector [34]. At higher implantation energy an increasing fraction of o-Ps annihilates during the diffusion through the pores via pick-off before reaching the surface and being emitted into the vacuum. These pick-off annihilations induce the observed decrease of $F_{3\gamma}$. The continuous line in Fig. 11.9 represents the best fit obtained by the diffusion model.

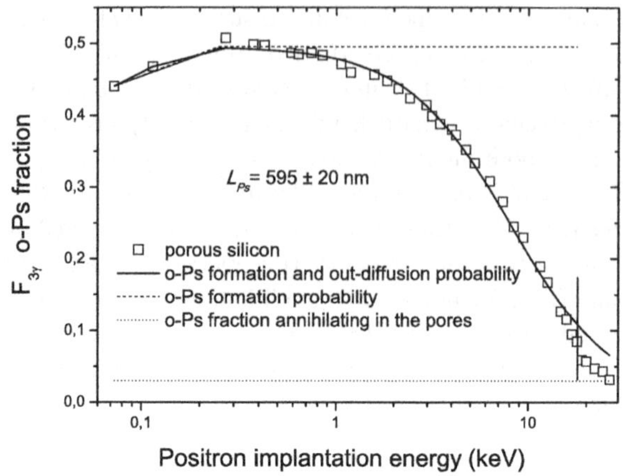

Fig. 11.9 Depth profile of the o-Ps fraction $F_{3\gamma}$ in a porous silicon sample with interconnected porosities. The continuous line represents the best fit obtained by the diffusion model described in the text. The dashed line is the related profile of o-Ps formation probability and the dotted line the estimated fraction of o-Ps annihilating via 3 gamma in the pores. The vertical line marks the border between the porous region and the silicon bulk. The extracted o-Ps diffusion length is also indicated

According to the fit the o-Ps formation probability in the bulk is around 50% (dashed line) while the fraction of o-Ps annihilating via 3γ in the pores has been found to be less than 5% (dotted line). The length of diffusion of o-Ps in the studied porous target has been estimated to be 595 ± 20 nm. By using the described technique, the interconnectivity of pores in low-k methyl-silsesquioxane films has been characterized and the porogen load threshold needed to get connected porosity has been identified [32]. Moreover o-Ps diffusion length up to 1.4 μm, corresponding to the thickness of the porous region, has been estimated in electrochemical etched silicon, indicating the presence of a deep, porous structure characterized by low tortuosity [34].

11.4.3 Pore Surface Characterization

In the last years, o-Ps has been also used as a probe to characterize the surface of pores where is formed or confined. The basis of the technique hinges on the fact that the presence of paramagnetic centers with unpaired spin or chemical groups on the pore surface affect the o-Ps behavior, in particular its annihilation rate and its cooling rate.

It is well known that Ps atoms react sensitively by spin exchange reaction with irradiation-induced paramagnetic centers of the surface (see for example [74–76]). In this process, o-Ps is quenched by spin conversion to p-Ps by an unpaired electron of the paramagnetic centers resulting in the shortening of the o-Ps lifetime.

The use of o-Ps PALS spectroscopy made possible to study the formation of paramagnetic centers due to ultraviolet and heavy ion irradiation on the pore surface of silica grains, aerogel samples [76] and zeolites [77]. The presence of paramagnetic centers on the surface of microvoids and pores in NiO/Al_2O_3 catalysts has been also investigated and the density of unpaired electrons has been correlated to the NiO continent [78]. The recent development of SSPALS technique allowed to study the time decay of transient paramagnetic centers in amorphous silica on a time scale of hundreds of nanoseconds [79]. An intense positron pulse has been used to create o-Ps atoms in a silica sample with closed pores. A short UV laser pulse (FWHM ~10 ns) with an energy of ~8 mJ was fired after the positron pulse to create paramagnetic centers. The production of these defects with unpaired electrons induces a transitory enhancement of the o-Ps decay rate resulting in an increase of o-Ps annihilations in the SSPALS spectrum starting in correspondence to the laser shot and decaying with a rate related to the paramagnetic center lifetime. Paramagnetic centers with a decay rate of ~40 μs^{-1} where observed with the described technique [79].

The use of different techniques—mainly electron spin resonance (ESR)—have demonstrated in the last half century to be solid tools for the study of paramagnetic centers [80], on the other hand recent studies have shown the usefulness of o-Ps lifetime techniques in this subject. In particular, they promise to be useful to investigate paramagnetic centers without well-defined ESR. Moreover, since o-Ps atoms emitted into the pores cannot penetrate back to the bulk, they interact only with the pore surface, making them specifically selective to paramagnetic centers on the pore surfaces [79].

The o-Ps lifetime is not the only parameter affected by the characteristics of the pore surface. As recently demonstrated, also o-Ps cooling in a pore depends on the pore peculiarities [42]. During the diffusion process in the pores, o-Ps loses energy and cools down through collisions with the walls (see Sect. 11.3.3). The Ps thermalization in open volumes has been modeled by Nagashima et al. [81]. According to this classical model, the average kinetic energy of o-Ps can be written as Eq. (11.14):

$$E_{av}(t) = \left(\frac{1+Ae^{-bt}}{1+Ae^{-bt}}\right)^2 \frac{3}{2} k_B T \tag{11.14}$$

Where t is the Ps diffusion time that depends mainly on the morphology and tortuosity of the pores. k_B is the Boltzmann's constant, T the target temperature and A is a term determined by initial conditions Eq. (11.15):

$$A = \frac{\sqrt{E_0} - \sqrt{\frac{3}{2} k_B T}}{\sqrt{E_0} + \sqrt{\frac{3}{2} k_B T}} \tag{11.15}$$

E_0 is the o-Ps energy at time $t = 0$. This corresponds to the energy of o-Ps when emitted into the pore, that in the case of silica is 1–3 eV [39]. Finally the term b is Eq. (11.16):

$$b = \frac{2}{lM}\sqrt{6mk_BT} \tag{11.16}$$

Where l is the Ps mean free path in the pores, m is the positron rest mass and M is the effective mass of the pore surface atoms and depends on the chemical groups decorating the walls. The Ps mean free path l can be typically estimated thanks to its correlation to the pore size [54].

In samples with pores connected to the surface, the o-Ps emission energy can be achieved by TOF measurements after its emission into the vacuum. o-Ps TOF measurements can be performed as a function of the positron implantation depth and, since in porous materials the time scale of positron diffusion is negligible with respect to the o-Ps diffusion, the positron implantation depth can be related to the o-Ps diffusion time t [42].

Using this approach He et al. [42] have studied the o-Ps emission energy from various mesoporous silica/hybrid silica films with similar tortuosity and with pore surfaces decorated by different groups, such as –OH, –H, and $-CH_3$. They have found that, after the same diffusion time, o-Ps emitted from pores with –OH and –H was much colder than Ps emitted from pores decorated with $-CH_3$. By analyzing the behavior of the o-Ps average emission energy $E_{av}(t)$ from the different samples they found that the effective mass (see Eqs. (11.14) and (11.16)) of the pore surface with $-CH_3$ is around three times higher than that of pore surface covered with –OH.

Although further studies are needed in order to investigate the correlation between o-Ps cooling rates and pore surface decorations, this preliminary experiment demonstrates that the presence of specific chemical groups can be distinguished on the basis of their effect on the o-Ps cooling.

11.5 Conclusion

The interaction between positronium and the surface of a pore is a complex subject where several parameters like pore dimension, shape, interconnectivity and surface characteristics concur to determine the fate of Ps atoms. In spite of this, much research conducted in the last decades have demonstrated that positronium is a powerful porosimetric and chemical probe for exploring the porosity and surface properties of various systems. In particular, beam-based o-Ps lifetime spectroscopy proved to give reliable and precise information about pore dimension of both open and closed porosity (Sect. 11.4.1). o-Ps lifetime techniques have also been applied to study the pore depth distribution and their interconnectivity (Sect. 11.4.2) and the formation of paramagnetic centers on the pore surface (Sect. 11.4.3). Other techniques, like o-Ps 3γ annihilation and o-Ps velocity measurements, have been successfully used to gain information about pore morphology (Sects. 11.4.1 and 11.4.2) and characteristics of the surface (Sect. 11.4.3).

While some of the described o-Ps porosimetry techniques are already widespread and frequently used, other techniques have been recently developed thanks to the recent technological improvements in positron manipulation and they are moving their first steps in the field of porosimetry. New porosimetry methods based on o-Ps could come into being in the next future taking advantage from the present effort to enhance positron storage, spatial and temporal compression. With continued deepening of our understanding of positronium formation, diffusion, cooling and quenching in different systems, the role of o-Ps porosimetry techniques promise to becomes wider in the next future in characterizing nanoporous materials.

Acknowledgements The author would like to thank R. S. Brusa for advice, counsel, and expertise about the laboratory activity and G. Nebbia for the useful discussions and suggestions.

References

1. Dirac PAM (1930) Proc Camb Philos Soc 26:361
2. Anderson CD (1932) Phys Rev 41:405
3. Anderson CD (1933) Phys Rev 43:491
4. Anderson CD (1933) Phys Rev 44:406
5. Deutsch M (1951) Phys Rev 82:455
6. Deutsch M (1951) Phys Rev 83:866
7. Schultz PJ, Lynn KG (1988) Rev Mod Phys 60:701
8. Coleman PG (2000) In: Coleman PG (ed), Positron Beams and their applications. World Scientific, Singapore-New Jersey-London-Hong Kong, pp 1–40
9. Krause-Rehberg R, Leipner HS (1999) Positron annihilation in semiconductors. Springer, Berlin
10. Asoka-Kumar P, Lynn KG, Welch DO (1994) J Appl Phys 76:4935
11. Surko CM, Leventhal M, Passner A (1989) Phys Rev Lett 62:901
12. Cassidy DB, Deng SHM, Greaves RG, Mills AP Jr (2006) Rev Sci Instrum 77:073106
13. Jorgensen LV et al (2005) (ATHENA collaboration). Phys Rev Lett 95:025002
14. Fitzakerley DW et al (2016) (ATRAP collaboration). J Phys B 49:064001
15. Aghion S et al (2015) (AEGIS collaboration). Nucl Instrum Method B 362:86
16. Valkealahti S, Nieminen RM (1984) Appl Phys A 35:51
17. Puska MJ, Nieminen RM (1994) Rev Mod Phys 66:841
18. Baker JA, Coleman PG (1988) J Phys C 21:L857
19. Massoumi GR, Hozhabri N, Lennard WN, Schultz PJ (1991) Phys Rev B 44:3486
20. Knights AP, Coleman PG (1995) J Phys Condens Matter 7:6477
21. Eldrup M, Vehanen A, Schultz PJ, Lynn KG (1983) Phys Rev B 51:2007
22. Mills AP, Crane WS (1984) Phys Rev Lett 53:2165
23. Tuomisaari M, Howell RH, McMullen T (1989) Phys Rev B 40:2060
24. Sferlazzo P, Berko S, Canter KF (1985) Phys Rev B 32:6067
25. Howell RH, Rosenberg IJ, Meyer P, Fluss MJ (1987) Phys Rev B 35:5303
26. Mills AP (1978) Phys Rev Lett 41:1828
27. Lynn KG, Welch DO (1980) Phys Rev B 22:99
28. Van Petegem S, Dauwe C, Van Hoecke T, De Baerdemaeker J, Segers D (2004) Phys Rev B 70:115410

29. Lynn KG (1979) Phys Rev Lett 43:391
30. Soininen E, Schwab A, Lynn KG (1991) Phys Rev B 43:10051
31. Eldrup M, Vehanen A, Schultz PJ, Lynn KG (1985) Phys Rev B 32:7048
32. Petkov MP, Weber MH, Lynn KG, Rodbell KP (2001) Appl Phys Lett 79:3884
33. Brusa RS, Macchi C, Mariazzi S, Karwasz GP (2005) Acta Phys Pol A 107:702
34. Mariazzi S, Bettotti P, Larcheri S, Toniutti L, Brusa RS (2010) Phys Rev B 81:235418
35. Mariazzi S, Bettotti P, Brusa RS (2010) Phys Rev Lett 104:243401
36. Bauer-Kugelmann W, Sperr P, Kögel G, Triftshäuser W (2001) Mater Sci Forum 363:529
37. Cassidy DB, Mills AP (2007) Nucl Inst Method B 580:1338
38. Cassidy DB, Deng SHM, Tanaka HKM, Mills AP (2006) Appl Phys Lett 88:194105
39. Nagashima Y, Morinaka Y, Kurihara T, Nagai Y, Hyodo T, Shidara T, Nakahara K (1998) Phys Rev B 58:12676
40. Vallery RS, Zitzewitz PW, Gidley DW (2003) Phys Rev Lett 90:203402
41. Ito K, Yu RS, Sato K, Hirata K, Kobayashi Y, Kurihara T, Egami M, Arao H, Nakashima A, Komatsu M (2005) J Appl Phys 98:094307
42. He C, Ohdaira T, Oshima N, Muramatsu M, Kinomura A, Suzuki R, Oka T, Kobayashi Y (2007) Phys Rev B 75:195404
43. Liszkay L, Barthe MF, Corbel C, Crivelli P, Desgardin P, Etienne M, Ohdaira T, Perez P, Suzuki R, Valtchev V, Walcarius A (2008) Appl Surf Sci 255:187
44. Tanaka HKM, Kurihara T, Mills AP (2006) J Phys: Condens Matter 18:858
45. Cassidy DB, Crivelli P, Hisakado TH, Liszkay L, Meligne VE, Perez P, Tom HWK, Mills AP Jr (2010) Phys Rev A 81:012715
46. Mills AP, Pfeiffer L (1979) Phys. Rev. B 43:1961
47. Mills AP, Pfeiffer L, Platzman PM (1983) Phys Rev Lett 51:1085
48. Mills AP, Pfeiffer L (1985) Phys. Rev. B 32:53
49. Aghion S et al (2016) (AEgIS collaboration). Phys Rev A 94:012507
50. Cassidy DB, Hisakado TH, Tom HWK, Mills AP (2011) Phys Rev Lett 106:173401
51. Goworek T, Jasińska B, Wawryszczuk J, Zaleski R, Suzuki T (2002) Chem Phys 280:295
52. Tao SJ (1972) J. Chem. Phys. 56:5499
53. Eldrup M, Lightbody D, Sherwood JN (1981) Chem Phys 63:51
54. Dull TL, Frieze WE, Gidley DW, Sun JN, Yee AF (2001) J Phys Chem B 105:4657
55. Gidley DW, Frieze WE, Dull TL, Yee AF, Ryan ET, Ho HM (1999) Phys Rev B 60:R5157
56. Gidley DW, Vallery RS, Liu M, Peng HG (2007) Phys Status Solidi (c) 4:3796
57. Sun JN, Hu YF, Frieze WE, Gidley DW (2003) Radiat Phys Chem 68:345
58. Zaleski R (2015) Nukleonica 60:795
59. Wang CL, Weber MH, Lynn KG (2006) J Appl Phys 99:113514
60. He CQ, Muramatsu M, Ohdaira T, Oshima N, Kinomura A, Suzuki R, Kobayashi Y (2007) Radiat Phys Chem 76:204
61. He CQ, Suzuki R, Ohdaira T, Oshima N, Kinomura A, Muramatsu M, Kobayashi Y (2007) Chem Phys 331:213
62. He CQ, Suzuki T, Hamada E, Kobayashi H, Kondo K, Shantarovich VP, Ito Y (2003) Mater Res Innovations **7**:37
63. Zanatta M, Baldi G, Brusa RS, Egger W, Fontana A, Gilioli E, Mariazzi S, Monaco G, Ravelli L, Sacchetti F (2014) Phys Rev Lett 112:04550
64. He C, Wang S, Kobayashi Y, Ohdaira T, Suzuki R (2012) Phys Rev B 86:075415
65. Mariazzi S, Salemi A, Brusa RS (2008) Phys Rev B 81:012715
66. Kajcsos Z, Liszkay L, Duplatre G, Varga L, Lohonyai L, Paszti F, Szilagyi E, Lazar K, Kotai E, Pal-Borbely G, Beyer HK, Caullet P, Patarin J, Azenha ME, Gordo PM, Lopes Gil C, de Lima AP, Ferreira Marques MF (2005) Acta Phys Pol A 107:729
67. Gidley DW, Peng H, Vallery Annu RS (2006) Rev Mater Res 36:49
68. Harms S, Raetzke K, Pakula C, Zaporojtchenko V, Strunskus T, Egger W, Sperr P, Faupel F (2011) J Pol Sci: Part B: Pol Phys 49:404
69. Sun J, Gidley DW, Hu Y, Frieze WE, Ryan ET (2002) Appl Phys Lett 81:19
70. Li S, Sun J, Li Z, Peng H, Gidley DW, Ryan ET, Yan Y (2004) J Phys Chem B 108:11689

71. Petkov MP, Wang CL, Weber MH, Lynn KG, Rodbell KP (2003) J Phys Chem B 107:2725
72. Brusa RS, Spagolla M, Karwasz GP, Zecca A, Ottaviani G, Corni F, Bacchetta M, Carollo E (2004) J Appl Phys 95:2348
73. Liszkay L, Corbel C, Perez P, Desgardin P, Barthe M-F, Ohdaira T, Suzuki R, Crivelli P, Gendotti U, Rubbia A, Etienne M, Walcarius A (2008) Appl Phys Lett 92:063114
74. Dauwe C, Tsumbu M (1992) Phys Rev B 45:9
75. Dauwe C, Consolati G (1993) J Phys IV 3:C4–161
76. Saito H, Hyodo T (1999) Phys Rev B 60:11070
77. Major P, Kajcos Z, Liszkay L, Zalan P, Kosanovic C, Bosnar S, Subotic B, Lazar K, Skuratov VA, Havancsak K, Gordo PM, Ferreira Marques MF (2008) Acta Phys. Pol. 113:1441
78. Zhang HJ, Chen ZQ, Wang SJ, Kawasuso A, Morishita N (2010) Phys Rev B 82:035439
79. Cassidy DB, Yokoyama KT, Deng SHM, Griscom DL, Miyadera H, Tom HWK, Varma CM, Mills AP (2007) Phys Rev B 75:085415
80. Griscom DL (2013) Phys Res Int 2013:379041
81. Nagashima Y, Kakimoto M, Hyodo T, Fujiwara K, Ichimura A, Chang T, Deng J, Akahane T, Chiba T, Suzuki K, McKee BTA, Stewart AT (1995) Phys Rev A 52:258

Index

P. Bettotti (ed.), *Submicron Porous Materials*,
DOI 10.1007/978-3-319-53035-2

The manufacturer's authorised representative in the EU is Springer Nature Customer Service Centre GmbH, Europaplatz 3, 69115 Heidelberg, Germany. If you have any concerns regarding our products, please contact ProductSafety@springernature.com

Printed and bound by CPI Group (UK) Ltd, Croydon, CR0 4YY
07/01/2026
02030952-0005